Metal Clusters

Metal Clusters

Proceedings of an International Symposium,
Heidelberg, April 7–11, 1986

Editors: F. Träger and G. zu Putlitz

With 219 Figures

Springer-Verlag Berlin Heidelberg New York
London Paris Tokyo

Professor Dr. *Frank Träger*
Professor Dr. *Gisbert zu Putlitz*

Physikalisches Institut der Universität Heidelberg, Philosophenweg 12, D-6900 Heidelberg

This book originally appeared as the journal
Zeitschrift für Physik D - Atoms, Molecules and Clusters, Volume 3, Number 2 and 3
(ISSN 0178-7683) © Springer-Verlag Berlin, Heidelberg 1986

ISBN-13: 978-3-642-71573-0 e-ISBN-13: 978-3-642-71571-6
DOI: 10.1007/978-3-642-71571-6

Offset printing: Weihert-Druck GmbH, D-6100 Darmstadt
Bookbinding: J. Schäffer OHG, D-6718 Grünstadt
2153/3150-543210

Preface

This volume contains papers which have been presented at the *International Symposium on Metal Clusters* in Heidelberg from April 7–11, 1986.

Clusters, and in particular metal clusters, have been the topic of fast growing scientific interest. Indeed, clusters constitute a field of interdisciplinary nature where both physical and chemical questions have to be addressed. Clusters are of fundamental importance for the deeper understanding of the transition from atoms via molecules and larger aggregates of particles to the properties of solid materials. Moreover, metal clusters and their characteristics are of vital significance for such applied topics as catalysis or photography.

Experimentally, the field exhibited rapid progress in the last years. Different sources for clusters have been developed. Intense beams made possible the investigation of free neutral clusters and cluster ions as well. Even though a number of issues concerning metal clusters is still discussed controversially, the present volume tries to give an overview of current work in this field and to illustrate the large variety of experiments as well as the advances made possible by modern theoretical methods. Looking at the many interesting questions still to be addressed it is fair to propose a rapid further growth of this field.

The International Symposium on Metal Clusters in Heidelberg was the inaugurate scientific meeting in the "Internationales Wissenschaftsforum Heidelberg", founded on the occasion of the 600th anniversary of the Ruprechts-Karls-Universität Heidelberg as an endavour to promote scientific discussion for all disciplines. Already in this very first symposium the basic goals of the "International Science Forum Heidelberg" became apparent: Exploitation of a special scientific topic or field with a limited number of participants ranging from experts to the advanced graduate student, promoting informal scientific discussions in a truly international atmosphere, and accessing the present knowledge as well as state of the art in the field of interest. Research has been and will be the keystone for the development of our universities.

The present meeting has benefited largely by the advice of the members of the International Advisory Board. Help in the organization and the local arrangements has been obtained by the directorate and staff of the Internationales Wissenschaftsforum Heidelberg and the Physics Institute of the University. Financial support has been provided by the Deutsche Forschungsgemeinschaft, by the State of Baden-Württemberg, and by the University of Heidelberg. In addition, some contributions have been made by private sponsors. To all of those we owe our gratitude.

July 1986

G. zu Putlitz
F. Träger

Contents

On the History of Cluster Beams

E.W. Becker

Kernforschungszentrum und Universität Karlsruhe,
Institut für Kernverfahrenstechnik, Karlsruhe, Federal Republic of Germany

Received April 23, 1986; final version May 15, 1986

The methods to produce and investigate cluster beams have been developed primarily with the use of permanent gases. A summary is given of related work carried out at Marburg and Karlsruhe. The report deals with the effect of carrier gases on cluster beam production; ionization, electrical acceleration and magnetic deflection of cluster beams; the retarding potential mass spectrometry of cluster beams; cluster size measurement by atomic beam attenuation; reflection of cluster beams at solid surfaces; scattering properties of ^4He and ^3He clusters; the application of cluster beams in plasma physics, and the reduction of space charge problems by acceleration of cluster ions.

PACS: 36.40

First Demonstration of Cluster Beams

The history of cluster beams dates back to 1956 when, at the Physics Institute of the University of Marburg, we tried to produce intense molecular beams at low temperature. We had built a nozzle source as proposed by Kantrowitz and Grey [1], and equipped it with a cooling system. To measure the velocity distribution of the beam, a time-of-flight analyser was applied as depicted in Fig. 1. Figure 2 shows time-of-flight distributions obtained with an argon beam produced at constant source pressure but at different source temperatures. Obviously, the single peak observed at 202 K splits up into two peaks at 187 K. At still lower temperatures the second peak becomes predominant. The temperature and pressure dependence of the effect suggested condensation of the argon atoms as the reason [2]. The lower velocity of the second peak exhibits a slip between the clusters and the uncondensed gas.

Condensation or cluster formation in a supersonic flow was already well known from wind-tunnel experiments. It was, however, not previously known that clusters formed under conditions of gas dynamics could be transferred into a high vacuum and separated almost completely from the non-condensed residual gas.

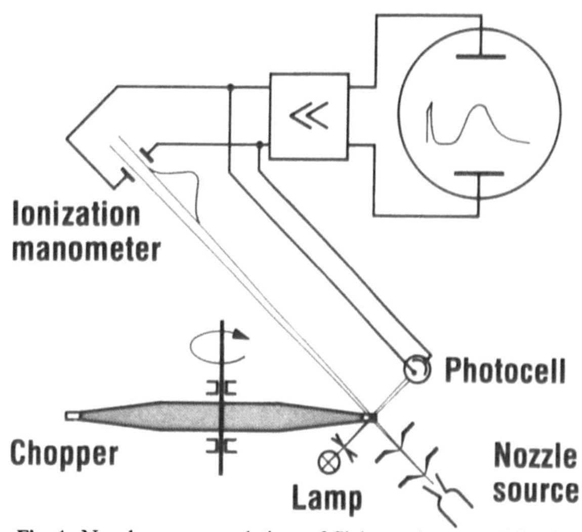

Fig. 1. Nozzle source and time-of-flight analyser used in the first demonstration of cluster beams in high vacuum [2]

When hydrogen was used as the feed gas and liquid hydrogen as the cooling fluid, the results were similar. In Fig. 3, which shows the dependence of intensity on source pressure of a hydrogen beam, cluster formation is indicated by the steep rise in intensity [2].

For our basic research, which aimed at measuring collision properties of single atoms or molecules, the

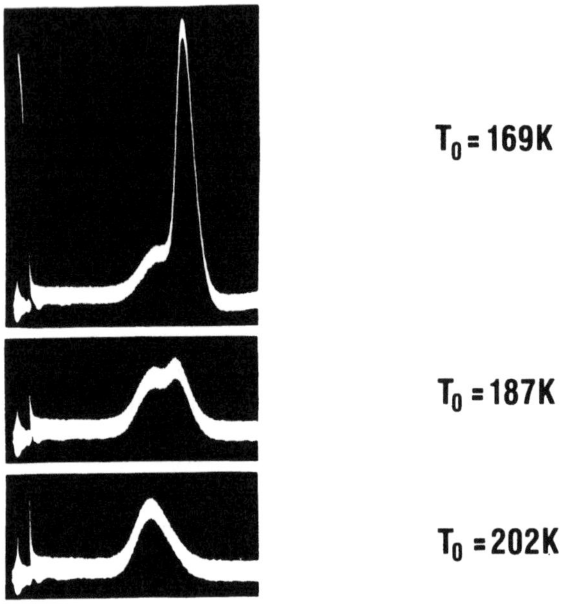

Fig. 2. Time-of-flight distribution of an argon beam at different temperatures. The second peak which appears at lower temperature indicates cluster formation [2]

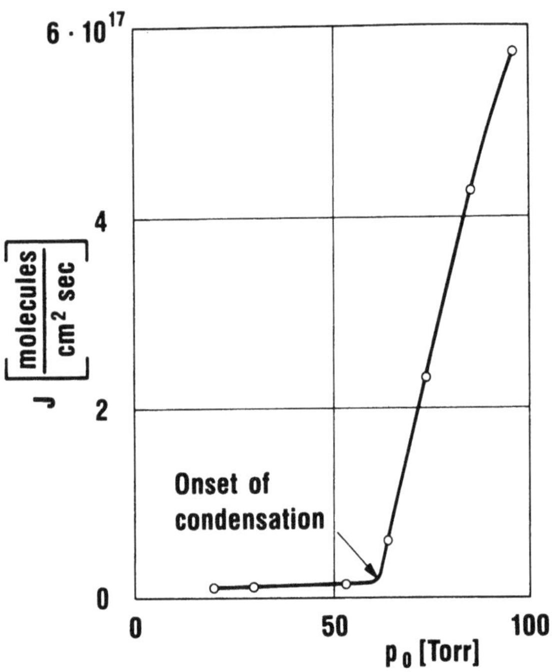

Fig. 3. Dependence of intensity on source pressure of a hydrogen beam at $T_0 = 20$ K. The steep rise in intensity indicates cluster formation [2]

formation of clusters was a severe drawback as it limited the attainable intensity of a beam consisting of single atoms or molecules. On the other hand, the possiblity of producing cluster beams in high vacuum was opening up interesting aspects in basic research as well as in practical application. Hence, work on cluster beams became one of the main research topics of the Institut für Kernverfahrenstechnik at Karlsruhe (founded in 1958[1]).

Considering the tremendous amount of work which has been done in so many laboratories with cluster beams since that time, it is impossible to discuss here all contributions. Therefore, as agreed with the chairman of the meeting, the report will be restricted to the contributions by the Karlsruhe institute[2]. As work there has been with permanent gases, the short history of cluster beams is a history of van-der-Waals cluster beams. A presentation of this at a meeting dealing mainly with metal clusters can be justified by the fact that the methods to produce and investigate metal cluster beams and van-der-Waals cluster beams are almost the same, and the boundary between metal clusters and van-der-Waals clusters is not always clear.

Effect of Carrier Gases on Cluster Beam Production

In an attempt to improve the intensity of cluster beams we investigated the influence of carrier gases, which, under the conditions applied, should not condense but help the feed gas to get rid of its heat of condensation [5]. It is obvious from the upper part of Fig. 4 that at liquid hydrogen temperature adding 300% He brings up the intensity of a hydrogen cluster beam by about a factor of three. For the case where nitrogen is the condensing gas and hydrogen the carrier gas, which is shown in the lower part of Fig. 4 with liquid nitrogen as the cooling fluid, the gain in intensity is more than a factor of thirty. In both cases, the cluster beam is almost free from the carrier gas. It should be noted that addition of a light carrier gas enhances the velocity of the clusters, an effect which may be of interest in special applications.

Ionization, Electrical Acceleration and Magnetic Deflection of Cluster Beams

In 1961 Henkes demonstrated with CO_2 [6] and a year later with H_2 [7] that cluster beams can be ionized by electron bombardment, so enabling mass spectra of cluster beams to be produced. Figure 5 shows the distribution of mass number M per unit

[1] It may be of interest that the strong separation of light and heavy particles, which we had observed in the course of the work on production of intense molecular beams, was the starting point of our development at the same institute of the so-called separation nozzle process for uranium enrichment [3]

[2] Hagena [4] has surveyed most of the work on cluster beams from nozzle sources done up until 1972

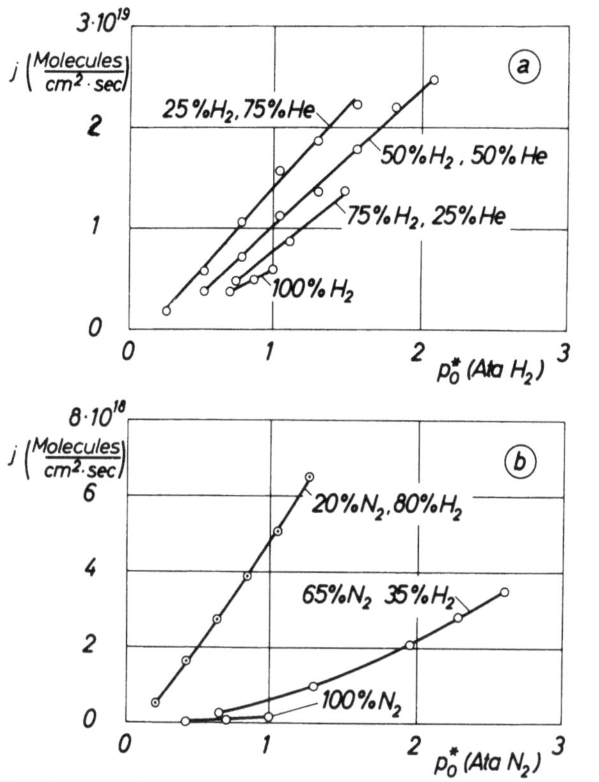

Fig. 4a and b. Enhancement of cluster beam intensity by a non-condensing carrier gas. **a** $T_0 \approx 20$ K; **b** $T_0 \approx 78$ K [5]. Ata H_2 and Ata N_2 = partial pressure of H_2 and N_2 respectively, given in atm

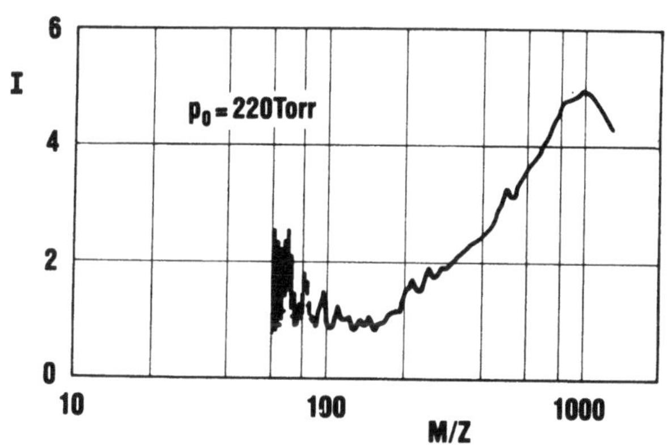

Fig. 5. Distribution of mass number M per unit charge Z of hydrogen cluster ions obtained by electron bombardment, electrical acceleration and magnetic deflection [7]

charge Z of hydrogen cluster ions obtained by electron bombardment, electrical acceleration and magnetic deflection.

The ionization process model that emerged from various experiments assumes that the loss in cluster mass by partial evaporation is negligible and that the observed changes in the cluster ion spectrum are due to multiple ionization of the cluster by one electron. For smaller clusters multiple ionization results in fragmentation [8].

On account of multiple charging and fragmentation, the distribution of the specific mass M/Z of the cluster ions cannot be expected to be identical with the original mass distribution of the cluster beam. However, it gives a good measure of the mean size of the original clusters.

Retarding Potential Mass Spectrometry of Cluster Beams

With higher values of the specific mass the magnetic deflection of electrically accelerated cluster ions becomes more and more ineffective. Therefore, Bauchert and Hagena introduced what is called retarding potential mass spectrometry [9]. It profits from the fact that all the cluster ions in an ionized cluster beam have practically the same velocity so that their kinetic energies are proportional to their masses. Figure 6 shows the mean number of molecules N per unit charge Z of ionized CO_2 cluster beams measured by a retarding electric field. It is obvious from the figure that the mean specific size N/Z increases considerably with increasing source pressure and increasing nozzle diameter. Hagena and co-workers derived similarity

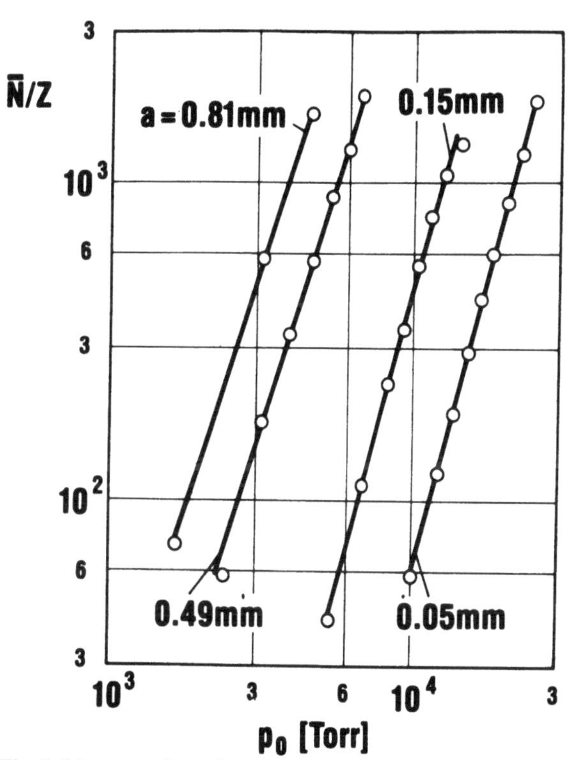

Fig. 6. Mean number of molecules N per unit charge Z of CO_2 clusters measured by a retarding electric field: a = nozzle diameter, p_0 = source pressure [9]

laws for the production of cluster beams [4] from such measurements.

Cluster Size Measurement by Atomic Beam Attenuation

From retarding potential mass spectrometry as well as from magnetic mass spectrometry only a mean size of the cluster *ions* can be derived. To obtain directly information about the mean number of molecules in the clusters of the original beam, Burghoff and Gspann measured the attenuation of a potassium oven beam by a nitrogen cluster beam [10]. Figure 7 shows that at a source temperature of 78 K, which is low enough for nitrogen to form clusters, the effective cross section of a nitrogen molecule in the beam drops by two orders of magnitude when the source pressure is brought up to 700 Torr. The effect is a consequence of the mutual screening of molecules within a cluster. Assuming the clusters to be spherical and having the density of solid nitrogen, Burghoff and Gspann calculated the mean number of molecules per cluster, as shown in the lower part of the figure. Later, Gspann repeated the size measurements with a time-of-flight mass spectrometer and obtained good agreement with the data based on potassium beam scattering [11], thus indicating the internal consistency of the various assumptions made.

Reflection of Cluster Beams at Solid Surfaces

One of the most striking effects with cluster beams is the enhancement of flux density attainable by reflection at a smooth surface, which we first demonstrated in 1967 [12]. The left part of Fig. 8 depicts the reflection of a nitrogen cluster beam of about 5 mm diameter at a polished steel plate at room temperature and an angle of incidence of 75°. The reflected beam is concentrated in a lobe of about 1 mm thickness at an angle of reflection of about 88°. The size of the beam perpendicular to the plane of incidence remains practically unaffected. It can be seen from the right part of the figure that the reflection results in an enhancement of flux density of up to a factor of 5. The increase in the reflected flux density, compared to the incident cluster beam, suggests the possibility of collimating a cluster beam with suitably shaped reflectors [12]. At a cooled target even He clusters can be reflected [12a].

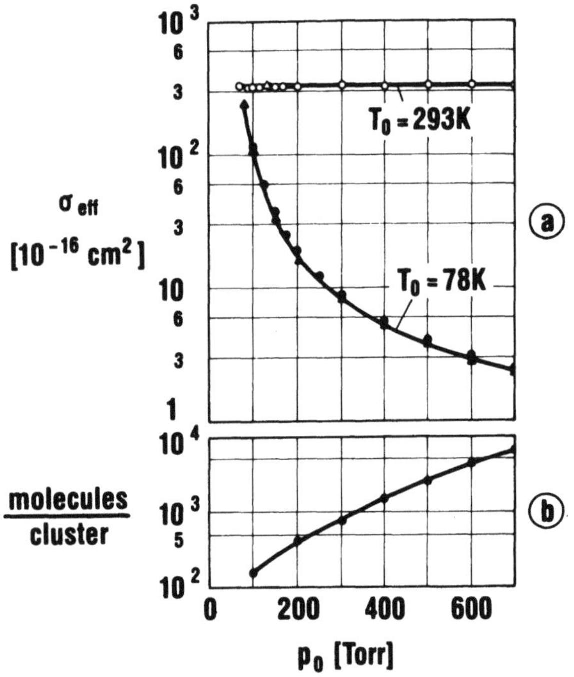

Fig. 7. a Effective cross section σ_{eff} of an N_2 molecule in a molecular beam ($T_0 = 293$ K) and in a cluster beam ($T_0 = 78$ K), varying with source pressure p_0. b Number of N_2 molecules per cluster calculated from the drop of σ_{eff} [10]

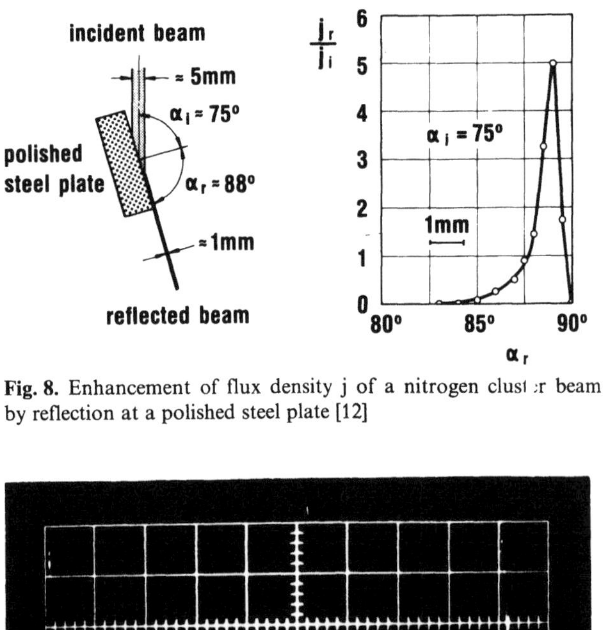

Fig. 8. Enhancement of flux density j of a nitrogen clust :r beam by reflection at a polished steel plate [12]

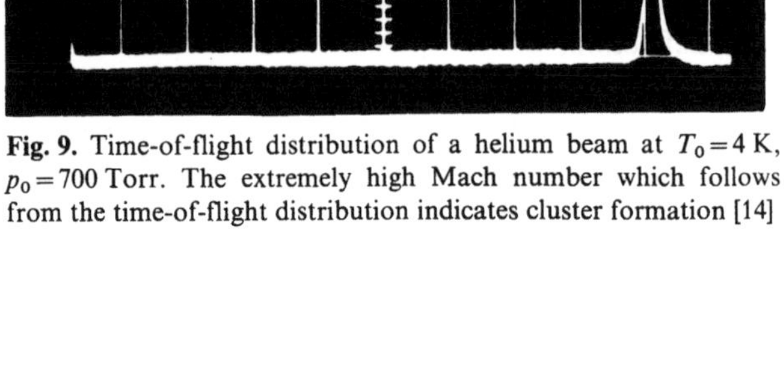

Fig. 9. Time-of-flight distribution of a helium beam at $T_0 = 4$ K, $p_0 = 700$ Torr. The extremely high Mach number which follows from the time-of-flight distribution indicates cluster formation [14]

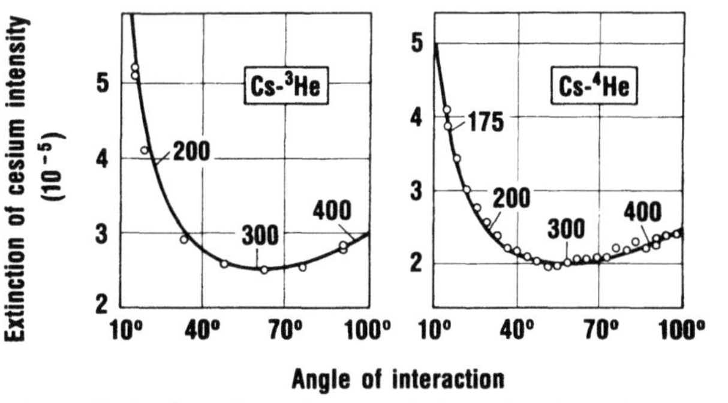

Fig. 10. Extinction of a cesium atomic beam by cluster beams of ^{3}He and ^{4}He, varying with the angle of interaction. The numbers indicate the mean relative velocities given in m/s. The curves are calculated under the assumption of a velocity-independent interaction [15b]

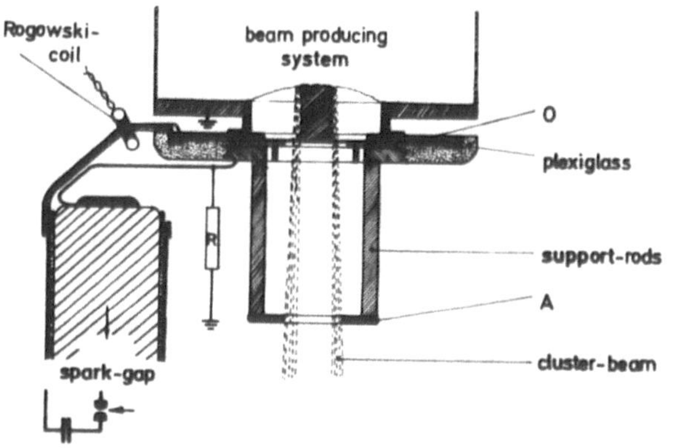

Fig. 11. Device used to produce a dense plasma column by electrical discharge along a hollow nitrogen cluster beam of about 4 cm diameter [17]

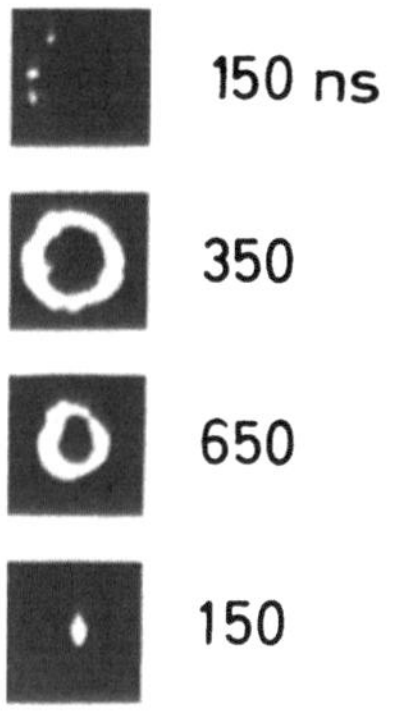

Fig. 12. End-on pictures of a hollow plasma cylinder produced by device depicted in Fig. 11 after several time intervals [17]

Scattering Properties of ^{4}He and ^{3}He Clusters

In 1959 we pointed out that ^{4}He cluster beams could provide a possibility of demonstrating superfluidity dependence on the total number of atoms involved [13]. In 1961 we produced the first helium cluster beam [14]. In 1975 we published first results on the scattering of cesium atoms by ^{4}He clusters [15]. Gspann and coworkers improved the scattering experiments and extended them to the rare and correspondingly expensive ^{3}He [15a]. Under the conditions applied, ^{3}He should exhibit no superfluidity at all.

Figure 10 shows the extinction of a cesium atomic beam by collision with cluster beams of the two helium isotopes, varying with the angle of interaction [156]. The number of atoms per cluster was of the order of 10^7. The curves are calculated under the assumption of a velocity-independent interaction. Fitting the curves to the experimental data at 90° was possible by assuming a content of some 10^{-3} of uncondensed gas within the cluster beams. The results still need to be explained theoretically.

Application of Cluster Beams in Plasma Physics

In 1959 we suggested using cluster beams as a fuel supply for thermonuclear plasmas [13]. The high mass-flux density obtainable in a cluster beam should allow the deuterium/tritium fuel to be injected to exactly that part of a high-vacuum chamber where the plasma is to be produced. So we started experiments with electrical discharges in cluster beams of hydrogen and nitrogen [16]. Figure 11 depicts a device we used in 1967 to produce a dense plasma column by electrical discharge along a hollow nitrogen cluster beam [17]. Figure 12 shows "end-on" pictures of the hollow plasma cylinder taken by an image converter camera after several time intervals. It can be seen from the photographs that within a microsecond the hollow plasma cylinder collapses without detectable instabilities, so that a dense plasma column is formed. Probe measurements demonstrated that most of the material of the cluster beam undergoes the pinch effect [18].

Reduction of Space Charge Problems by Acceleration of Cluster Ions

The possibility of reducing space charge problems in the production of intense beams of comparatively low energy per atom by high voltage acceleration of cluster ions was first proposed in 1962 by Henkes in connection with nuclear fusion [7, 19, 20]. The principle is demonstrated in Fig. 13. Obviously the acceleration by a voltage U of cluster ions of N atoms and the total charge Q results in an energy per atom of $U \cdot Q/N$. For a given energy per atom and constant beam geometry, the space charge limited beam current in-

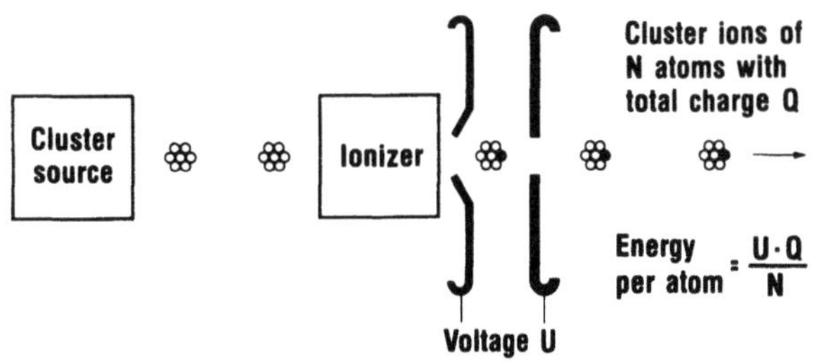

Fig. 13. Principle of reducing space charge problems in the production of intense beams of comparatively low energy per atom by high voltage acceleration of cluster ions [7, 19, 20]

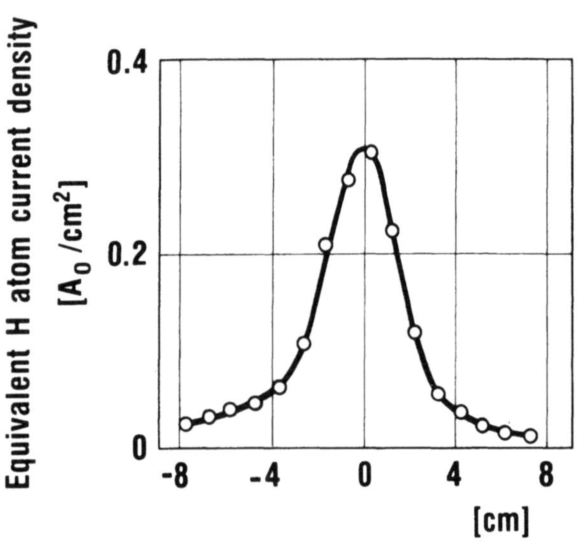

Fig. 14. Profile of an accelerated hydrogen cluster ion beam, 215 cm downstream of the earth electrode of the accelerator. Acceleration voltage 350 kV; average energy per H atom 7 eV; total equivalent H atom current $6A_0 = 3.6 \times 10^{19}$ atoms s^{-1} [21]

creases proportionally to the square of N/Q [20]. So the bigger the clusters are, the more effective the method is. On the other hand, for a given energy per atom, bigger clusters require higher acceleration voltage.

A cluster ion accelerator built at Karlsruhe with an acceleration voltage of up to 1 MV was devoted mainly to producing intensive cluster ion beams of hydrogen with some eV per atom. Such beams are of interest in nuclear fuel injection into fusion devices as well as in the production of negative hydrogen ions in connection with plasma heating. Figure 14 depicts the profile of an accelerated hydrogen cluster ion beam of 6 A_0 equivalent H atom current ($1\ A_0 = 6 \cdot 10^{18}$ H atoms$\cdot s^{-1} = 1$ amp for full ionization) 215 cm downstream of the earth electrode of the accelerator. Considering the desired low energy of 7 eV per atom and the long distance from the accelerator, the current density achieved is remarkable [21].

Conclusions

The methods to produce and investigate cluster beams have been developed primarily with the use of permanent gases. The present meeting demonstrates that cluster beam techniques have been successfully extended to materials which are solid under normal conditions. In addition to many interesting theoretical aspects, a lot of practical applications can be anticipated. For example, cluster beam techniques allow high quality thin film deposition and formation of new materials [22].

Although the conditions for producing cluster beams from materials with high boiling temperature, especially metals, are rather different from those typical for permanent gases, the two classes of material seem to exhibit similar dependences of cluster formation on source conditions. Therefore the extended work being done with cluster beams of permanent gases can be used as a guideline in the further development of the cluster beam technique for materials with a high boiling temperature.

References

1. Kantrowitz, A., Grey, J.: Rev. Sci. Instrum. **22**, 328 (1951)
2. Becker, E.W., Bier, K., Henkes, W.: Z. Phys. **146**, 333 (1956)
3. See. e.g. Becker, E.W., Bier, W., Ehrfeld, W., Schubert, K., Schütte, R., Seidel, D.: Naturwiss. **63**, 407 (1976)
4. Hagena, O.F.: Cluster beams from nozzle sources. In: Molecular beams and low density gas dynamics. Wegener, P.P. (ed), pp. 95–181. New York: M. Dekker Inc. 1974. See also Hagena, O.F.: Surf. Sci. **106**, 101 (1981)
5. Becker, E.W., Klingelhöfer, R., Lohse, P.: Z. Naturforsch. **17a**, 432 (1962)
6. Henkes, W.: Z. Naturforsch. **16a**, 842 (1961)
7. Henkes, W.: Z. Naturforsch. **17a**, 786 (1962); see also Becker, E.W., Henkes, W.: D.P. 1178152 (appl. 12 may 1962)
8. Falter, H., Hagena, O.F., Henkes, W., v. Wedel, H.: Int. J. Mass Spectrom. Ion Phys. **4**, 145 (1970). See also Henkes, W., Isenberg, G.: Int. J. Mass Spectrom. Ion Phys. **5**, 249 (1970)
9. Bauchert, J., Hagena, O.F.: Z. Naturforsch. **20a**, 1135 (1965)
10. Burghoff, H., Gspann, J.: Z. Naturforsch. **22a**, 684 (1967)
11. Gspann, J.: Entropie **42**, 129 (1971)
12. Becker, E.W., Klingelhöfer, R., Mayer, H.: Z. Naturforsch. **23a**, 274 (1967)

12a. Becker, E.W., Gspann, J., Krieg, G.: Entropie **30**, 59 (1969)

13. Becker, E.W.: Beams of Condensed Matter in High Vacuum; paper presented at the Brookhaven Conference on Molecular Beams at Heidelberg, 11.6.1959
14. Becker, E.W., Klingelhöfer, R., Lohse, P.: Z. Naturforsch. **16a**, 1259 (1961)
15. Becker, E.W., Gspann, J., Krieg, G.: Proceedings of the 14th International Conference on Low Temperature Physics, Otaniemi 1975. Krusius, M. (ed.), p. 426. Amsterdam, Oxford, New York: North-Holland 1975

15a. Gspann, J., Krieg, G., Vollmar, H.: J. Phys. Suppl. No. 7, Tome **38**, C2–171 (1977)

15b. Gspann, J., Ries, R.: Ergebnisbericht über Forschung und Entwicklung 1984, Kernforschungszentrum Karlsruhe, S. 320

16. Becker, E.W., Klingelhöfer, R.: Z. Naturforsch. **19a**, 813 (1964)
17. Becker, E.W., Burghoff, H., Klingelhöfer, R.: Z. Naturforsch. **22a**, 589 (1967)
18. Becker, E.W., Klingelhöfer, R., Wüst, J.: Naturforsch. **27a**, 1406 (1972)
19. Henkes, W.: Phys. Lett. **12**, 322 (1964)
20. Henkes, W.: KfK-Nachrichten **3**, 16 (1/1972)
21. Becker, E.W., Hagena, O.F., Henkes, P.R.W., Keller, W., Klingelhöfer, R., Krevet, B., Moser, H.O.: Nucl. Eng. Design **73**, 187 (1982)
22. See e.g. Takagi, T.: Ionized-cluster beam technique for thin film deposition,: paper presented at this meeting

E.W. Becker
Institut für Kernverfahrenstechnik
Kernforschungszentrum Karlsruhe GmbH
Postfach 3640
D-7500 Karlsruhe 1
Federal Republic of Germany

Shell Structure and Response Properties of Metal Clusters

W. D. Knight, Walt A. de Heer, and Winston A. Saunders
Department of Physics. University of California, Berkeley, California, USA

Received April 7, 1986

Electronic shell structure, which was first recognized in sodium clusters, has been observed in alkali and noble metals, as well as in divalent and trivalent metals. Shell structure with modifications is expected to be broadly applicable to most metals. Features in the cluster abundance spectra and in the experimental dipole polarizabilities and ionization potentials correlate well with predictions of electronic level filling in spherical and spheroidal potential wells. The lack of precise quantitative agreement between experiment and theory for the response properties indicates necessary refinements in the self-consistent uniform background jellium model for clusters.

PACS: 36.40+d; 35.20.My; 35.20.Vf

1. Electronic Shell Structure

The application of angular momentum conditions in systems with spherical symmetry, such as atoms and nuclei [1], results in the prediction of energy levels and shell structure which explain satisfactorily the observed properties. The shell models which have worked well for atoms and nuclei are also applicable to metal clusters [2].

All three systems may be described in terms of particles confined in spherically symmetric potentials on the respective scales of 10^{-12}, 10^{-8} cm for nuclei and atoms, and $10^{-7}-10^{-6}$ cm for clusters. The details, the scale, and the interactions governing the potentials are different. Nevertheless the systems have common features based on the behavior of fermions in a spherical potential, and reasonable agreement with experiment is obtained for stabilities and response properties.

The patterns of stability over the respective ranges of homologous species reflect the basic symmetries and energy level structures. For metal clusters the abundance spectra reflect the underlying spherical symmetries appropriate to the shell closings, and also the spheroidal symmetries of distorted geometries produce subshells which result in fine structure patterns in the abundances [3, 4]. The order of level filling depends somewhat on the shape of the well, of which the following have been used: three-dimensional harmonic oscillator; infinite square well; or intermediate between these the finite square well with rounded corners [5]. The filling of shells in the intermediate case correlates quite well with the abundance patterns in the cluster mass spectra [2].

2. The Spheroidal Model

The shell modell of the nucleus included, in addition to a strong spin-orbit coupling [1] which has not yet been identified in metal clusters, spheroidal distortions for the open shell systems between spherical shell closings. A similar treatment of clusters has been successful in accounting for energy levels, abundances, static electric polarizabilities, and ionization potentials. Following the Nilsson model [6], clusters are allowed to undergo spheroidal distortions at constant volume, and total electronic energy is minimized with respect to the distortion parameter η, in order to determine the latter for each cluster [3].

The Nilsson single particle hamiltonian [6] for the spheroidal harmonic oscillator is written

$$H=\frac{p^2}{2m}+\frac{1}{2}m\omega_0^2(\Omega_\perp^2\rho^2+\Omega_z^2z^2)-U\hbar\omega_0(L^2-\langle L^2\rangle_n) \quad (1)$$

where z is the axis of symmetry, $\rho^2 = x^2 + y^2$, and harmonic oscillator frequencies are assumed proportional to Ω_z and $\Omega_\perp$ parallel and perpendicular to the z axis. These become equal in the spherical limit. The anharmonic term in L^2, where L is the angular momentum operator, reshapes the harmonic potential well to resemble the rounded corner square well potential. The term in $\langle L^2 \rangle_n$, which is an average over all states with the same n, and is equal to $n(n+3)/2$, is included in order to preserve a constant energy spacing between oscillator shells of different n. The constant U is an adjustable parameter which determines the strength of the anharmonic correction, and hence the effective shape of the potential well.
The distortion parameter is defined as

$$\eta = 2(z_0 - \rho_0)/(z_0 + \rho_0) \tag{2}$$

where z_0 and ρ_0 are the major and minor semi-axes of the spheroid. The total electronic energy for each cluster is minimized [3, 4] as a function of the distortion parameter, and each cluster is assigned a distortion parameter corresponding to the minimum. The Nilsson diagram is plotted [4], see Fig. 1, to give energy levels for the clusters as a function of the distortion parameter η. The number of oscillator quanta is n, the number of quanta along the symmetry axis is n_z, and the projection of the electronic orbital angular momentum along the z axis is Λ. The levels are either two ($\Lambda = 0$) or fourfold ($\Lambda \neq 0$) degenerate. The energy levels for a given cluster may be found from the intersections of a vertical line on the Nilsson diagram with the plotted eigenvalues. The effect of the anharmonic term removes degeneracies in the spherical harmonic oscillator levels, giving for example the $1d-2s$ and $1f-2p$ splittings for $\eta = 0$, see Fig. 1.
The spherical shell closings $\eta = 0$ occur, see Fig. 2a, for occupation numbers $N = 2$, 8, 20, 40, 58, and 92. Subshell closings occur at 18, 26, 30, 34, 36, 38, 50, 54, etc. Conspicuous sequences of four ending for example at 18, 26 and 30 are associated with the fourfold degeneracies ($\Lambda \neq 0$), and subshell closings at for example 36 and 38 correspond to the twofold degeneracies ($\Lambda = 0$). The calculated second differences of the electronic energies of successive clusters for sodium are shown for comparison in Fig. 2b. The agreement of experiment and theory is good, except for 12, which according to recent calculations [7, 8], appears to be ellipsoidally distorted which lowers the energy and increases stability. With this provision the overall agreement is excellent.
The agreement with experiment, including both spherical and subshell closings, reflects the powerful role played by the symmetry of the problem, which is basically governed by the single adjustable parame-

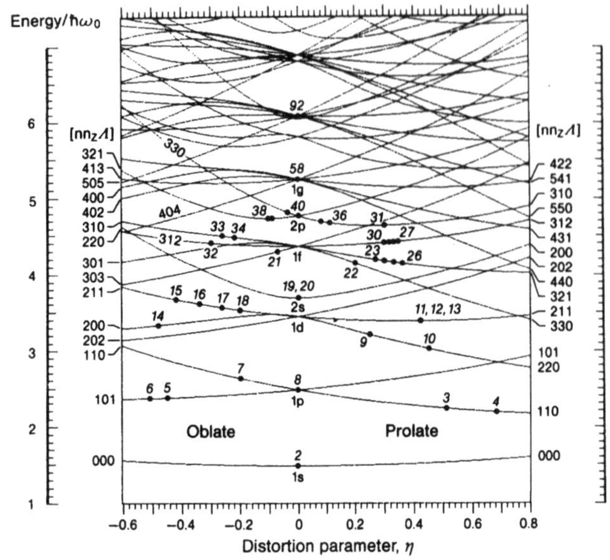

Fig. 1. Nilsson diagram of cluster energy vs distortion parameter η, for $U = 0.04$. The numbered dots indicate the highest occupied level in the ground state for each clusters. [After K. Clemenger, Ref. 4]

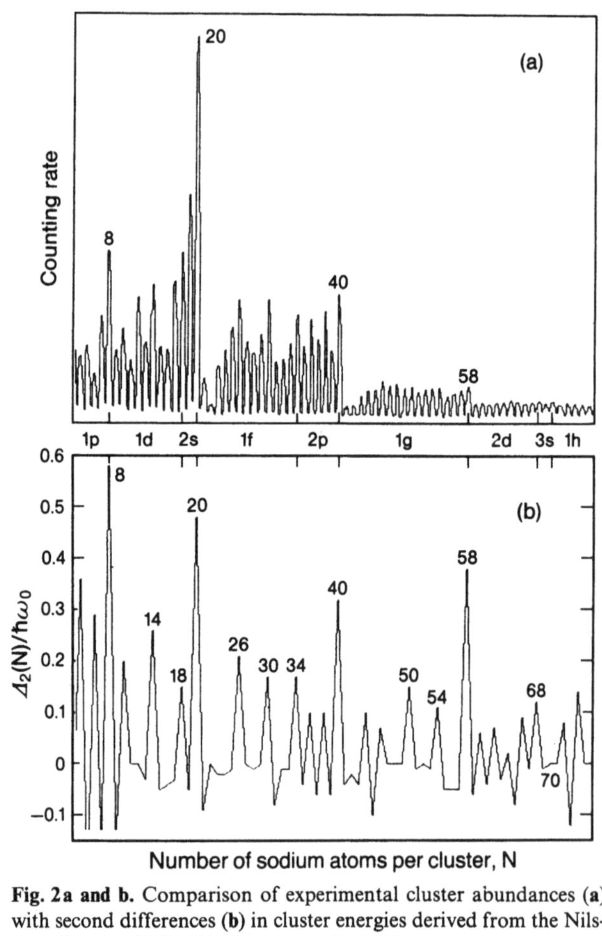

Fig. 2a and b. Comparison of experimental cluster abundances (**a**) with second differences (**b**) in cluster energies derived from the Nilsson diagram. [After K. Clemenger, Ref. 4.]

ter U, and hence the shape of the potential. The Nilsson diagram in Fig. 1, is plotted for the constant value $U=0.04$. A major aim in this paper is to show that the electronic energies for the respective clusters as determined by the Nilsson theory are directly related to the stabilities, the polarizabilities, and the ionization potentials.

3. Application of the Jellium Model to Clusters

The success of the harmonic oscillator type model despite the neglect of the structural arrangement of the ion cores suggests that further theoretical refinements may follow the techniques of the self consistent jellium background model (SCUJB) [9], in which the lattice structure of the ion cores is replaced by a uniform positive background. The application of this model to simple metals in calculating properties such as cohesive energy, dielectric functions, and surface work functions gives fair agreement with experiment and with other theories which specifically include the lattice structures [10, 11].

The jellium model is applicable when the electronic energies are dominant. The existence of a residual lattice structure is accounted for [12] by introducing appropriate pseudopotentials. An important result of this type of calculation is the electron density spillout beyond the jellium edge for both infinite solids and clusters [10]. The SCUJB calculations are, except in a few cases [13], not yet applicable to excited states of systems. The calculations for energy levels leading to predicted abundance spectra for sodium in Ref. 2 first used a Woods-Saxon type potential [5]. Further analysis of the problem using the SCUJB model with the local density functional approach [9] improved the ageement with experiment.

Evidence for shell structure has been observed in the cluster abundance spectra for sodium [2], potassium [14], copper, silver, gold [15], zinc and cadmium [16], and aluminum [17]. In view of the extent of present observations the model is expected to be useful for wider applications.

4. Static Electric Dipole Polarizability

We now compare the measured response properties with theoretical predictions. The screening of external electric fields from the interior of metals is related to the delocalized nature of the conduction electrons. An important question is whether metal clusters also exhibit screening [18]. This question is answered by measuring the dipole polarizability. Lang and Kohn found that the response of bulk metal surfaces includes strong screening, and Friedel oscillations in electron density, with appreciable charge spillout beyond the jellium boundary [10]. Several theoretical calculations [18–20] find analogous behavior for clusters, and show that the static dipole polarizability may be expressed

$$\alpha=(R+\delta)^3 \tag{3}$$

where $\delta \sim 1$ au is the charge spillout parameter. The result implies that a small cluster responds to electric fields like a metal sphere whose effective radius is larger than the classical value. In the jellium model the sphere radius is given by

$$R=r_s N^{1/3} \tag{4}$$

where r_s is the Wigner Seitz radius, which is usually taken to be the value for the bulk solid at 0 K. The δ parameter is relatively more important for smaller clusters, and gives a 30% enhancement for the sodium 20 atom cluster, whose classical radius is 11 au. Accordingly it is expected that the polarizability will fall toward the bulk value as the cluster size increases.

The SCUBJ calculations [18–20] for the polarizability of closed shell clusters agree among themselves and follow the expected trend toward the bulk value with increasing cluster size. A calculation of polarizability for open shell clusters [21] has been carried out, using fractional occupation numbers to obtain spherically symmetric wavefunctions. These calculations show definite shell effects, with minima in the polarizability curve near 2, 8, and 20 atom clusters. The deviations from a smooth curve are enhanced when depolarizing factors based on the distortion parameters given by the Nilsson diagram are included, see Fig. 3.

The experimental polarizabilities were measured [22] by deflection of the cluster beam in an inhomogeneous electric field. The deflection z is given by

$$z=K\alpha E^2/Mv^2, \tag{5}$$

where α is the polarizability, E is the applied electric field, M is the cluster mass (number of atoms N times atomic mass m), v is the beam velocity, and K is a constant depending on the configuration of the apparatus. Since the polarizability is roughly porportional to cluster volume, the deflection should be approximately independent of cluster size, with larger deflections for smaller clusters because of the spillout factor. The experimental deflection profiles were combined with time of flight profiles to give values for polarizability.

Experimental polarizability per atom normalized to the experimental atomic polarizability is shown in Fig. 3 along with calculated values corrected for depolarizing effects. Both the experimental and theoretical curves show correlations with shell structure. Although the above trends are reasonable, the quantita-

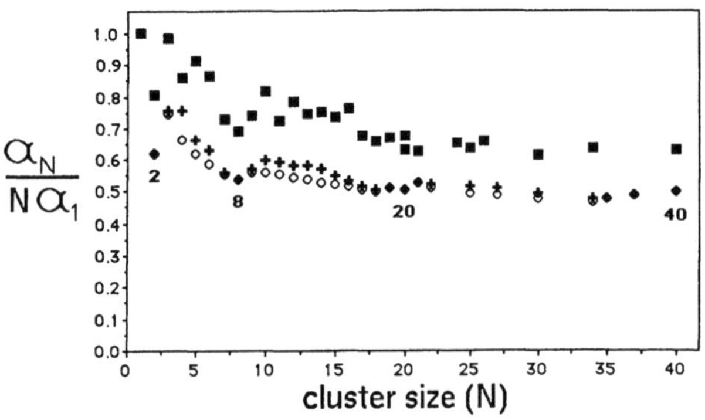

Fig. 3. Sodium cluster polarizabilities, per atom: upper curve, experimental; lower curve, unpublished calculations after Reference 21 (open dots), corrected (crosses) for depolarization factors, using distortion parameters from Fig. 1

tive agreement between experiment and theory is not as good as might be hoped, the latter values being some 15–20% lower. Several factors [4] might account for the discrepancy, including for example centrifugal distortions, ion core polarizability, and thermal expansion. It is not clear whether correction for these factors would account for the discrepancy. Furthermore, the use of the bulk value for r_s could be incorrect. If r_s is smaller in clusters than in the bulk [23], the discrepancy would be even worse.

It is not surprising that SCUJB theory and experiment disagree. The inclusion of gap narrowing effects would increase the calculated polarizability, and the weak pseudopotential structure associated with a residual ion lattice, would produce crystal field splittings, reduce the energy gaps, and improve agreement with experiment, as was found by Lang and Kohn [10] in their calculations of the work function. The lack of agreement indicates the directions for needed improvement of the SCUJB, which provides a good basis for the calculations. The SCUJB model should include a self-consistent treatment of the spheroidal clusters, a proper accounting of the ion background crystal field splitting, and a calculation of excited state wavefunctions

5. Ionization Potential

Calculations of the cluster ionization potentials by the SCUBJ model have been given explicitly for the alkali metals [9], based on the framework developed by Lang and Kohn for the bulk work function. This involves a self-consistent treatment of the inhomogeneous electron gas at the edge of the jellium boundary. For the infinite plane surface [10] this involves three terms, representing electrostatic, exchange-correlation, and kinetic energies

$$W = E_{es} + E_{xc} - E_F \tag{6}$$

where E_{es} represents the surface barrier associated with the electron spillout, E_{xc} represents the exchange and correlation interactions among the delocalized electrons, and E_F is the Fermi energy. Lang and Kohn calculate the work function for bulk sodium according to Eq. (6), $W = 0.91 + 5.28 - 3.13 = 3.06$ eV compared to the experimental value 2.7 eV. For potassium the similarly calculated value is 2.74 eV compared to the experimental 2.39 eV.

Calculations of the ionization potential (IP) for metal clusters find variations with N showing direct correlation with shell structure. Ekardt [18] finds that total work function and the electrostatic part approach the corresponding bulk limits [10]. Thus the theory of ionization potential for clusters is consistent with the analogous calculations for the bulk material by the same method.

It has proved convenient [24, 25] to relate the cluster ionization potentials (IP) to the bulk work function plus a term inversely proportional to the cluster radius R

$$\mathrm{IP} = W + A(e^2/R) \tag{7}$$

where W is the bulk work function and A is a constant, usually found to be in the range between 0.3 and 0.5. The second term has been called the electrostatic term, and the constant A has been variously identified in connection with cluster charging or image potential. The process of ionization is in fact more complicated, and (7) is to be regarded as a convenient empirical description rather than an explicit statement of the true quantum mechanical behavior.

The experimental ionization potentials have been determined [26] for potassium clusters containing between 2 and 101 atoms. The trends, and reflections of shell structure follow the theoretical predictions [9, 18, 27] qualitatively. If we consider only the IPs at shell closings and relate the results to (7) we find $W = 2.37$, compared to the bulk value 2.4, and the constant $A = 0.33$. In view of the fact that the open shell clusters vary in a known consistent manner, an average of the constant A for all clusters is hardly meaningful.

The measured ionization potentials for potassium, Fig. 4, are in reasonable agreement with other published data for $N < 9$ clusters [28, 29]. The appearance of jumps in the IP at shell closings is another striking example of the general manifestation of shell structure over a wide range of cluster sizes. Nevertheless, it is significant that the magnitude of these jumps is less than 1/2 of that predicted by SCUJB models [9, 18, 27]. As was noted in the discussion of polarizability, the jellium model requires some refinements in order to deal with response properties which depend on excited state wavefunctions, and to account

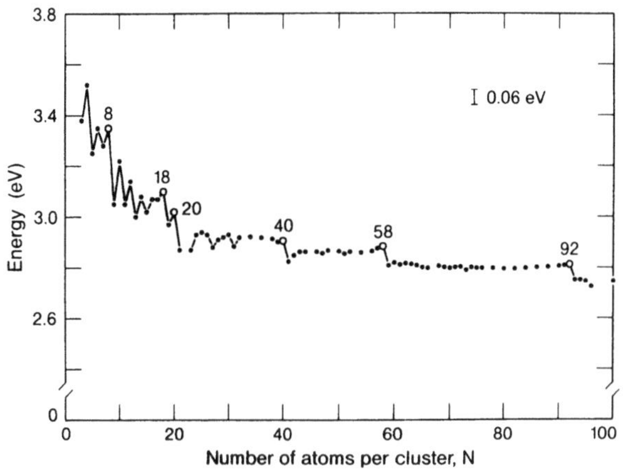

Fig. 4. Ionization potentials for potassium clusters, showing steps at shell closings. The bulk work function for potassium is 2.4 eV

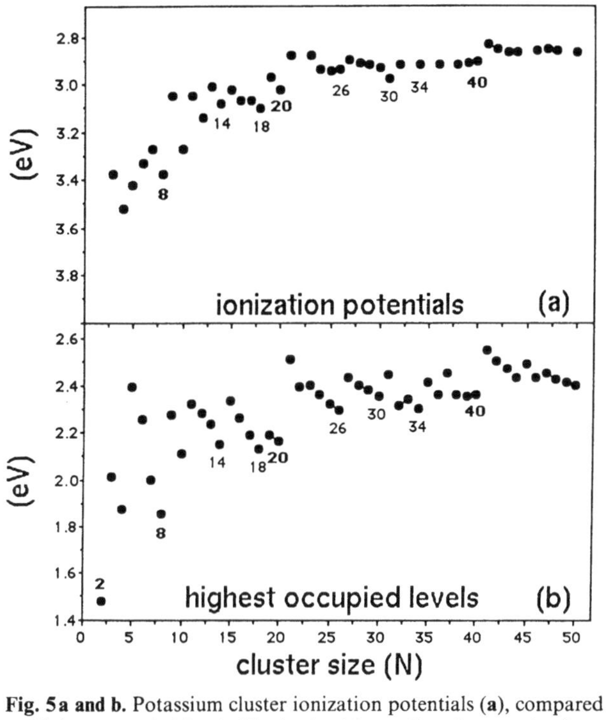

Fig. 5a and b. Potassium cluster ionization potentials (**a**), compared to highest occupied levels (**b**) obtained by scaling the energies from Fig. 1, using (8)

properly for crystal field splitting effects, the inclusion of which would reduce the energy gaps and related IP jumps for better agreement with experiment. It is probable furthermore that the exchange-correlation energy term, which makes an important contribution to the potential well depth, needs to be refined, as does the application of the local density approximation.

The spheroidal distortions which give rise to definite patterns in the abundance spectra, and in the polarizability parameter, are also reflected in the ionization potentials. Scaling the Nilsson diagram according to the relation

$$\hbar\omega_N \sim E_F/N^{1/3} \quad (8)$$

and assuming that the well depth is constant, the highest occupied levels for the clusters as found in the Nilsson diagram may be correlated with the measured ionization potentials. Figure 5a shows the experimental ionization potentials for potassium, to be compared with Fig. 5b, which gives the highest occupied levels derived from Fig. 1 for a series of potassium clusters. The odd-even alternations associated with the filling of the two-fold degenerate levels may be seen between 6 and 14, providing the energy for 12 is lowered. It is probable that the odd-even patterns arise not only from symmetry effects on the levels in the Nilsson diagram but from spin effects as well [30]. Finally, the fourfold sequences ending at 18, 26, 30 can be seen. Although the quantitative comparisons of the results in Fig. 5 are far from precise, the good correspondence of the patterns gives further strength to the role of symmetry, and the validity of the spheroidal model for predicting energy levels [31].

6. Conclusions

The main features of the alkali metal cluster abundance patterns, as well as the polarizabilities and ionization potentials, may be shown to result from the symmetry of the potential wells. The ground state energy levels are given to first order in the Nilsson diagram. Further development of the SCUBJ model is expected to give accurate values for the energies, and to account for further details in the abundances and the response properties of the alkali and other simple metals.

We thank the following colleagues for stimulating and productive conversations: Prof. Marvin L. Cohen, Dr. M.Y. Chou, Dr. Keith Clemenger, Dr. W. Ekardt, and Dr. J.L. Martins. This work has been supported by the Materials Research Division of the U.S. National Science Foundation under Grant No. DMR 84-17823.

References

1. Mayer, M., Jensen, J.: Elementary theory of nuclear shell structure. New York: Wiley 1955
2. Knight, W., Clemenger, K., Heer, W. de, Saunders, W., Chou, M., Cohen, M.L.: Phys. Rev. Lett. **52**, 2141 (1984)
3. Clemenger, K.: Phys. Rev. **B32**, 1359 (1985)
4. Clemenger, K.: PhD Thesis, University of California, Berkeley (1985)
5. Woods, R., Saxon, D.: Phys. Rev. **95**, 577 (1954)
6. Nilsson, S.: Kgl. Dan. Videns. Selsk. Mat.-Fys. Medd. **29**, No. 16 (1955)
7. Heer, W. de: (unpublished 1985)
8. Saunders, W.: (unpublished 1985)

9. Chou, M., Cleland, A., Cohen, M.L.: Solid State Commun. **52**, 645 (1984)
10. Lang, N.: Solid state physics. Vol. 28, p. 225. New York: Academic 1970
11. Mahen, G.: Many particle physics. Chap. 5. New York: Plenum 1981
12. See for example Harrison, W.: Pseudopotentials in the theory of metals. New York: Benjamin 1966; Cohen, M.L., Heine, V.: Solid state physics. Vol. 24, p. 37. New York: Academic 1970
13. Hybertsen, M., Louie, S.: Phys. Rev. Lett. **55**, 1418 (1985)
14. Knight, W., de Heer, W., Clemenger, K., Saunders, W.: Solid State Commun. **53**, 445 (1985); see also Kappes, M., Radi, P., Schär, and Schumacher, E.: Chem. Phys. Lett. **119**, 11 (1985)
15. Katakuse, I., Ichihara, I., Fujita, Y., Matsuo, T., Sakurai, T., Matsuda, H.: Int. J. Mass Spectrosc. Ion Proc. **67**, 229 (1985)
16. ibid, (in press, 1986). We are grateful to the authors for communicating their results prior to publication.
17. Devienne, F., Roustan, J.-C.: Organic Mass Spectrosc. **17**, 173 (1982)
18. Ekardt, W.: Surf. Sci. **152**, 180 (1985)
19. Snider, D., Sorbello, R.: Phys. Rev. **B28**, 5702 (1983)
20. Beck, D.: Phys. Rev. **B30**, 6935 (1984)
21. Puska, M., Nieminen, R., Manninen, M.: Phys. Rev. **B31**, 3487 (1985)
22. Knight, W., Clemenger, K., Heer, W. de, Saunders, W.: Phys. Rev. **B31**, 2539 (1985)
23. Martins, J., Car, R., Buttet, J.: Surf. Sci. **106**, 265 (1981)
24. Robbins, E., Leckenby, R., Willis, P.: Adv. Phys. **16**, 739 (1967)
25. Schumacher, E., Kappes, M., Marti, K., Radi, P., Schär, and Schmidhalter, B.: Ber. Bunsenges Phys. Chem. **88**, 220 (1984)
26. Saunders, W.: PhD Thesis, University of California, Berkeley (1986)
27. Beck, D.: Solid State Commun. **49**, 381 (1984)
28. Hermann, A., Schumacher, E., Wöste, L.: J. Chem. Phys. **68**, 2327 (1978)
29. Bréchignac, C., Cahuzac, Ph.: Chem. Phys. Lett. **117**, 477 (1985)
30. Snider, D., Sorbello, R.: Solid State Commun. **47**, 845 (1983)
31. de Heer, W., Knight, W., Chou, M., Cohen, M.L.: Solid state physics. New York: Academic (to be published 1986)

W.D. Knight
W.A. de Heer
W.A. Saunders
Department of Physics
University of California
Berkeley, CA 94720
USA

Shell Closings and Geometric Structure Effects. A Systematic Approach to the Interpretation of Abundance Distributions Observed in Photoionisation Mass Spectra for Alkali Cluster Beams

M.M. Kappes, P. Radi, M. Schär, C. Yeretzian, and E. Schumacher
Institut für Anorganische, Analytische und Physikalische Chemie, Universität Bern, Switzerland

Received May 2, 1986

Recent mass spectroscopic studies of continuous cluster beams resulting from supersonic expansions of alkali metal vapor have led to the postulation of islands of enhanced thermodynamic stability among the clusters produced. We discuss the various assumptions being made in converting ion abundances measured in these mass spectra into information about neutral stabilities. In this connection a number of experiments are described which allow insight into unimolecular dissociation of alkali cluster ions and into neutral cluster growth.

PACS: 35.20.V; 35.20.W; 36.40

I. Introduction

Theoretical interest in the electronic structure of metal clusters has been stimulated by recent measurements of broad band photoionisation mass spectra obtained for homonuclear alkali cluster beams which show multimodal cluster ion size distributions [1–4]. These ion distributions appear to be independent of the alkali metal studied and have also been observed in similar form for mixtures of alkali metals [3, 5]. Many indirect pieces of evidence suggest that the ion distributions measured for these beams approximate the actual neutral distributions present [4]. On this experimental basis there have been several attempts to explain particularly abundant neutrals in terms of higher relative thermodynamic stability. One moderately successful approach has been the jellium model which produces a set of occupation numbers for closed spherical electron shells which partly coincide with observed abundance maxima [2]. Initially applied to explain dominant abundance maxima in alkali beams, then to rationalize mass spectra obtained for other *s*-electron metals such as Cu, Ag and Au, the jellium model has recently found its way into interpretations of mass spectra measured for clusters of more complicated metals [6]. In a further development the *fine structure details* of alkali mass spectra have been interpreted in terms of a perturbed jellium approach – with limited success [7]. In view of this interpretative involvement, it is of interest to consider both the assumptions being made in the reduction of mass spectral data to neutral abundances as well as in the various theoretical approaches which have so far been proposed to deal with the observations. We will present several aspects of the problem together with new experimental data which casts some light on several of the assumptions being made.

II. Correspondence between Ion Abundances and Neutral Abundances

There are three factors which need to be taken into account before broad band photoionisation mass spectra can be mapped onto neutral abundances.

(*a*) Variations in active photon flux. Measurements on Na_x ($x \leq 65$) and K_x ($x \leq 39$) have demonstrated that ionisation potentials of simple metals can to first order be described in terms of classical electrostatics with I.P. = W.F. + $3/8\, e^2/R$ where W.F. is the bulk polycrystalline work function and R is the radius of a sphere with the same dimensions as an

x-atomic metal cluster [4]. Of course there are deviations from this simple dependence, particularly for very small clusters due to quantum size effects. However, in general, cluster I.P.'s decrease with increasing size. As a result irradiation of an alkali cluster beam with broad band UV light will give rise to cluster ion signals that are a function of "active" photon flux; i.e. that portion of the incident spectrum capable of ionising the cluster in question [8].

(*b*) Ionisation cross section and its wavelength dependence. We know very little about ionisation cross section in these species. It is likely that the cross section at threshold (where fragmentation effects are at a minimum) will to first order go as $n^{2/3}$ (i.e. scale with the geometric cross section). However, it is possible, particularly in the light of recent photodissociation experiments on sodium clusters which show strong size dependence, that there is a non monotonic development of threshold photoionisation cross section with particle size [9]. This is a problem which needs to be investigated further. For present purposes we assume a monotonic size dependence for ionisation cross section at threshold.

Beyond threshold, the experimentally determined photoionisation cross section for a specific particle size is a combination of direct ionisation and fragmentation terms. Fragmentation will be treated below. The direct ionisation cross section appears to decrease beyond threshold with greater effects being observed for larger clusters [8].

(*c*) Fragmentation. Many of the neutral "magic numbers" proposed for van der Waals clusters on the basis of electron impact ionisation mass spectra were in fact reflections of enhanced ion stabilities (due to excessive ionisation induced fragmentation) [10]. Unimolecular dissociation of alkali cluster ions *near* their appearance threshold is thought to be insignificant on the basis of a set of internally consistent ionisation potential determinations. We have recently investigated this problem further by measuring pairs of photoionisation efficiency (PIE) curves for Na_x generated by two types of expansion in which the resulting neutral composition differs. As we can determine neutral abundances unequivocally from threshold ion signals, it then becomes possible to separate direct ionisation and fragmentation contributions to the post threshold ion signals (it appears that in a typical fragmentation only one atom is lost – consistent with the energetics involved). This in turn makes it possible to gauge the validity of the assumed correspondence between cluster ion intensity and neutral abundance – simply by integrating this direct ionisation term over the extent of the PIE curve while correcting for fragmentation losses to smaller cluster ions. It turns out that fragmentation occurs for nascent cluster ions with as little as 0.5 eV excess energy above threshold [8]. However, the effects of unimolecular dissociation are only significant for those ions formed in part by ionisation induced fragmentation of particularly abundant neutrals (M_8, M_{20}, M_{40} ...).

Consideration of factors (*a*) and (*c*) prior to mapping ion abundances observed in broad band measurements onto associated precursor neutral distributions then requires deconvolution. The effect of this deconvolution is to *accentuate* the discontinuities inferred in the neutral distributions [8].

It is clear that the method of choice for the determination of relative neutral abundances involves ionisation at threshold with subsequent correction for active photon flux. At low count rates this is not feasible, consequently broad band measurements become necessary. This in turn requires careful deconvolution before neutral abundances can be deduced. Given the above measurements on Na_x, it is likely that for previous uncorrected broad band measurements, all conclusions concerning neutral stabilities based on $<30\%$ changes in ion signal betwenn M_n^+ and M_{n+1}^+ are suspect.

III. Neutral Cluster Distributions as Indications of Relative Thermodynamic Stabilities

Keeping in mind that we have not ruled out discontinuities in ionisation cross section, we conclude that the *dominant* abundance maxima observed in broad band photoionisation mass spectra obtained for alkali cluster beams (Li_x, Na_x, K_x, Li_xNa_y, NaK_x, and LiK_x) correspond to particularly abundant neutral clusters with the same composition (M_2, M_8, M_{20} ...).

We therefore need to establish a connection between observed enhancements and some kinetically/thermodynamically driven mechanism which leads to these neutral abundance maxima.

Cluster growth in supersonic expansions of pure metal vapor may occur either by aggregation (of metal atoms) or coalescence (with other clusters). Which of the two mechanisms prevails has not yet been established. It is likely that for either process collisional cooling is insufficient to fully remove all heat of condensation and therefore a growing cluster can be expected to go through numerous accretion/evaporation cycles – prior to reaching a collisionless regime [11]. Clearly the result of these processes is *not* an equilibrium distribution, but it seems plausible that relative differences in cluster stability or collision cross section may lead to multimodal neutral distributions via variations in sticking and fragmentation cross sections.

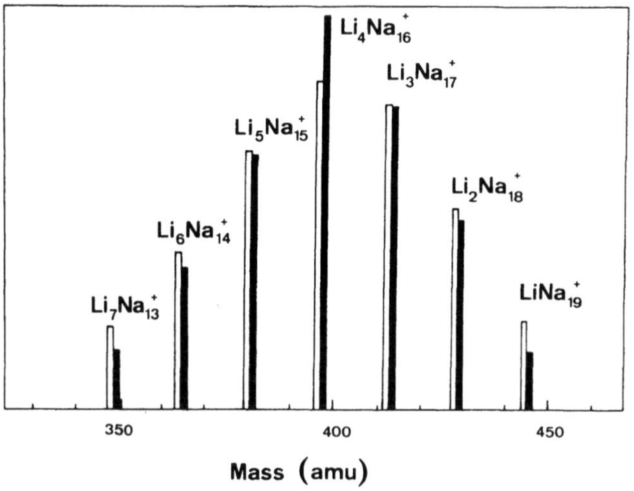

Fig. 1. Plotted (shaded) are representations of the integrated ion signals obtained for all 20 atom cluster ions in a photoionisation study of Li_xNa_y clusters. Experimental parameters are described in the text. Also shown (not shaded) is the result of a fit to the data of a binomial distribution in which we assume that Li and Na are chemically equivalent and that the only parameter determining the intensity of a particular $Li_xNa_y^+$ signal is the statistical weight of the two atoms. This analysis yields a Li to Na vapor pressure ratio of 0.25 which compares with a vapor pressure ratio of 6×10^{-3} present in the high temperature oven. The observed enrichment in lithium content can perhaps be explained in terms of a cluster production mechanism involving multiple growth/evaporation cycles

Directly probing the expansion zone as a means of determining the growth mechanism presents a number of technical difficulties. An alternative line of attack may involve studying mixed alkali cluster beams.

Figure 1 shows a stick plot deriving from a photoionisation mass spectrum obtained for an expansion of Li and Na metal vapors from an oven at a temperature of 770 C. Clusters were generated by expansion from a 0.6 mm throat diameter, 45 degree conical nozzle. The initial lithium mole fraction was 4.6%. As the oven utilized was one stage, the lithium to sodium ratio changed with time. However, at the time at which the data shown was obtained, the lithium mole fraction present within the oven had not changed appreciably. To minimize fragmentation, clusters were ionized near their appearance thresholds using 280 nm radiation from an arc lamp monochromator (5 mm slits, 1,200 lines/mm blazed at 300 nm) configuration. Plotted in Fig. 1 are the integrated ion signals obtained for all observed $Li_xNa_y^+$ having $x+y=20$. As expected from other alkali studies, the 20 atom cluster ion is dominant for each constant x series (e.g. $Li_3Na_{15}^+$, $Li_3Na_{16}^+$, $Li_3Na_{17}^+$ ...). There are of course other minor species (with different total atom numbers) present within the mass range shown. We have chosen not to include them in Fig. 1 for reasons of clarity. Figure 1 also contains the result of a fit to the data of a binomial distribution. We are assuming for the purpose of discussion that Li and Na are chemically equivalent and therefore that the only factor which determines the relative abundances of mixed cluster species is the partial pressure ratio of the respective atoms. Therefore the relative intensity of $Li_xNa_y^+$ is given by:

$$\binom{x+y}{x}\alpha^x(1-\alpha)^y \quad (x+y=\text{const})$$

where $\alpha/(1-\alpha)$ is the partial pressure ratio. The best fit to the data results in a Li/Na ratio of 0.25. How does this compare to the vapor pressure ratio present within the high temperature oven prior to expansion? The thermodynamics of mixtures of lithium and sodium vapors at elevated temperatures is not well known. Nevertheless it is possible to provide a reasonable guess of the actual pressure ratio. At the oven temperature used in these experiments, lithium and sodium are soluble in all proportions. If we assume the applicability of Raoults law – on the basis of the low (4.6%) lithium ratio present – we obtain a Li to Na pressure ratio of $3.1*10^{-4}$. Thermodynamic studies at lower temperatures indicate small positive deviation from ideal solution behaviour for Li/Na liquid alloys, implying destabilization due to an atomic size mismatch [12]. Consequently it is possible that Raoults law is no longer valid at 4.6% lithium. In this case we can obtain an upper limit for the lithium content of the expanding vapor by assuming complete immiscibility of the two liquids. Pure metal vapor pressures then being additive we obtain a Li to Na ratio of 0.006, which is still more than an order of magnitude *smaller* than the ratio obtained from analysis of the statistical distribution of 20 atom mixed clusters.

Can the lithium enrichment observed be understood in terms of the cluster production mechanisms discussed previously? The relative thermodynamic stabilities of Li_xNa_y clusters have not been determined. We know that $D_e(\text{LiNa})=0.8763$ and $D_e(Na_2)=0.7423$ eV [13]. Furthermore cohesive energies are 1.63 and 1.113 eV/atom for bulk lithium and sodium respectively [14]. Therefore it is clear that in a mixed lithium/sodium cluster, lithium will on average be bound more strongly than sodium. Dissipation of excess energy via fragmentation during cluster growth (with elimination of atoms) would lead to preferential loss of sodium. The observation of lithium enrichment in Li_xNa_y species is therefore consistent with a cluster production mechanism involving atom evaporative cooling. Given the enrichment factor observed, it is likely that the number of growth/

evaporation cycles involved in the generation of a 20 atom cluster is very large ($>10^3$). Correspondingly we would expect sticking coefficients to be quite small ($<10^{-3}$).

We have determined the extent of lithium enrichment in Li_xNa_y for all $x+y<23$. Above dimers the statistical ratio shows only minor (but significant) variations with cluster size. This suggests that the cluster distribution observed may be quite close to thermodynamic (rough equilibrium constant determinations provide reasonable bond energy differences). Variations in nozzle geometry have a significant influence on the statistical ratio observed (under otherwise identical expansion conditions). We will present a detailed account of these experiments in a future publication [15].

Interestingly, a similar photoionisation study has independently established sodium enrichment in Na_xK_y ($x+y<10$) generated by coexpansion of Na and K vapor from a cylindrical nozzle [16].

IV. Structure and Bonding in Alkali Clusters

In light of the above statements it is not unreasonable to assume that the dominant neutral abundance maxima observed in alkali cluster beams correspond to species with *enhanced relative thermodynamic stability*. Of the various theoretical approaches used to rationalize these conclusions, two have been particularly illuminating.

(a) Jellium Model. This treatment assumes that an alkali cluster can be considered as a uniform spherically symmetric positive potential well into which valence electrons are to be filled (one electron per atom). The quantum mechanics of this problem is basically that of a particle in a three dimensional box. Solution is straightforward and a series of partially degenerate energy levels results. The filling order of these levels (or shells) generates a set of numbers which can be mapped onto the atom numbers of observed dominant abundance maxima. Agreement of the two series is satisfactory but not perfect for sodium clusters and becomes less so for heavier alkalis [2, 17].

The model has several merits. It describes an effect in terms of a simple, familiar picture (shell models have been applied to stability islands in atomic nuclei [18], to various classes of chemical compounds [19] and also – before the advent of good data – to alkali clusters [20]). Furthermore the model suggests that in contrast to van der Waals clusters with localized bonds and optimized packing, metal cluster stability is determined by delocalized bonding interactions. In fact according to the model – in which outside parameters are atomic size and well depth – the *only* quantity determining relative thermodynamic stability is the total number of valence electrons present in the cluster.

Apart from moderate success as a number generating device [21], the jellium model fails to quantitatively predict any dynamic system response properties yet experimentally determined. For example, predicted and measured ionisation potentials are not in agreement [4, 22]. Experimentally determined polarisabilities are systematically higher than those predicted by the first order model [23]. Experiments in which alkali atoms are replaced by Mg, Ca, Sr, Ba, Zn, Eu and Yb, respectively, lead to dominant abundance maxima for heterocluster species which in many cases have the "wrong" number of valence electrons (e.g. K_8Mg instead of K_6Mg) [5, 24]. We conclude from these measurements that the jellium model is a good first approach to phenomena dependent on electronic structure. However, calculations with predictive power need to include a rigorous consideration of localized core structure – i.e. geometries.

There has been an attempt to introduce residual geometric structure into jellium calculations by means of ellipsoidal distortion of the spherical potential [7]. Reportedly, this reduces but still does not entirely eliminate the discrepancy between predicted ionisation potentials and their experimental values.

(b) Ab-Initio Calculations. There have been a significant number of higher level calculations on small alkali clusters. These have resulted in ground state equilibrium geometries and total energies for clusters with up to about nine atoms [25]. There have recently been calculations for Li_{13} and Li_{14}, in which a set of reasonable geometries were studied but in which geometry was not actually optimized [26]. Interestingly calculations on Li_x carried through to $x=9$ show a local stability maximum at Li_8 (bond energy per atom), consistent with experimental data. Ground state geometries are planar up to and including Li_6. Larger clusters have high symmetry more closely approximating spheres with increasing size. Calculations are reportedly under way for Li_x ($x<21$) [27].

In addition to the neutral ground state, many calculations also include the corresponding ionic ground state. Consequently it becomes possible to predict ionisation potentials to high precision. The numbers obtained (up to about M_9) are in very good agreement with experiment, confirming that core structure is important [28].

A problem common to these calculations is the neglect of zero point energy. Reportedly there are many local minima on the respective potential hypersurfaces which are only separated from one another by small amounts of energy. Barrier heights are typically not well characterized, but it is conceivable that in many of the clusters considered, inclusion of zero point energy might already be enough to surmount some of these barriers. We have very little experimental data concerning internal energy. It is not implausible that significant vibrational excitation exists for certain cluster sizes. Consequently we cannot rule out the possibility that many of the clusters studied are characterized by large amplitude motion on the time scale of the experiment. In this case, the assumption of spherical structure for *larger* clusters may be quite reasonable. In the absence of unequivocal experimental data, it is debatable to what lower limit of cluster size "sphericity" is tenable.

We thank the Swiss National Science Foundation for financial support of this work (grant No. 2.431.84).

References

1. Kappes, M., Kunz, R., Schumacher, E.: Chem. Phys. Lett. **91**, 413 (1982)
2. Knight, W., Clemenger, K., Heer, W. de, Saunders, W., Chou, M., Cohen, M.: Phys. Rev. Lett. **52**, 2141 (1984)
3. Knight, W., Heer, W. de, Clemenger, K., Saunders, W.: Solid State Commun. **53**, 445 (1985)
4. Kappes, M., Schär, M., Radi, P., Schumacher, E.: J. Chem. Phys. **84**, 1863 (1986)
5. Kappes, M., Schär, M., Radi, P., Schumacher, E.: Chem. Phys. Lett. **119**, 11 (1985)
6. Knight, W., Heer, A. de, Saunders, W.A.: Z. Phys. D – Atoms, Molecules and Clusters **3**, 109 (1986); Begemann, W., Dreihöfer, S., Meiwes-Broer, K., Lutz, H.O.: Z. Phys. D – Atoms, Molecules and Clusters **3**, 183 (1986)
7. Clemenger, K.: Phys. Rev. B **32**, 1359 (1985)
8. Kappes, M., Schär, M., Schumacher, E.: (to be published)
9. Kappes, M., Schär, M., Schumacher, E.: (to be published)
10. Mühlbach, J., Recknagel, E., Sattler, K.: Surf. Sci. **106**, 188 (1981)
11. Evaporation may of course continue in the collisionless regime of the cluster beam. Typical cluster flight times between source and detector are on the order of 500 μs. An upper limit for cluster internal temperature is about 500 C. Consequently one would not expect to lose more than about one atom per cluster by evaporation – after collisions have ceased
12. Down, M., Hubberstey, P., Pulham, R.: J.C.S. Dalton 1490 (1975)
13. Kappes, M., Schär, M., Schumacher, E.: J. Phys. Chem. **89**, 1499 (1985)
14. Kittel, C.: Introduction to solid state physics. New York: Wiley 1976
15. Kappes, M., Schumacher, E.: (to be published)
16. Bréchignac, C., Cahuzac, P.: Z. Phys. D – Atoms, Molecules and Clusters **3**, 121 (1986)
17. Chou, M., Cleland, A., Cohen, M.: Solid State Commun. **52**, 645 (1984)
18. Mayer-Kuckuk, T.: Physik der Atomkerne Stuttgart: Teubner 1970
19. Shell structure pictures have been prevalent in all theories of chemical bonding since Lewis, G.N.: J. Am. Chem. Soc. **32**, 762 (1916)
20. Herrmann, A., Schumacher, E., Wöste, L.: J. Chem. Phys. **68**, 2327 (1978)
21. An analogous number generating device applied to islands of enhanced stability in atomic nuclei was a great advance because interactive forces were poorly characterized. This is not the case for molecules containing alkali atoms, to which the full arsenal of quantum chemical methods can and should be applied
22. Saunders, W., Clemenger, K., Heer, W. de, Knight, W.: Phys. Rev. B **32**, 1366 (1985)
23. Knight, W., Clemenger, K., Heer, W. de, Saunders, W.: Phys. Rev. B **31**, 2539 (1985)
24. Kappes, M., Yeretzian, C., Schumacher, E.: (to be published)
25. Koutecký, J., Fantucci, P.: Chem. Rev. (in press)
26. Koutecký, J., Fantucci, P.: Z. Phys. D – Atoms, Molecules and Clusters **3**, 147 (1986)
27. Koutecký, J.: Personal communication
28. Martins, J., Buttet, J., Car, R.: Phys. Rev. Lett. **53**, 655 (1984); Phys. Rev. B **31**, 1804 (1985)

M.M. Kappes
P. Radi
M. Schär
C. Yeretzian
E. Schumacher
Institut Anorganische, Analytische und Physikalische Chemie
Universität Bern
CH-3012 Bern
Switzerland

Evolution of Photoionization Spectra of Metal Clusters as a Function of Size

C. Bréchignac and Ph. Cahuzac
Laboratoire Aimé-Cotton, CNRS II, Orsay, France

Received April 7, 1986; final version May 7, 1986

Electronic structure effect in small metallic clusters up to $n=15$ are investigated through three series of experiments performed on one-valence-electron-atom clusters and two-valence-electron ones. In these experiments the cluster molecular beam is probed by photoionization mass spectroscopy, either by using a tunable laser source for alkali clusters or by synchrotron radiation for mercury ones. With alkali clusters the results are related to fragmentation effects, ionization potential measurements and photoionization efficiency curve profile analysis in the threshold region. The similar behavior of the homogeneous and heterogeneous clusters and the comparison with theoretical models suggest that for $n \geqq 3$ the valence electrons are partially delocalized. This similarity against the electronic structure is not found in the nucleation process which generates homogeneous and heterogeneous clusters with a strong difference in their respective abundances. For mercury clusters the evolution with size for excitation spectra of two autoionizing lines is obtained up to $n=8$. Results show that they do not have a metallic character. This is also supported by the observation of small doubly charged mercury clusters for $n \geqq 5$ which are stable against the Coulomb explosion.

PACS: 36.40

Introduction

As opposed to organic clusters which opened up the channel of macromolecule-chemistry, the inorganic clusters have been studied more recently. These aggregates, from a few to a about 10^4 atoms or molecules constitute a specific domain which differs both from the gas phase and from macroscopic condensed matter. In addition, how fast the properties in the clusters approach the bulk value may depend on the property being investigated. For example, as the metal cluster size increases the metallic nature of the bonds appears before the lattice properties. The geometrical structures of metal clusters are certainly, in many cases, different from the bulk and it is useful to understand the link between the equilibrium geometry and its electronic structure. The general trends differ according to the nature of the metallic element. For alkali or noble clusters, built up with atoms of group I, the equilibrium distance of the dimer is smaller than the nearest-neighbour distance in the corresponding metal. On the opposite for clusters built up with atoms of group II the reverse evolution prevails: the shrinkage of the nearest-neighbour distance has to take place when going from the dimer, bound with a weak Van der Waals interaction, to the bulk with a metallic bond [1–2].

Recent advances in experimental techniques (supersonic beam, laser vaporisation ...) now make possible to generate clusters of any element. Since the species are isolated and cold they are suitable for studies of unsupported clusters. Although no direct experimental results are available yet on geometrical structures, some results have been recently obtained for electronic structures [3–9]. The experimental methods which involve photoionization spectroscopy

* Laboratoire associé à l'Université de Paris-Sud

allow both the probing of valence electrons and the ionization of the clusters which is necessary for mass spectroscopy.

On the other hand any calculation of physical properties of microclusters must either assume a reasonable geometrical structure or try to predict the equilibrium geometry from the minimization of the calculated total energy. A dominant part of the theoretical activities deals with quantum chemistry [10–11], but also other local methods originally developed in solid state physics (electronic density fonctional) are transferred to the clusters [12–14]. It is clear that both theory and experiment are necessary to understand the link between the equilibrium geometry and electronic structure.

In this paper are emphasized recent results that we obtained in the search of the stability and electronic structure for alkali clusters K_n and NaK_{n-1} and for mercury clusters Hg_n. For these two kinds of clusters, containing either atoms with one valence-electron or atoms with two-valence-electrons, electronic behavior are quite different for small clusters and can be used to understand how the localized electrons in an atom transform to delocalized electrons in metal.

Experimental Set-Up

The basic elements of the set-up are as follows. The cluster beam is formed in the adiabatic expansion of the metallic vapor through a small diameter nozzle. The stagnation pressure is of the order of few hundred torr for alkali atom clusters and two atmospheres for mercury ones. Higher pressure is necessary for the observation of the Hg_n due to the weakness of the Van der Waals bond. Once formed the clusters enter the ionizing region of a mass-spectrometer. In order to minimize the fragmentation processes we used a one-photon ionization in the near threshold region. For alkali-atom clusters this requires a photon energy tunable in the 3–4 eV range. The source includes a Nd-YAG pumped dye laser (15 Hz repetition rate, 15 ns pulse duration). The dye laser output is frequency doubled or mixed with the fundamental frequency of the pump laser. This provides the tunable U.V. line in the appropriate domain. The photoions are then selected by using a time-of-flight (TOF) mass spectrometer in connection with a synchronous gate detection. For mercury-atom clusters more than 10 eV are necessary for one-photon ionization. Therefore we used the facilities of the synchrotron radiation available at Orsay (LURE). With such a source having a high repetition rate we used a quadrupolar mass-spectrometer which is more suitable in this case, in connection with a ion-counting system. Whatever the mass analysis is, the ion signal is recorded at a given mass while the photon energy is varied.

Let us mention here some advantages offered by a time of flight mass spectrometer. A simple three-grid-system (Fig. 1) allows time and space focusing and provides high mass-resolution. For example isotopic components of alkali-clusters are easily resolved (Fig. 6, inset a). Such a high resolution alone without either loss of intensity or mass limitation makes this quite an attractive design. Even more it makes possible the study of fragmentation processes by measuring the kinetic energy of the photoions [15]. When fragmentation occurs during the ionization, fragment-ion-peaks appear at the daughter ion mass. They can be detected only as a broadening in the corresponding time-profile due to the small change in the kinetic energy of the fragment ions. The shape of the signal reflects the anisotropy of the fragmentation process. In the simple case involving only one dissociative channel, the dissociation energy can be deduced from the time-broadening value [16]. When fragmentation occurs within the drift-region, the velocity of the fragment-ions corresponds to that of

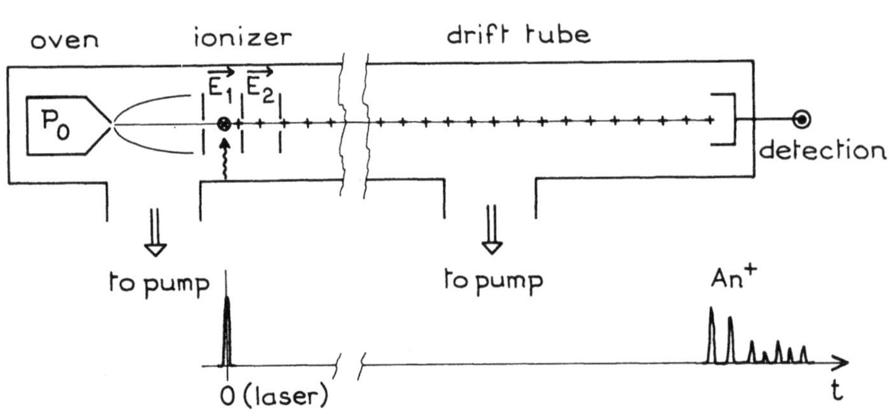

Fig. 1. Experimental set-up used in the study of alkali clusters. Ionization is provided by a pulsed and tunable UV line (wavy arrow) focused in a low electrostatic field region ($\mathbf{E}_1$) whereas the second field $\mathbf{E}_2 \gg \mathbf{E}_1$ provides the main acceleration of the photoions. The drift tube length is 1.70 m. In typical conditions the total accelerating voltage is 2 KV and the time-of-flight is 18 μs for K^+

the parent ion from which it originates, hence the broadening occurs on the parent ion. When fragmentation takes place in the transient region, within the accelerating grids, the fragment-ions give rise to spikes distributed over a large time-interval and shifted toward the larger mass of the parent ion. The versatility of this conventional system can be improved by using retarding field techniques or a reflectron-type TOF mass spectrometer. A connection with a multichannel transient digitizer offers to benefit from the "multiplex" effect.

Results and Discussion

Before drawing any conclusion from mass spectrum pattern and appearance potential of ions it is important to realize how each ion-peak-intensity I_n^+ is obtained. The abundance of cluster ion A_n^+ depends on the abundances a_{n+p} of neutral clusters A_{n+p} $(p \geqq 0)$, on the photoionization cross-section $\sigma(n+p, \nu)$ of A_{n+p} and on the probabilities of fragmentation $\rho_{n,p}$ (ν, T) for the processes $A_{n+p} + h\nu \rightarrow A_n^+ + A_p$.

$$I_n^+(\nu) = \sum_p a_{n+p}\, \sigma(n+p, \nu)\, \rho_{n,p}(\nu, T) - \sum_{p<n} a_n\, \sigma(n, \nu)\, \rho_{p,n-p}(\nu, T).$$

The first factors a_n depend on the nucleation conditions i.e. stagnation pressure and temperature which are related to each other in the case of metallic vapors. The last two factors $\sigma(n, \nu)$ and $\rho_{n,p}(\nu, T)$ depend on both ionization and dissociation thresholds.

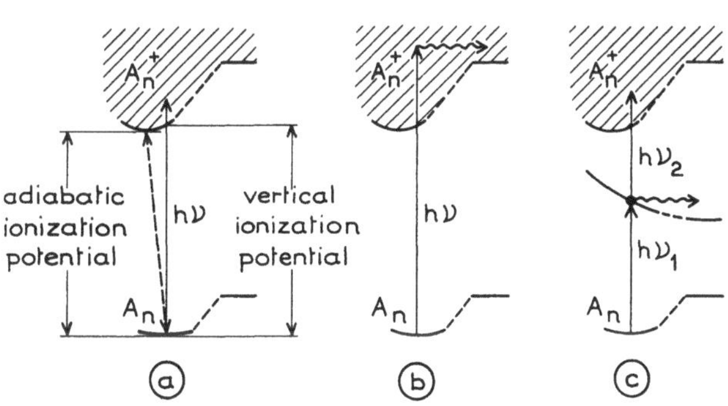

Fig. 2. Three typical situations for photoionization of clusters. The one-photoionization near the threshold – case a – minimizes the fragmentation probability provided that A_n^+ is stable against $A_p^+ + A_{n-p}$. When the photon energy $h\nu$ reaches the region of the ion dissociation limit – case b – fragmentation occurs and the rate depends on the height of an eventual dissociation barrier. In the case of two-photon ionization – case c – the probability is high to reach a dissociative level in the neutral cluster. This leads to a competition between the two-photon ionization and fragmentation paths [16]. It may be remarked that the ionization is always a vertical process. In most cases the measured appearance potential has a larger value than the adiabatic one

The probability of fragmentation is also a function of cluster-temperature.

At a given ionization-photon frequency, nucleation processes can be investigated by varying stagnation pressure, whereas stabilities and electronic structure of clusters can be studied by varying the laser frequency for a given stagnation pressure. In Fig. 2 are shown three typical situations for one or two-photon-ionization. In the case a the cluster is photoionized in the near threshold region, the probability of dissociation is very weak and $I_n^+ = a_n\, \sigma(n, \nu)$. In the case b the energy of photoionization is above the dissociation-limit of the parent ion and fragmentation processes can lower the ion signal. In the case c two-photons are necessary to photoionize the cluster and fragmentation can occur after the first step of excitation in the neutral [16–17]. In this last case competition between dissociation and ionization can takes place depending on the ionizing laser intensity.

When a beam of clusters is photoionized, a given cluster is in a situation a, b, or c depending on the value of ionization potential as compared to the photon energy. Mass spectra obtained in various conditions are necessary to obtain reliable informations on stabilities and ionization potential of clusters.

A. Photoionization Spectroscopy of Alkali Clusters

1. Stability of Potassium Clusters. In Fig. 3 are shown three mass spectra of potassium clusters obtained in the same stagnation pressure conditions for different laser wavelengths and for different laser flux density.

At $\lambda_1 = 392$ nm the energy of the photon $\sigma_1 = 25{,}500\ \mathrm{cm}^{-1}$ is lower than the appearance potential of light ions K_3^+ to K_8^+ and higher than the appearance potential of ions K_n^+ with $n > 8$. At low laser flux density the lack of light ions in mass spectrum indicates that fragmentation processes do not affect seriously ion peak intensities in the near-threshold-ionization region. This situation corresponds to case a of Fig. 2. For the same laser wavelength but by increasing the laser flux density we studied the influence of multiphoton ionization processes on the mass spectrum. Ion peaks appear for low masses $(n \leqq 8)$ and their intensities increase faster than the laser intensity whereas for large masses the mass spectrum pattern is not too much affected by varying the laser intensity. This last point is not surprising since for these clusters $(n > 8)$ the photon energy is above the ionization energy (case a of Fig. 2) so the one-photon ionization is predominant against the two photon process which involves a bound-free transition in the ion. We have to notice that for low masses, which are in a situation corresponding to case c of Fig. 2,

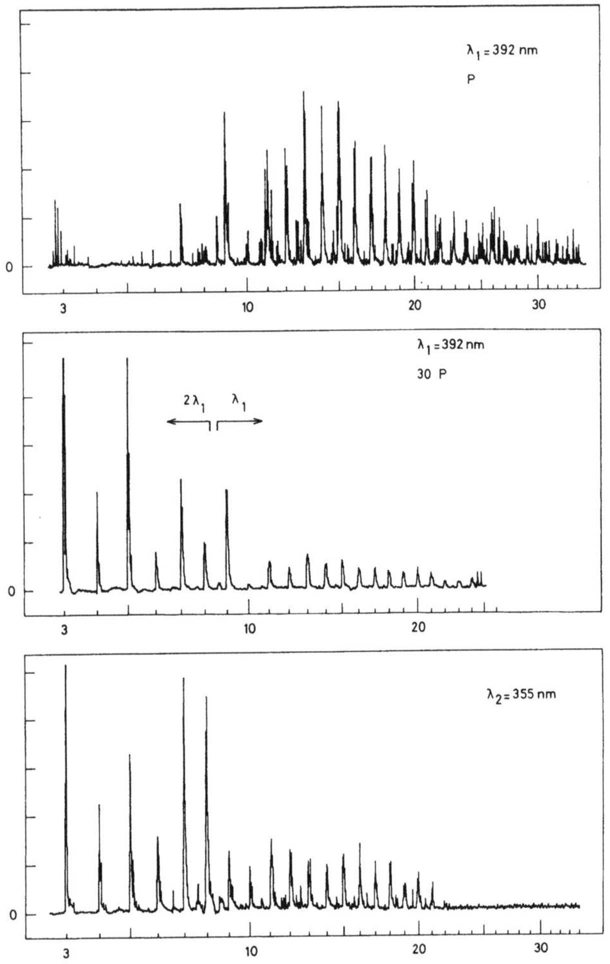

Fig. 3. Evolution of mass spectrum of potassium clusters with laser flux density P and ionizing photon energy. These spectra have been obtained at a stagnation pressure $P_0 = 480$ torr (720 °C). The small peaks visible between the main K_n^+ peaks correspond to the NaK_{n-1}^+ species, the sodium being present as impurity in potassium metal. The comparison between the spectra at $\lambda_1 = 392$ nm for two different values of the laser flux shows the effect of the two-photon ionization which manifests itself as a shift in the mass spectrum intensities toward low masses. Also the low masses predominate in the spectrum obtained at $\lambda_2 = 355$ nm despite the difference with regard to the ionization situation

the two-photon processes preferentially produce odd-ionized clusters. In this case two-photon ionization passes through Rydberg states and the ion signals result from the competition between ionization and dissociation of these intermediate levels. Assuming that the potential hypersurfaces of Rydberg states are almost parallel with that of the ground state of corresponding ion, the relative two-photon ion signals reflect the relative stabilities of K_n^+ [18]. The observed stabilities which are larger for K_{2n+1}^+ than for K_{2n}^+ are in agreement with Martins's et al. calculations for sodium [13]. These stabilities can be easily understood in therm of monoelectronic model in which the alkali clusters having an even number of electrons are more stable than the clusters having an odd number of electrons and then present an antibonding molecular orbital.

At $\lambda_2 = 355$ nm the ionizing energy $\sigma_2 = 28{,}169$ cm^{-1} is just above the ionization threshold of small clusters K_n with $3 \leqq n \leqq 8$. The lack of potassium cluster ions above K_{21}^+ indicates that dissociation occurs during ionization. These large clusters ($n > 20$) are in a situation which is relevant to case b of Fig. 2. For small masses ($n \leqq 8$) the ion peak intensities are strongly dependant on the photoionization cross sections. The odd-even alternation that we found in ion peak intensities is a consequence of the odd-even alternation that we observed in photoionization efficiency curves [8].

Note that for all these spectra the ion peaks remained localized. This points out that if fragmentation exists it occurs either during ionization (< 10 ns) or a long time after ionization, during the time of flight (> 1 μs). When an I.R. photon ($\lambda = 1.06$ μm) is added the situation is different: spikes appear randomly distributed in time which is an indication for fragmentation taking place in the accelerating region.

2. Comparison between Ionization Potential Values of K_n *and* NaK_{n-1}. To obtain reliable informations on electronic structure of alkali clusters from photoionization measurements it is essential to get rid of fragmentation processes which mix the photoionization contributions of daughter and parent ions. In order to minimize fragmentation processes we recorded photoion signals against photon energy at low laser flux density and in the near threshold ionization-region. We have checked a linear dependence of the ion signals against the laser flux density in the whole frequency range [8]. Under these conditions we compared the results obtained for homogeneous K_n clusters and heterogeneous NaK_{n-1} in order to obtain deeper insight into the link between geometrical and electronic structures, since sodium and potassium are isoelectronic but the sodium atom may introduce an asymmetry in the potassium cluster.

The results are shown in Fig. 4. The most remarkable feature is the similarity in the behavior of photoionization profiles of homogeneous and heterogeneous clusters. Both display a linear behavior versus photon energy for odd-numbered-clusters while the even-numbered-cluster-yields show a more pronounced curvature. This regular odd-even alternation in the shapes of the yield-curves, whatever their compounds are, indicates a general distinction between clusters with paired-electrons and those with an unpaired-electron. If we assume that photoionization efficiency curves are related to geometrical structures

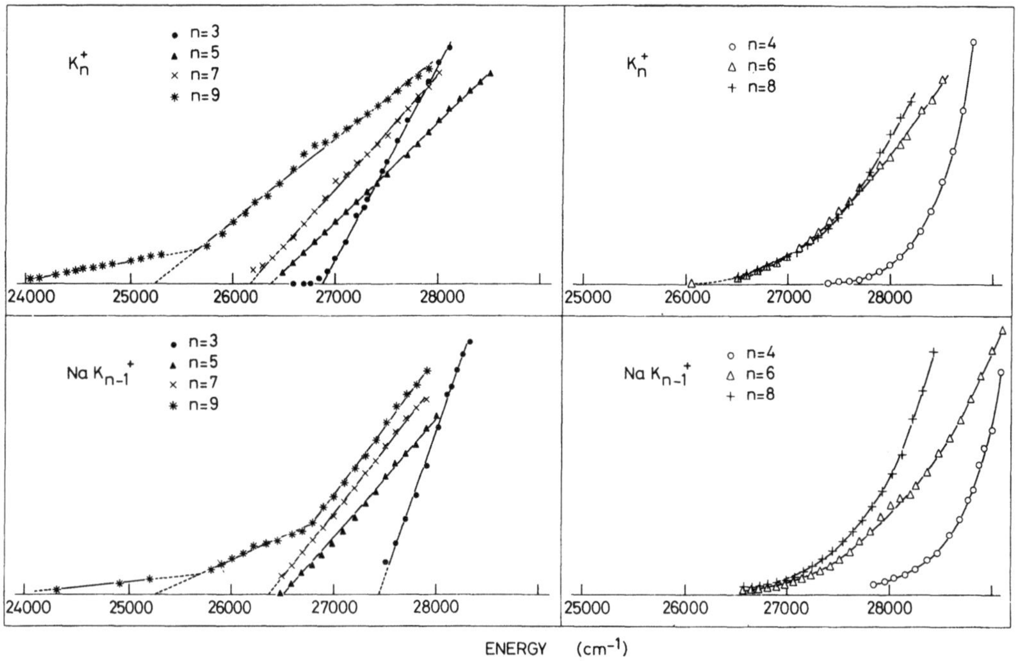

Fig. 4. Photoionization efficiency curves of homogeneous and heterogeneous K_n, NaK_{n-1} clusters until $n=9$ at $P=5.10^{-3}$ mJ. The 24,000–29,000 cm^{-1} tuning range has been obtained by using four different dyes. The relative ion intensities are not to scale. Notice the peculiar behavior at $n=9$ for which two extrapolations are possible

via Franck-Condon factors [19], this similarity suggests that the geometrical structures of K_n and NaK_{n-1} should be comparable. However the recent calculations on sodium clusters [13], which show the existence of multiple minima, separated by potential barriers in the potential surfaces, make difficult the interpretation of the photoionization curves.

From the slowly rising shapes of photoionization efficiency curves, the adiabatic ionization potential cannot be accurately deduced. For odd-numbered potassium clusters significant values of appearance potential can be deduced from linear extrapolation. These values are in agreement with the values obtained by E. Schumacher et al. [3, 20] and W. Knight et al. [4]. On the other hand, for even-numbered clusters the strong curvature of the yield curves makes more difficult the determination of the appearance ionization potential. We have systematically plotted the values corresponding to the beginning of these curves. This may explain the difference between our values and the values given after deconvolution of lamp photoionization profiles [3, 4, 20]. In Fig. 5 are plotted the appearance potential values for K_n and NaK_{n-1} with $n=1$ to 9, 11, 15 and for Na_2K_5 as an example of a cluster in which two potassium atoms are substituted with sodium atoms. For $n=9$ a red wing occurs at high temperature in the photoionization profile. Under these conditions fragmentation

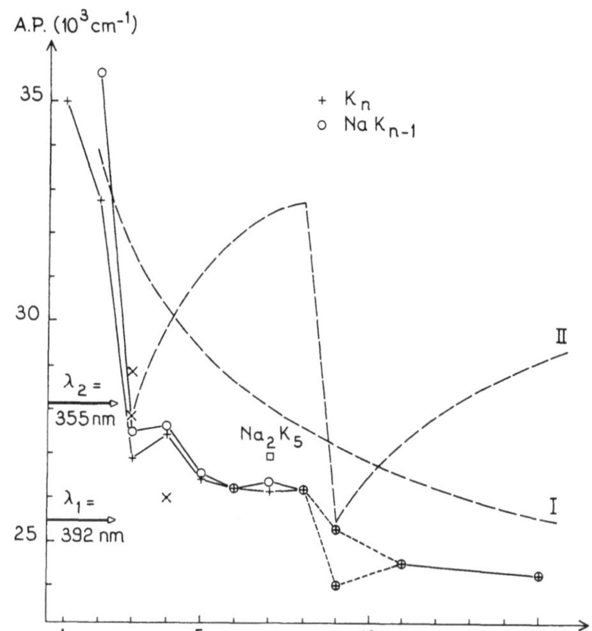

Fig. 5. Measured appearance ionization potentials (A.P.) of K_n, NaK_{n-1} and Na_2K_5 clusters. The dashed curves correspond to the result of a simple electrostatic model for the work function of a metallic droplet (curve I) and to the ionization potential calculated for a jellium sphere (curve II). The crosses indicate the calculated values taken from Ref. 24 for the upper value and from Ref. 25 for the lower one at $n=3$ and from Ref. 10 at $n=4$

of high masses $n \geqq 11$ cannot be totaly ruled out. So two extrapolated values are proposed. We have to notice that in lamp photoionization experiment the presence of a low-energy tail is observed in the same extending energy range [4]. The remarkable feature of Fig. 5 is that both homogeneous and heterogeneous clusters display the same behavior as the cluster size increases. However a significant blue shift up to $n=7$ is in evidence when a potassium atom is substituted with a sodium one in a given numbered cluster. This blue shift results partially from an increase of the stability of NaK_{n-1} as compared to K_n. The more the sodium is dilute in potassium cluster the less the shift is. The general evolution of the appearance potentials of both NaK_{n-1}^+ and K_n^+ as a function of cluster size presents a general decrease which drops for $n=3$ and $n=9$. This could be understood in term of electronic shell structure in one electron shell model [14, 21, 22]. In Fig. 5 we compare our experimental results to the size dependance calculation of the work function of a small metal sphere [23]

$$W(R)=W_{\infty}+\frac{5.4}{r_s\, n^{1/3}},$$

where r_s is the radius of Wigner -Seitz cell in the bulk (2.43 Å) and W_{∞} the work function of the planar metal (2.22 eV), as well as to the ionization potential values calculated for jellium spheres taking into account the exchange-correlation-energies [22]. We can find some correlation between our results and these two simplified models which is in favor of metallic character of small alkali clusters. However the quantitative disagreement between experiment and these models is not surprising since atomic arrangement plays a role in small alkali clusters. To go further we have compared our results with more refined calculations taking into account geometrical structure using either configuration-interaction ab-initio calculations [10] or pseudopotential and local spin density approximation [24, 25].

B. Heterogeneous Nucleation in Mixed Clusters

The stability properties of clusters as well as the kinetics of cluster formation are governed by the interaction potential. This correlation between the interaction potential and the formation processes can be reached by comparing the relative peak intensities of homogeneous and mixed clusters obtained under the same conditions for various concentrations. Only a few experiments on mixed clusters have been reported earlier for example: [26–29]. We have done an experiment on alloy clusters of sodium and potassium. As it has been seen these two species have equivalent electronic structure but slightly different stabilities.

The mixed clusters were formed as follows. A liquid mixture of sodium and potassium was prepared and placed in the same oven. The two components were evaporated congruently forming a mixed vapor. A disadvantage of this simple method is that the thermodynamics of liquid alloy evaporation is not obvious. The partial vapor pressures of NaK liquid alloy are rather close to the ideal solution values only for small fractional abundance of one of the components. Therefore in our experiments we used either a small fractional abundance of sodium in potassium or a small fractional abundance of potassium in sodium. Our studies were done by following the evolution of the mass spectrum pattern with binary composition at a given ionizing photon energy, chosen to minimize the fragmentation processes. First results put into evidence that for low sodium concentration mixed clusters are more probable than expected from statistical considerations, whereas for low potassium concentration homogeneous sodium clusters dominate the mass spectrum more than expected. Similar results have been recently obtained in a Li-Na mixture [30]. As an example in Fig. 6 is shown a mass spectrum of Na_pK_{n-p} clusters produced from a 8% atomic fraction of sodium.

It is easy to calculate the probability that a cluster containing n atoms has the composition A_pB_{n-p} if it is assumed that A and B are chemically equivalent. At each step of the growth either one A atom can be added with the probability α or one B atom with the probability $1-\alpha$. For n steps the probability is $C_n^p\, \alpha^p(1-\alpha)^{n-p}$, where C_n^p is the binomial coefficient. Isotopes offer a typical case of chemically equivalent elements. In the inset a of Fig. 6 is displayed the isotopic composition of K_8. The calculated binomial distribution indicated by the vertical lines assuming for α the natural fractionnal abundance of ^{41}K ($\alpha=0.07$) fits very well the isotopic experimental distribution. The inset b of Fig. 6 shows experimental intensities of mixed Na_pK_{n-p} clusters for $n=8$ obtained from a 8% atomic fraction of sodium in the initial liquid sodium-potassium-alloy. Using α as a parameter to give the best fit to the observed distribution we found the value $\alpha_{exp}=0.30$. If the sodium and the potassium atoms play equivalent roles in the nucleation processes, the probability α could be set equal to the ratio of the density of sodium to the total density in the vapor, which is equivalent to the ratio of the partial pressure of sodium to the total pressure. The equilibrium partial vapor pressures over liquid binary alloys is given by $p_A=p_A^0\, a_A$ where p_A^0 is the vapor pressure of component A in its pure state and a_A is the activity of component A in liquid

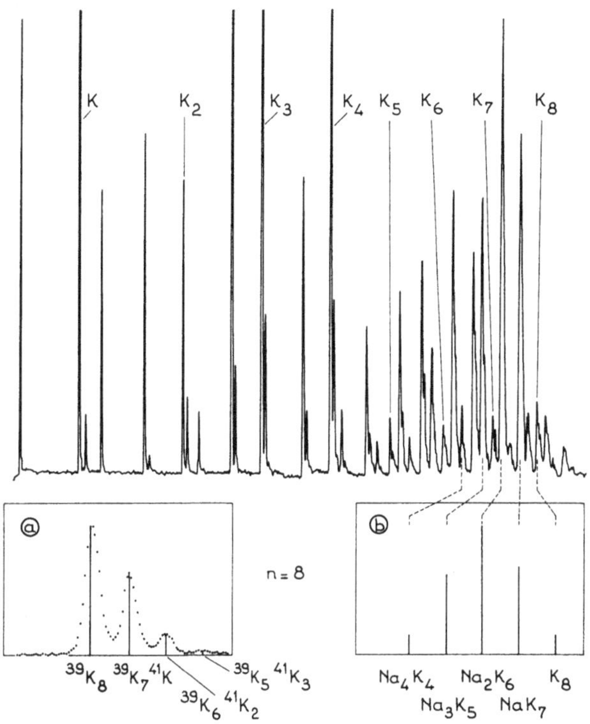

Fig. 6. Mass spectrum of K_n and Na_pK_{n-p} clusters with 8% atomic fraction of Na obtained at $P_0 = 290$ torr (675 °C) with the $\lambda = 337$ nm N_2 radiation. In the inset *a* is displayed the isotopic structure of K_8 as obtained by using a fast multichannel transient digitizer [35]. The inset *b* displays the observed components of the $n = 8$ mixed clusters. The relative intensities of the Na_3K_5 and Na_4K_4 components are probably lowered because their situation against the ionization process is closer to the case b of the Fig. 2

alloy. For NaK these activities are not known in the whole temperature range [31]. However with a fairly good approximation we obtained $\alpha = 0.05$ in our conditions. The discrepancy between this value and that derived from the mass spectrum, which exists for all cluster sizes, indicates a tendancy for sodium atoms to be easily dissolved in potassium clusters. It is surprising that the small difference between the binding energies of Na_pK_{n-p} and K_n can be responsible for such a large amount of mixed clusters. However the facility for a potassium atom to be substituted with a sodium atom does not depend only on the total binding energy but also on geometrical parameters (relative radii of Na and K atoms, equilibrium distances ...). In fact for atoms of unequal size a single small atom is more easily built in a cluster than a single large one. It should be interesting to know if sodium atoms take place in central positions in potassium clusters in order to maximize the number of Na – K bonds. It should finally be noted that the discrepancy between the composition of mixed clusters and a simple random mixing model is much more pronounced for mixed alkali clusters than for mixed alkali halide clusters [27], and was not found in the case of Ni – Cr and Ni – Al alloys [26].

C. Photoionization Spectroscopy of Mercury Clusters

As we have just seen a physically measurable quantity which can explore the electronic structure of clusters is the ionization potential. However for aggregates which are formed with atoms of group II the adia-

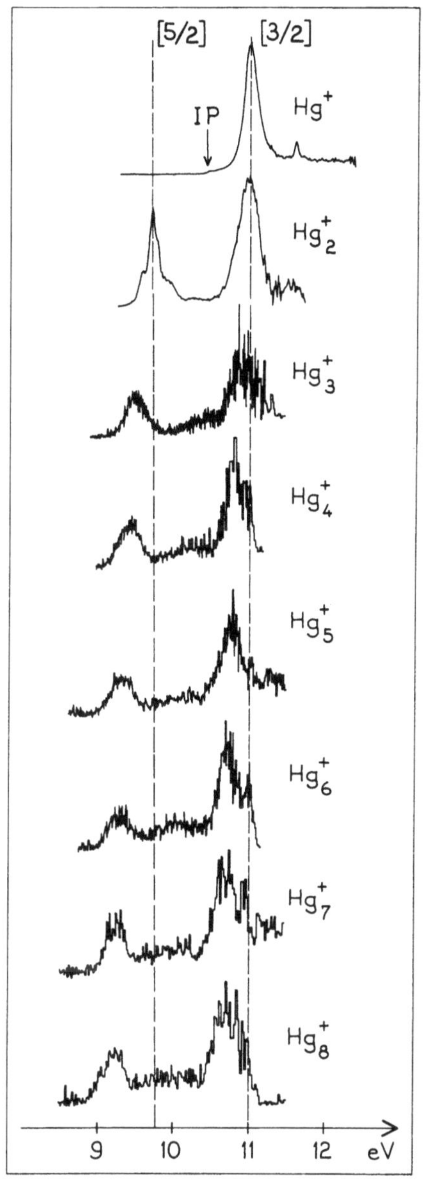

Fig. 7. Excitation spectra of autoionizing lines in mercury clusters correlated to the atomic transitions $5d^{10}\ 6s^2\ {}^1S_0 \rightarrow (5d^9\ 6s^2\ {}^2D_{5/2,\ 3/2})$ 6p as observed by using the synchrotron radiation. I.P. corresponds to the atomic ionization potential at 10.43 eV. Notice that in the atomic spectrum only the ${}^2D_{3/2}$ line is situated above the ionization potential

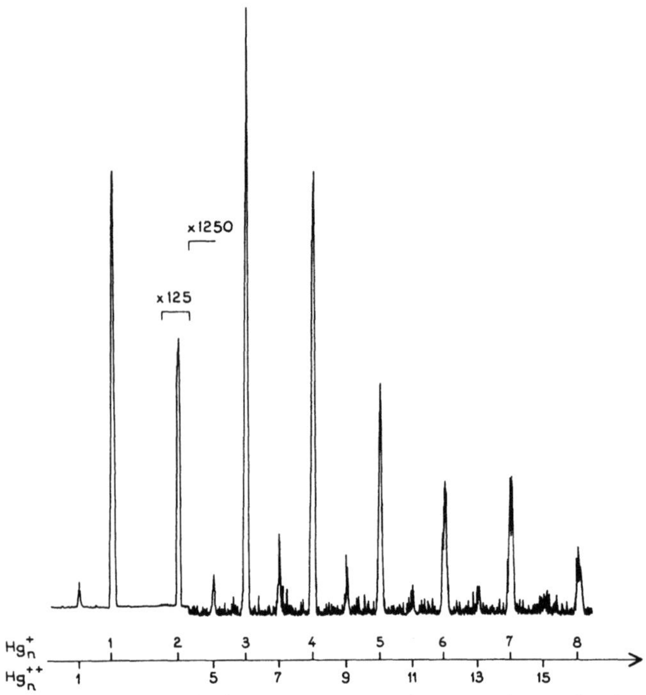

Fig. 8. Photoionization mass spectrum of mercury clusters showing odd-numbered doubly charged species for $n \geqq 5$. The even numbered ones cannot be discriminated from the single ionized clusters. In this experiment the spectrum is limited to Hg_8^+ and Hg_{15}^{++} due to the limited mass range of the quadrupole mass spectrometer

batic ionization potential is difficult to reach. In the neutral ground state, each atom essentially keeps an (s^2) configuration leading to a cluster which is essentially bound by Van der Waals forces. In the positive ion the unpaired s electron (or the positive hole delocalized on the different sites of the cluster) induces a strong binding energy. It follows that the minima of the two potential hypersurfaces correspond to different geometries and the Franck-Condon factors are unfavorable for vertical ionization. This problem can be overcome via spectroscopic studies of autoionizing lines which offer the two advantages of probing the electronic structure and of ionizing the clusters for mass spectrometry. In Fig. 7 is shown the excitation spectra of autoionization lines correlated to the atomic transitions $5d^{10}\,6s^2\,{}^1S_0 \rightarrow (5d^9\,6s^2\,{}^2D_{5/2,\,3/2})\,6p$. These autoionization peaks are observed up to $n=8$ with comparable widths mainly due to autoionization broadening. Moreover with increasing cluster size beyond the dimer, the two autoionization lines appear to be red-shifted with respect to the corresponding atomic transitions and their corresponding energies are inversely proportional to n [9]. This functional dependance can be compared with the dependence of ionization potential for a regular chain or cycle of n atoms which decrease as n^{-1} [2]. In our case as n increases, mercury clusters become more compact; increasing the number of nearest neighbours generates a frequency shift for the atomic transition analogous to a pressure effect in gas phase. The existence of these autoionization lines which are not as broad as one would expect for delocalized valence electrons and their shifts which do not converge to the bulk values [32] show that the excited $6p$ electron still remains localized. These two results concerning both the shifts and the widths of the line show that the small mercury clusters do not yet have a metallic character.

This conclusion is also supported by the observation of small doubly charged Hg_n^{++} clusters for $n \geqq 5$ up to $n=15$ in a one photon double ionization experiment (Fig. 8). We have shown that their surprising stabilities against the Coulomb explosion, invoked to explain the absence of doubly charged species in lead cluster experiments [33], may be explained by the polarisation effects which balance the Coulomb explosion [34].

Conclusion

As we have seen the photoionization experiments can be used as a tool to explore electronic structure of clusters. The simplest examples of metal are the alkalis. Built up with atoms having one-valence electron, the alkali clusters present a monotonic evolution toward the metal as predicted by very simple models. However for the very small clusters the link between geometrical and electronic structure cannot be neglected and it manifests itself as a deviation from this regular trend. With two electrons per primitive cell mercury could in principle be insulator. The metallic character of the bulk is due to the overlap between the empty p and the filled s band. Our results show that the small mercury clusters have not yet a metallic character and the question of the existence of a transition to a metallic phase at a given size is open.

References

1. Huber, K., Herzberg, G.: Molecular spectra and molecular structure. Vol. IV. New York: Van Nostrand 1979
2. Ashcroft, N., Mermin, D.: Solid State physics. Vol. 391. Philadelphia, Tokyo: Holt-Saunders international editions 1981
3. Kappes, M., Schär, M., Radi, P., Schumacher, E.: J. Chem. Phys. **84**, 1863 (1985)
4. Saunders, W., Clemenger, K., De Heer, W., Knight, W.: Phys. Rev. B **32**, 1366 (1985)
5. Peterson, K., Dao, P., Farley, R., Castleman, Jr., A.: J. Chem. Phys. **80**, 1780 (1984)
6. Rohlfing, E., Cox, D., Kaldor, A., Johnson, K.: J. Chem. Phys. **81**, 3846 (1984)

7. Delacrétaz, G., Wöste, L.: Surf. Sci. **156**, 770 (1985)
8. Bréchignac, G., Cahuzac, Ph.: Chem. Phys. Lett. **117**, 365 (1985)
9. Bréchignac, C., Broyer, M., Cahuzac, Ph., Delacrétaz, G., Labastie, P., Wöste, L.: Chem. Phys. Lett. **120**, 559 (1985)
10. Paccioni, G., Beckmann, H., Koutecký, J.: Chem. Phys. Lett. **87**, 151 (1982)
11. Rao, B.K., Jena, P.: Phys. Rev. B **32**, 2058 (1985)
12. Flad, J., Stoll, H., Preuss, H.: J. Chem. Phys. **71**, 3042 (1979)
13. Martins, J.L., Buttet, J., Car, R.: Phys. Rev. B **31**, 1804 (1985)
14. Ekardt, W.: Phys. Rev. B **29**, 1558 (1984)
15. Wiley, W., Mac Laren, I.: Rev. Sci. Instrum. **26**, 1150 (1955)
16. Bréchignac, C., Cahuzac, Ph.: Chem. Phys. Lett. **112**, 20 (1984)
17. Broyer, M., Delacrétaz, G., Labastie, P., Whetten, R.L., Wolf, J.P., Wöste, L.: Z. Phys. D – Atoms, Molecules and Clusters **3**, 131 (1986)
18. Bréchignac, C., Cahuzac, Ph., Roux, J.Ph.: Chem. Phys. Lett. (to be published)
19. Guyon, P.M., Berkowitz, J.: J. Chem. Phys. **54**, 1814 (1971)
20. Hermann, A., Schumacher, E., Wöste, L.: J. Chem. Phys. **68**, 2327 (1978)
21. Beck, D.E.: Solid State Commun. **49**, 381 (1984)
22. Chou, M.Y., Cleland, A., Cohen, M.: Solid State Commun. **52**, 645 (1984)
23. Wood, D.. Phys. Rev. Lett. **46**, 749 (1981)
24. Stoll, H., Flad, J., Golka, E, Krüger, Th.: Surf. Sci. **106**, 251 (1981)
25. Martins, J.L., Car, R, Buttet, J.: J. Chem. Phys. **78**, 5646 (1983)
26. Rohfing, E.A., Cox, D.M., Petkovic-Luton, R., Kaldor, A.: J. Phys. Chem. **88**, 6227 (1984)
27. Diefenbach, J., Martin, T.P.: J. Chem. Phys. **83**, 2238 (1985)
28. Knight, W.D., Heer, W.A. de, Clemenger, K.: Solid state Commun. **53**, 445 (1985)
29. Kappes, M., Radi, P., Schär, H., Schumacher, E.: Chem. Phys. Lett. **119**, 11 (1985)
30. Kappes, M., Radi, P., Schär, M., Yeretzian, C., Schumacher, E.: Z. Phys. D – Atoms, Molecules and Clusters **3**, 115 (1986)
31. Waseda, Y., Jacob, K.T., Tsuchiya, Y., Tamaki, S.: Z. Naturforsch. **33**, 940 (1978)
32. Svensson, S., Martensson, N., Basilier, E., Malmqvist, P., Gelius, V., Siegbahn, K.: J. Electron Spectrosc. **9**, 51 (1976)
33. Sattler, K., Mühlbach, J., Echt, O., Pfau, P., Recknagel, E.: Phys. Rev. Lett. **47**, 160 (1981)
34. Bréchignac, C., Broyer, M., Cahuzac, Ph., Delacrétaz, G., Labastie, P., Wöste, L.: Chem. Phys. Lett. **118**, 174 (1985)
35. Bréchignac, C., Cahuzac, Ph.: Laser Chem. **5**, 321 (1986)

C. Bréchignac
Ph. Cahuzac
Laboratoire Aimé-Cotton
Université de Paris-Sud
CNRS II
Bâtiment 505
F-91405 Orsay Cedex
France

Spectroscopy of Na_3

M. Broyer[1], G. Delacrétaz[2]*, P. Labastie[1], R.L. Whetten[3], J.P. Wolf[2], and L. Wöste[2]

[1] Laboratoire de Spectrométrie Ionique et Moléculaire (associé au CNRS n° 171), Université Lyon I, Villeurbanne, France
[2] Institut de Physique Expérimentale, Ecole Polytechnique Fédérale de Lausanne PHB-Ecublens, Lausanne, Switzerland
[3] Department of Chemistry and Biochemistry, Solid State Science Center, University of California, Los Angeles, California, USA

Received April 7, 1986; final version May 16, 1986

The excitation spectrum of Na_3 was systematically investigated from 700 to 330 nm. Four excited states were observed. One of them exhibits fractional quantization of the vibronic pseudorotation which constitutes the first direct verification of the adiabatic sign-change theorem. Photofragmentation is studied by depletion spectroscopy and the 420 nm system is found to be completely predissociated. The structure of the ground state and the ionization potential are also measured.

PACS: 36.40

I. Introduction

The electronic properties of small metallic clusters are of great importance for the understanding of the fundamental mechanism of catalysis and surface chemistry. Informations concerning the electronic structure can be carried out by means of molecular spectroscopy. However, despite intensive activity in the research area, only the gas phase spectra of clusters larger than dimers have been reported for Na_3 [1–3] and Cu_3 [4]. We describe in this paper an investigation of the gas phase spectra of Na_3 in the region of 330–700 nm. Four excited states were observed. Among these the 625 nm system exhibits the first ever observed spectroscopic example of vibronic pseudorotation that follows a series of non-integer quantum numbers [3]. This fractional quantization constitutes a direct verification of the adiabatic sign change theorem [5]. Moreover photofragmentation processes are studied and the excited state in the region of 420 nm is found to be fastly predissociated. For the four spectroscopic systems, hot bands are observed leading to the ground state structure. Finally the Na_3 ionization potential is measured by means of resonant two-photon ionization.

II. Experiment

The basic part of the experimental setup is the seeded supersonic molecular beam [6]. The metal vapor is expanded through a small orifice (typically 50 or 100 μm) into the vacuum using argon as carrier gas. The argon pressure which is mainly limited by the 1,500 l/s diffusion pump can reach 10 bar, when the 50 μ nozzle is used. The metal pressure ranges modestly between 10 and 100 mbar. Neutral beam intensities are permanently monitored with a surface ionization detector. With this expansion system very cold Na clusters were obtained. A hot band analysis of

* Present address: Central Analytical Department, Ciba Geigy LTD, CH-4002 Basel, Switzerland

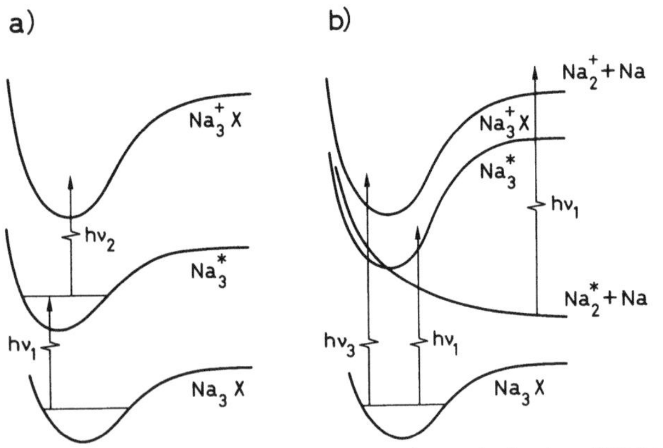

Fig. 1a and b. Principle of the **a** two-photon ionization (T.P.I.) experiment, and **b** the depletion experiment. The indicated potential curves are only meant to be schematic

the obtained spectra showed vibrational temperatures of Na_3 down to 20 K [7] and rotational temperatures of Na_2 down to 7 K [2].

The principles of our experiments are illustrated in Fig. 1a and 1b. The Na_3 molecules are electronically excited with a tunable laser $h\nu_1$. In the resonant two-photon ionization (T.P.I.) scheme (Fig. 1a) the excited molecules are ionized with a second laser $h\nu_2$. The photoions formed are detected with a quadrupole mass spectrometer. In the depletion spectroscopy experiment (Fig. 1b), which is applied for probing the Na_3 dissociative states, an ultraviolet laser $h\nu_3$ (308 nm) directly ionizes Na_3 and monitors the remaining population in the molecular beam. Whenever a predissociated transition $h\nu_1$ is simultaneously irradiated, a depletion of the Na_3^+ intensity is observed. The effect is most favourably detected, when the predissociated transition $h\nu_1$ is saturated. A similar method was used by Smalley and coworkers for detecting predissociated states of Cu_3 [4]. In our experiment the resulting photofragments Na_2 and Na were also detected. This was performed by increasing the laser power $h\nu_1$ to a degree which was sufficiently high to ionize the excited Na_2- or Na-fragments by means of two- or three-photon ionization.

The experiments were carried out with a copper vapor laser (CVL)-pumped dye laser as excitation source (ranging from 600–700 nm), while the 510 nm line of the pump laser or a second CVL-pumped dye laser, was used for the ionization step (two-color experiment). The laser system operated at 30 ns pulses with a 6 kHz repetition rate. The average power of the dye lasers at 4 GHz bandwidth was about 800 mW. Detailed informations about the specifications of this dye laser system are published elsewhere [8]. For excitation wavelengths below 600 nm an excimerpumped dye laser (system Quanta Ray) was used for simultaneously exciting and ionizing resonant particles (one color experiment). The depletion experiments in this wavelength range, however, needed to be carried out as two color experiments as well, which required the dye laser and 20% of the excimer laser output power to be irradiated simultaneously. The excimer laser operated at 14 ns pulses with a 200 Hz repetition rate. The average output power was about 2 W, the dye laser conversion efficiency was about 10%.

In order to obtain a mass-selective detection of the produced photoions, the entrance of a quadrupole mass spectrometer was directed to the interaction area, which extracted the ions perpendicular to both the particle and the laser beam. Measurements were performed by setting the mass spectrometer on the mass peak of Na_3^+, or the corresponding Na_2^+- or Na^+-fragment, while the ion signal was recorded as a function of the excitation wavelength λ_1. During the measurement of the Na_3-ionization potentials, $h\nu_1$ was kept on a fixed transition, and $h\nu_2$ was tuned.

III. Results

The excitation spectrum of Na_3 was investigated from 700 to 330 nm. Figure 2 shows the two-photon ionization (T.P.I.) spectrum observed in this region. This spectrum exhibits three systems: the A-system in the 675 nm region, the B- and B'-system which corresponds to the same excited state (see below), located between 550 and 625 nm, and the C-system in the 475 nm region. No other system is observed by T.P.I. between the C-system and the ionization threshold. A fourth predissociated excited state is found by depletion spectroscopy in the 420 nm region (see below).

1. The 675 nm System (A-State)

This system has been observed first by Hermann et al. [1]. Only a very short vibrational progression is seen.

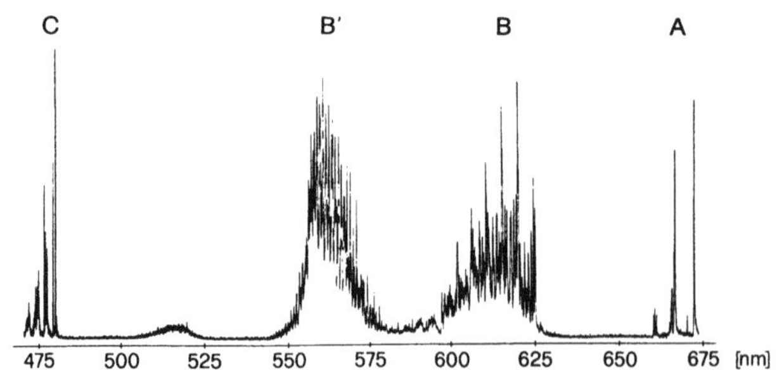

Fig. 2. Compressed spectrum showing the electronically excited bands (A, B, B′ and C) of Na_3 obtained by means of resonant two-photon ionization (T.P.I.)

This is most probably due to Franck-Condon factors because we have found no evidence for the predissociation of this state in depletion experiments. This seems to indicate that the A-state and the ground state have very similar geometric structures. The structure of the bands is complicated most probably because of Jahn-Teller perturbations. However two vibrational frequencies can be determined: $\omega_1' = 128.5\ \text{cm}^{-1}$ which represents the energy difference between the main bands and $\omega_2' = 47\ \text{cm}^{-1}$ the energy between the two first cold bands. These frequencies are close to the measured hot bands (see paragraph 2) $\omega_1'' = 139\ \text{cm}^{-1}$ and $\omega_2'' = 49\ \text{cm}^{-1}$, which affirms our assumption that the A-state and the ground state have similar structures.

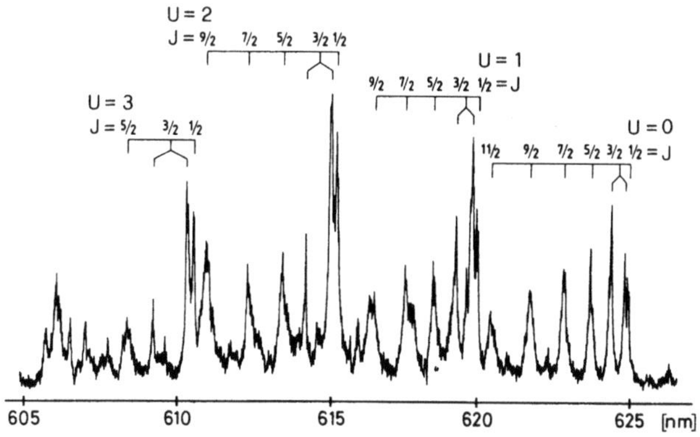

Fig. 3. Vibronic pseudorotation sequence appearing in the B-system of Na_3. The U-values represent the vibrational distortion amplitudes, and the J-values the non-integer pseudorotation quantum numbers

2. The 510 nm and 560 nm System (B- and B′-State)

The B-system has also been observed by Herrmann et al. [1] in a pure metal expansion where the vibrational temperature of Na_3 is in the order of 100 K [7]. The rotational temperature of Na_2 at similar conditions is about 50 K [2]. Under these conditions the spectrum appears as a quasi-continuum with a superimposed vibrational structure. With our seeded beam source, however, a very rich structure emerges at lower beam temperatures. Figure 3 shows the details of this structure on the first band of the B-state. The salient features of this bands system are the following: (*1*) There is a long progression of nearly equally spaced bands ($\omega_0 = 128\ \text{cm}^{-1}$) which appear to be split in doublets. (*2*) Accompanying each member of the main progression there is a series of closely spaced bands fanning out from the doublet.

It can be shown [3] that this spectrum corresponds to the vibronic pseudorotation spectrum predicted by Longuet-Higgins [9] in the case of large Jahn-Teller distortions. In that case a simple approximate pattern emerges

$E(u, j) = (u + 1/2)\,\omega_0 + Aj^2$

where $u = 0, 1, 2, \ldots$ corresponds to the distortion amplitude and $|j| = 1/2, 3/2, \ldots$ to the internal pseudorotation. j is half integer because the electronic wavefuntion changes sign in a degenerate state (adiabatic sign-change theorem).

A detailed analysis of the spectrum by solving the dynamical Jahn-Teller problem shows that the B-state has E′ symmetry in the D_{3h} symmetry group. The Jahn-Teller stabilization energy is $1{,}050\ \text{cm}^{-1}$ and the localization energy $26\ \text{cm}^{-1}$ [3].

The B′-system appears very complicated and complex. A great number of vibrational bands is observed with a small vibrational spacing. A slow transition seems to exist between the structure of the B-state and that of the B′-state. This suggests that the observed spectra of the B′-system correspond to transitions toward the second Born-Oppenheimer surface of the same electronic state as the B-system. It has E′ symmetry and is located beyond the conic intersection. This interpretation is in agreement with the Jahn-Teller stabilization energy deduced from the B-state analysis [3].

3. The 475 nm System (C-State)

This system has a short vibrational progression of four bands. The rather complex structure of each band is not yet interpreted. Spectral features in the same region were also observed by Gole et al. [10]: In their experiment the Na_3 photodissociation was studied through the detection of the Na resonance *D* line. The rotational dimer temperature in their beam was about 30 K. Under these experimental conditions they only observed very broad bands. It is difficult to correlate these results with our observed C-state structure. If their measurement corresponds to the Na_3 C-state, it means that the C-state is at least partially predissociated. We plan to perform in the near future depletion spectroscopy on this electronic state in order to arrive to a definitive conclusion.

4. The Predissociated 420 nm System (D-State)

No signal was recorded by T.P.I. on the Na_3^+-channel between the C-state and the ionization threshold. During the systematic investigation, however, the fragmentation channels of Na^+ and Na_2^+ were always recorded as well. In the 420 nm region (see Fig. 4a) we found structure on Na_2^+, which could not be attributed to Na_2. The spectrum exhibits a regular pro-

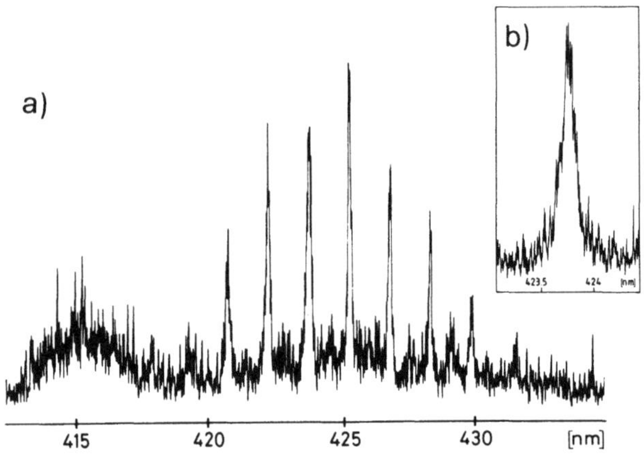

Fig. 4a and b. The predissociated D-state of Na_3 recorded on the Na_2^+-channel. Despite the harmonic character of the spectrum (**a**), the observed progression cannot be attributed to Na_2. This is also affirmed by the symmetric shape of the observed lines (**b**)

gression corresponding to vibrational levels of the excited states. Despite the harmonic character of this progression, the rather symmetric structure of each band does not look like the usual band structure of dimers (see Fig. 4b). Moreover the observed excited states cannot be attributed to Na_2 for which all states in this region are known [11]. This suggests that the observed spectrum corresponds to the predissociation of larger clusters. The depletion spectrum observed on Na_3^+ (Fig. 5) affirms Na_3 as the parent cluster. On Na_5^+, on the other hand, a strong dissociative absorption continuum without any lines was detected by means of depletion spectroscopy. This confirms the importance of dissociation processes in the excited states of clusters.

We also recorded the Na_3 predissociation spectrum as a function of the cluster temperature by varying the carrier gas pressure. The intensity variation of small peaks, which we observed between the intense bands of the main vibrational progression, have the characteristic behavior of Na_3-hot bands.

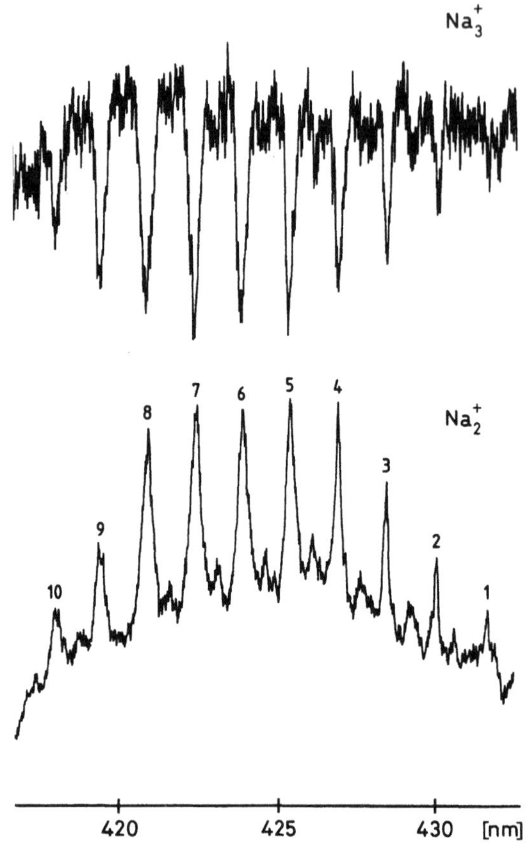

Fig. 5. Simultaneous representation of the Na_3-depletion spectrum and the Na_2-fragment spectrum. Contrary to Fig. 4a, hot bands appear between the main band peaks. This is due to the highly saturated character of the progression

5. The Electronic Ground State

During our systematic spectroscopy of Na_3, hot bands have been assigned by recording the spectra as a function of the carrier gas pressure. Figure 6 shows typical spectra obtained for the bandhead of the B-system. The carrier gas pressure of the experiment was 9 bar (top), 1,9 bar (middle) and 1 bar (bottom), which corresponds to vibrational Na_3 temperatures of 20 K, 60 K and 100 K [7]. Three vibrational frequencies were measured by this method: 49, 87, and 139 cm^{-1}. In the A- and C-systems we observed further hot bands at 49 and 139 cm^{-1}. The hot bands of the D-system could only be determined with less precision, but the result (45 ± 3 cm^{-1}) is close to 49 cm^{-1}. The observed values correlate quite well with theoretical calculations of the Na_3 ground state [12], which are based on a potential surface that was calculated by Martins et al. [13]. These theoretical values are: 58, 94, and 143 cm^{-1}.

6. The Na_3*-Ionization Potential*

The T.P.I. method allows also very precise measurements of the ionization potential: The laser $h\nu_1$ is set to a resonant absorption line, while the second laser $h\nu_2$ is tuned across the ionization threshold. Results, which were obtained for potassium dimers by Herrmann, Leutwyler, Schumacher, Wöste [18], showed very distinct ionization thresholds. This was due to the onset of vibrational autoionization. Figure 7 shows results obtained for Na_3 by using various bands of the B′-system as intermediate levels. Con-

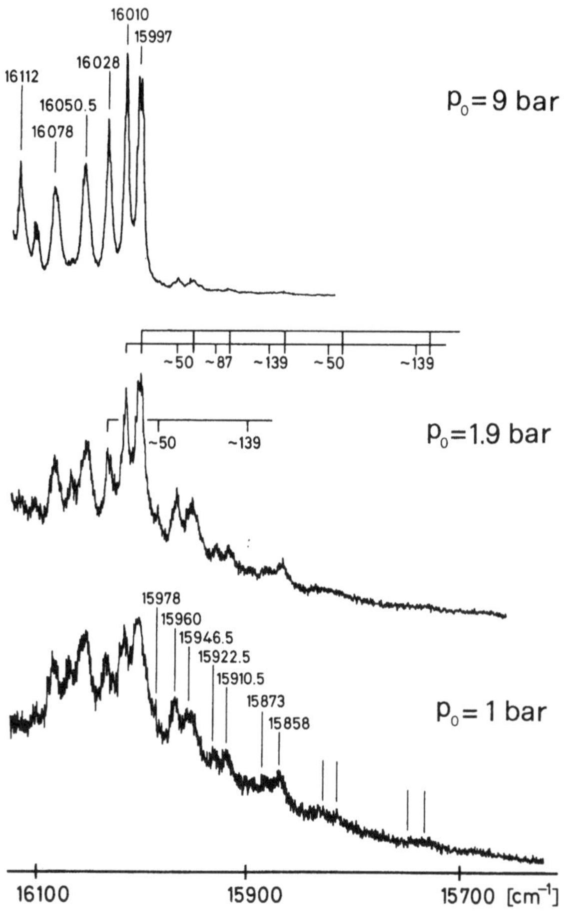

Fig. 6. Hot band progressions recorded at the band head of the B-system. The experiment was carried out at a carrier gas pressure of 9 bar (top), 1.9 bar (middle) and 1 bar (bottom), which corresponds to vibrational Na_3-temperatures of 20 K, 60 K and 100 K [7]

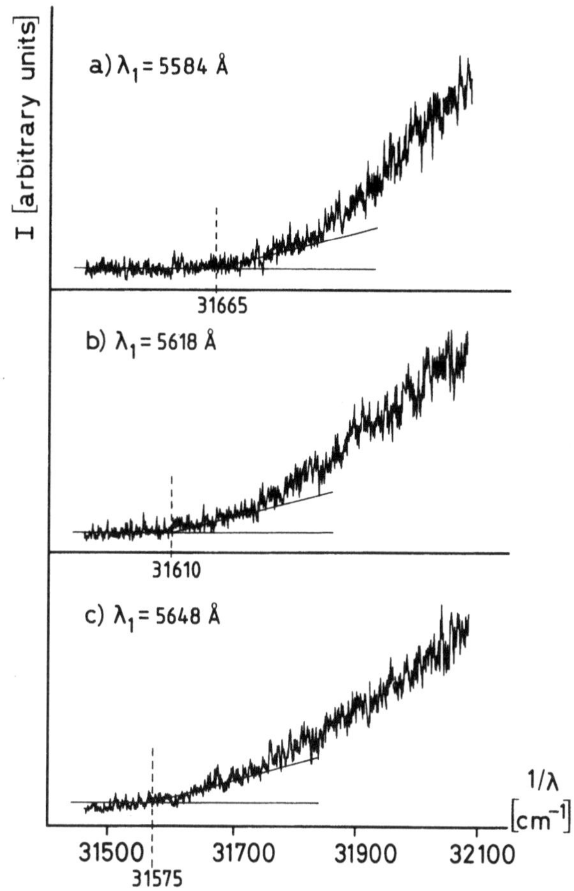

Fig. 7. Na_3-appearance potential measurements obtained by resonant two-photon ionization, using different excited states in the B′-system as intermediate level

trary to results obtained with alkali dimers, there is no expressed step function behaviour, which means that autoionization features are not seen in our spectrum. The gradual increase in the photoionization efficiency indicates that the observed curves correspond to a direct photoionization process [15]. Predissociation processes are obviously faster than the autoionization of Na_3-Rydberg states. Similar features have been observed for larger clusters as well [16]. In the case of dimers, on the other hand, autoionization processes are faster than predissociation, which leads to a step-like ionization threshold [17, 18]. The result illustrates once more the importance of photofragmentation processes in clusters.

Figure 7 shows slight shifts of the appearance potential as a function of the vibrational intermediate levels. This is due to Franck Condon factors between the intermediate levels and Na_3^+-ground state levels. The result, illustrates a difficulty for obtaining precise values for ionization potentials, even when the T.P.I. method is applied. We excited six different vibrational levels of the B- and B′-system as intermediate states. The appearance potentials, which we determined this way are all scattered betwen 3.915 eV and 3.935 eV. This result is in agreement with the appearance potential measurements obtained by one photon ionization [14, 15, 19].

IV. Conclusion

The sodium trimer appears now as the best known small metal cluster. The comparison of our experimental results with theory will constitute an interesting test of quantum chemistry calculations. It must be noticed that Na_3- and Na_2-electronic excited states have very different behaviours: In Na_2 more than one hundred electronic states are known to be stable with respect to dissociation, in Na_3 only three electronically excited states are found not to be predissociated. In larger alkaline clusters no structure was observed and all excited electronic states are suspected to be strongly coupled with dissociative states. In very large metal clusters (as in the bulk) the electronic excitation undergoes rapid relaxation by vibration (phonon). Therefore with regard to the stability of the excited states, the sodium trimer can be considered as the frontier between the molecular and metallic behavior.

References

1. Herrmann, A., Hoffmann, M., Leutwyler, S., Schumacher, E., Wöste, L.: Chem. Phys. Lett. **62**, 216 (1979)
2. Delacrétaz, G., Wöste, L.: Surf. Sci. **156**, 770 (1985)
3. Delacrétaz, G., Grant, E.R., Whetten, R.L., Wöste, L., Zwanziger, J.W.: Phys. Rev. Lett. **56**, 2598 (1986)
4. Morse, M.D., Hopkins, J.B., Langridge-Smith, P.R.R., Smalley, R.E.: J. Chem. Phys. **79**, 5316 (1983)
5. Herzberg, G., Longuet-Higgins, H.C.: Disc. Faraday Soc. **35**, 77 (1963); Longuet-Higgins, H.C.: Proc. R. Soc. London Ser. A **344**, 147 (1975)
6. Delacrétaz, G., Stein, G., Wöste, L.: J. Chem. Phys. (submitted for publication)
7. Broyer, M., Delacrétaz, G., Labastie, P., Wolf, J.P., Wöste, L.: J. Phys. Chem. (submitted for publication)
8. Broyer, M., Chevaleyre, J., Delacrétaz, G., Wöste, L.: Appl. Phys. B **35**, 31 (1984)
9. Longuet-Higgins, H.C.: Adv. Spectrosc. **2**, 429 (1961)
10. Gole, J.L., Green G.J., Pace, S.A., Preuss, D.R.: J. Chem. Phys. **76**, 2247 (1982)
11. Brechignac, C., Cahuzac, Ph.: Chem. Phys. Lett. **112**, 20 (1984)
12. Thompson, T.C., Izmirlian, G., Lemon, S.J., Truhlar, D.G., Mead, C.A.: J. Chem. Phys. **82**, 5597 (1985)
13. Martins, J.L., Car, R., Buttet, J.: J. Chem. Phys. **78**, 5646 (1983)
14. Herrmann, A., Schumacher, E., Wöste, L.: J. Chem. Phys. **68**, 2327 (1977)
15. Broyer, M., Delacrétaz, G., Chevaleyre, J., Fayet, P., Wöste, L.: Chem. Phys. Lett. **114**, 477 (1985)
16. Brechignac, C., Cahuzac, Ph.: Chem. Phys. Lett. **117**, 365 (1985)
17. Broyer, M., Chevaleyre, J., Delacrétaz, G., Martin, S., Wöste, L.: Chem Phys. Lett. **99**, 206 (1983)
18. Leutwyler, S., Herrmann, A., Wöste, L., Schumacher, E.: Chem. Phys. **48**, 253 (1980)
19. Peterson, K., Das, P.D., Farley, R., Castleman, A.W., Jr.: Chem. Phys. **80**, 1780 (1984)

M. Broyer
P. Labastie
Laboratoire de Spectrométrie
Ionique et Moléculaire
(associé au CNRS n° 171)
Université Lyon I
Bât. 205
43, Bd du 11 Novembre 1918
F-68622 Villeurbanne Cedex
France

G. Delacrétaz
Central Analytical Department
Ciba Geigy Ltd
CH-4002 Basel
Switzerland

J.P. Wolf
L. Wöste
Institut de Physique Expérimentale
Ecole Polytechnique Fédérale de Lausanne
PHB-Ecublens
CH-1015 Lausanne
Switzerland

R.L. Whetten
Department of Chemistry and Biochemistry
Solid State Science Center
University of California
Los Angeles, CA 90024
USA

The Formation and Kinetics of Ionized Cluster Beams

I. Yamada, H. Usui, and T. Takagi
Ion Beam Engineering Experimental Laboratory, Kyoto University, Japan

Received April 7, 1986

The formation and kinetics of large vapourized-material cluster beams (large size metal clusters) are discussed. The clusters are formed by injecting the vapour of solid state materials into a high vacuum region through a nozzle of a heated crucible. The conditions under which metal clusters form are analysed using nucleation theory. Computer simulation by combining the nucleation and flow equations has also been made. The results show that the theory can be useful in predicting qualitative dependences of metal cluster formation on operation conditions. Several experimental results are also presented, which support the finding that a large size metal cluster is formed by homogeneous nucleation and growth. The advantageous characteristics of ionized cluster beam for thin film formation are also discussed.

PACS: 36.40; 47.55E; 68.55.tb

I. Introduction

Takagi et al. postulated in 1972 that it should be possible to form high quality thin films and functional devices using the Ionized Cluster Beam (ICB) technique [1]. Various kinds of metal, semiconductor and insulating films have since been prepared. These results show that the technique allows high quality film deposition on to a wide variety of substrate surfaces at low temperature and even permits the formation of thin film materials not previously possible. Film formation mechanism and recent novel results of film deposition are reviewed elsewhere in this volume [2].

Clusters can be created by condensation of super saturated vapour atoms produced by an adiabatic expansion through a small nozzle into a high vacuum region. The source of atoms which form the clusters can be solid state material heated to a sufficiently high temperature to give rise to the pressure differential causing expansion through the nozzle [1]. Due to the enormous difference in the nucleation tendencies of metals compared to gases, cluster formation conditions of metals must necessarily be quite different from those of gases. Clusters from gaseous material have been studied by Becker et al. for other purposes [3].

Characteristic of our cluster formations is that clusters of 100–2,000 atoms are formed purely by expansion of the vapourized solid state material into the high vacuum through a nozzle. In cluster formation, it was commonly believed that metals would have little tendency to condense due to their high surface tension [4]. However, we have pointed out elsewhere [5] that the barrier height and nucleation rates for metals are similar to those of gases, even though metals are characterized by high surface tension. These results have also been confirmed by Yang and Lu for many different metals and semiconductors [6].

Experimental evaluations of cluster size have been made so far by using a time-of-flight (TOF) [1, 7], electrostatic retarding field energy analyser [8], a 127° electrostatic energy analyser [9] and a transmission electron microscope [10]. These results show that clusters of the order of 10^2–10^3 atoms can be formed.

This paper describes the formation and kinetics

of metal clusters during expansion through the nozzle. The theoretical background regarding if metals can nucleate efficiently is discussed using numerical calculations from condensation theory, taking a function of the surface tension-temperature ratio σ/T instead of σ. By computer simulation of expansion in the nozzle, dependences of the kinetic parameters and the cluster size on the operation conditions can be predicted qualitatively. Experimental results of the translational velocity and temperature, the cluster size and other important characteristics related to film formation are compared.

II. Nucleation Rates

According to classical nucleation theory, the change in the Gibbs free energy ΔG for a cluster of radius r is given by

$$(\Delta G)_{\max} \equiv \Delta G^* = \frac{16\pi\sigma^3}{3\left\{\frac{kT}{v_c}\ln S\right\}^2}, \tag{1}$$

where σ is the surface tension, v_c is the molecular volume in the cluster, and S is the saturation ratio P/P_∞, P_∞ being the equilibrium vapour pressure. Assuming thermally stable equilibrium, the nucleation rate J of the clusters is expressed by

$$J = K\exp\left(-\frac{\Delta G^*}{kT}\right), \tag{2}$$

where K is a factor which varies much slower with P and T than the exponential term. It is given by Frankel [11] as

$$K = (P/kT)^2 v_c (2\sigma/\pi m)^{1/2}. \tag{3}$$

Since the formation energy of the cluster ΔG^* is a function of the surface tension σ^3 in Eq. (1) and appears in the exponent of the nucleation rate expression J, the value J is very sensitive to variation in σ. This information had for many years been responsible for the misunderstanding that metal clusters are hard to form because of high surface tension [4].

The situation is, however, more subtle. Figure 1 shows the calculated results of the change of the nucleation energy barrier $\Delta G/kT$ required to assemble a cluster of radius r for several materials. The results showed that no drastic difference in the nucleation barrier is found for metals as compared to gases. But Hg is an exception. In this calculation, no correction of the surface tension from the flat plane value was made. In Table 1, numerical com-

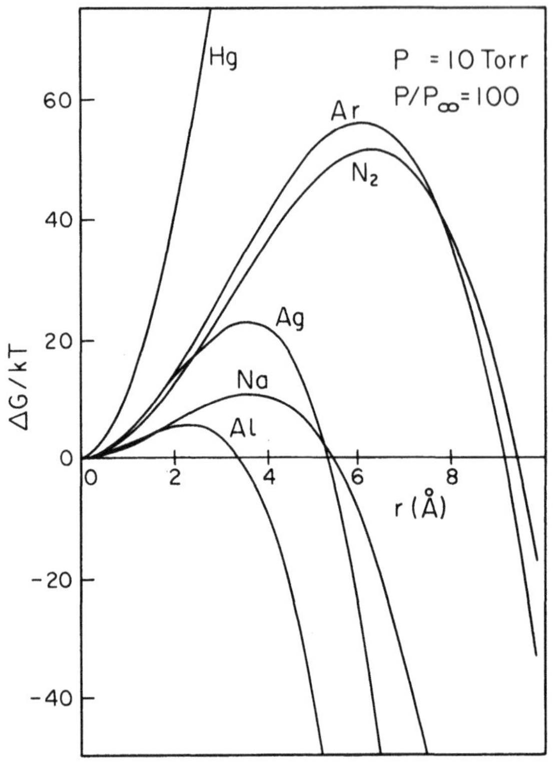

Fig. 1. $\Delta G/kT$ as a function of cluster radius for several materials. Pressure and supersaturation ratio are fixed to $P = 10$ Torr and $P/P_\infty = 100$, respectively

Table 1. Comparison of important kinetic parameters for different materials at $P = 10$ Torr and $P/P_\infty = 100$

	Al	Ag	Ar	N_2	Hg
$T(K)$	1,640	1,435	47.1	41.1	353
σ(mN/m)	577	839	23	17.4	450
$\Delta G^*/kT$	9.5	27.3	55.9	51.4	854

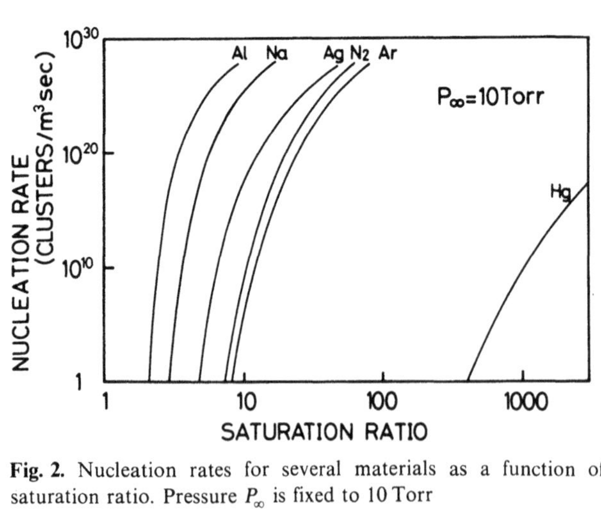

Fig. 2. Nucleation rates for several materials as a function of saturation ratio. Pressure P_∞ is fixed to 10 Torr

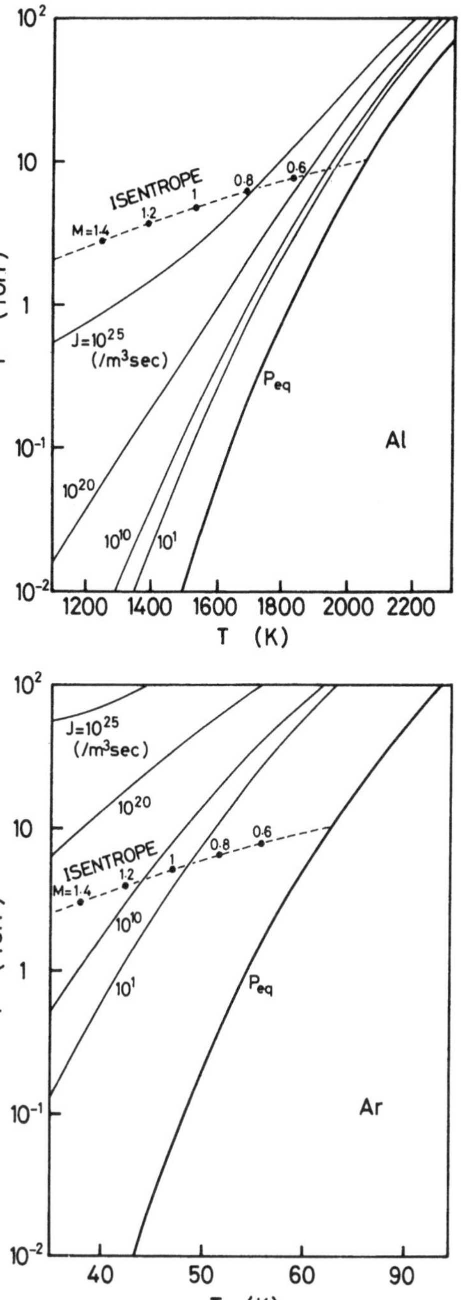

Fig. 3a and b. Equi-nucleation rate lines of Al **a** and Ar **b** on pressure-temperature (P-T) diagrams

parisons of T, σ, and $\Delta G^*/kT$ are shown for different materials for $P=10$ Torr and $S=100$ after the isentropic expansion from the saturated vapour at $P_0=20$ Torr. Supersaturation S, pressure P and temperature T were calculated by assuming that the expansion was adiabatic. They clearly show that the energy barriers of metal cluster formation are comparable, even though the surface tension is high. This is due to the fact that metals have to be heated to a high temperature to get a sufficiently high vapour pressure to cause expansion through the nozzle.

Figure 2 shows the nucleation rates of different materials as a function of saturation ratios. Since the nucleation rate of 10^{20}–$10^{24}/m^3$ s is generally taken as a criterion for cluster formation, the result shows that sufficiently high nucleation rates can be expected for metals, as in the case of gases. But the nucleation rate of mercury is exceptionally low, which is in agreement with the report by Merritt and Weatherston [12]. Figures 3a, b show the equi-nucleation rate lines for Al and Ar in Pressure-Temperature (P-T) diagrams, respectively. The P-T relations under conditions of isentropic expansion are also shown by dotted lines in those figures. The curves show that a sufficient nucleation rate is possible in Al, even with weak expansion, compared to Ar. When the expansion starts from a saturated vapour of $P_0=10$ Torr, J of 10^{25} clusters/cm^3 s can be obtained at Mach number $M=0.8$ for Al, whereas a much higher M or S is required for Ar to obtain the same nucleation rate. These results show that Al vapour condenses much more easily than Ar. Similar results for other metals have been obtained.

III. Computer Simulation

The process of metal cluster formation during supersonic expansion through the nozzle has been simulated by applying classical nucleation theory and one-dimensional flow equations. A cluster is assumed to start to grow from the critical radius with maximum value of ΔG^*. The nuclei grow through a process whereby impinging vapour atoms are balanced out against re-evaporation at a rate given by

$$\frac{\mathrm{d}t}{\mathrm{d}t}=\left(\frac{\xi}{\rho_c}\right)(P/\sqrt{2\pi RT}-P_c/\sqrt{2\pi RT_c}), \tag{4}$$

where ξ is the sticking coefficient, ρ_c the density of the cluster and T_c the temperature of the cluster. Here P_c is the saturated vapour pressure at the surface of the cluster and is given by Thomson [13] as

$$P_c=P_\infty \exp(2\sigma/\rho_c R T_c r). \tag{5}$$

The one-dimensional flow equations for the conservation of mass, momentum and energy are

$$\frac{\mathrm{d}}{\mathrm{d}x}\frac{\rho A u}{1-\mu}=0; \tag{6}$$

$$u\frac{\mathrm{d}u}{\mathrm{d}x}+\frac{1-\mu}{\rho}\frac{\mathrm{d}P}{\mathrm{d}x}=0; \tag{7}$$

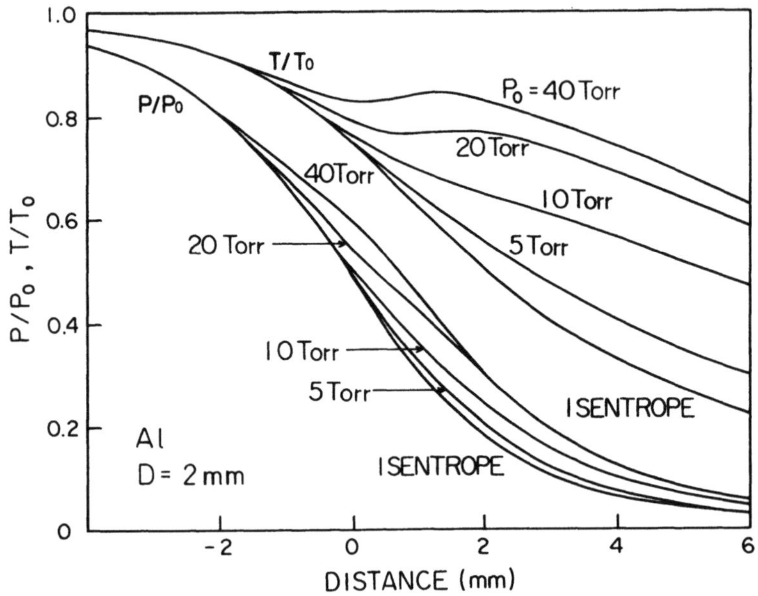

Fig. 4. Variations of relative flow temperature T/T_0 and pressure P/P_0 of Al cluster beam along the x axis

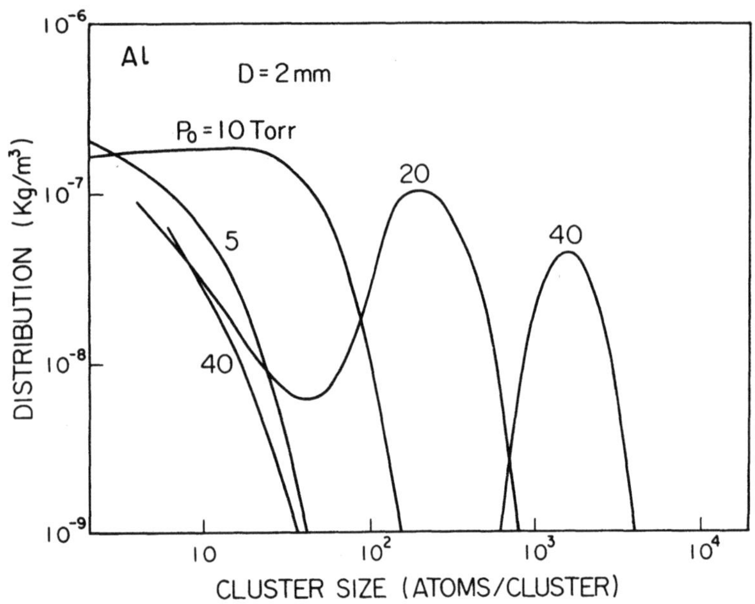

Fig. 5. Cluster size distribution of Al clusters obtained by simulation under different source pressures P_0

$$u\frac{\mathrm{d}u}{\mathrm{d}x}+(1-\mu)C_p\frac{\mathrm{d}T}{\mathrm{d}x}=h_{fg}\frac{\mathrm{d}u}{\mathrm{d}x}, \quad (8)$$

where ρ is the vapour density, A the cross-sectional area of the flow, u the velocity, μ the ratio by mass of the condensed and vapour phase, C_p the specific heat and h_{fg} the latent heat. These equations, together with the equation of state for the monomer vapour, were solved simultaneously by means of Adams method. The source condition was set to a saturated vapour and the flat plane surface tension was used in the calculation. The area of the nozzle as a function of the x axis was taken as

$$A=(\pi D^2/4)\left[1+\left(2x\tan\left(\frac{\theta}{2}\right)\Big/D\right)^2\right], \quad (9)$$

where D is the throat diameter and θ is the nozzle opening angle. These are fixed at 2 mm and 30°, respectively.

Figure 4 shows the variation of relative flow temperature T/T_0 and relative flow pressure P/P_0 for Al. An increase of P_0 over several Torr promotes cluster formation, which can be seen by the increase in relative flow temperature caused by the release of the latent heat of condensation. The calculation shows that the condensation of Al vapour starts upstream of the throat ($x<0$). Figure 5 shows the cluster size distribution for Al 6 mm from the throat for different P_0. The cluster size increased with increasing P_0. The lower tail of the size distribution is attributed to those clusters that were formed near the region downstream from the nozzle throat, while the large clusters are generated upstream. The simulation shows that a considerable portion of large clusters is generated upstream of the throat.

IV. Experimental Evaluations

The translational velocity and temperature of an Ag cluster beam as a function of crucible temperature have been measured by the TOF method [14]. The nozzle geometry was cylindrical and three different diameters were used. Figure 6 shows the translational velocity of the cluster beam. The measured velocity was much higher than the most probable thermal velocity (broken line in the figure). The velocity increases rapidly with increasing crucible temperature. A higher velocity was obtained by increasing the nozzle diameter. The translational temperature of the cluster beam is shown in Fig. 7: the temperature decreased with increasing vapour pressure in the crucible. The observed temperature value was much lower than the crucible temperature T_0. This means that the vapour was efficiently supercooled by adiabatic expansion. A velocity analysis on simply evaporated Ag vapour from a Knudsen cell was also made to obtain the translational temperature. An apparent contrast was observed by comparing the temperature of the beams from a cluster source with those from the Knudsen cell. The temperature of the evaporated vapour was nearly equal to the crucible temperature, and increased with increasing crucible temperature. The results demonstrate that efficient adiabatic expansion enables the onset of vapour condensation; that is, cluster formation is taking place.

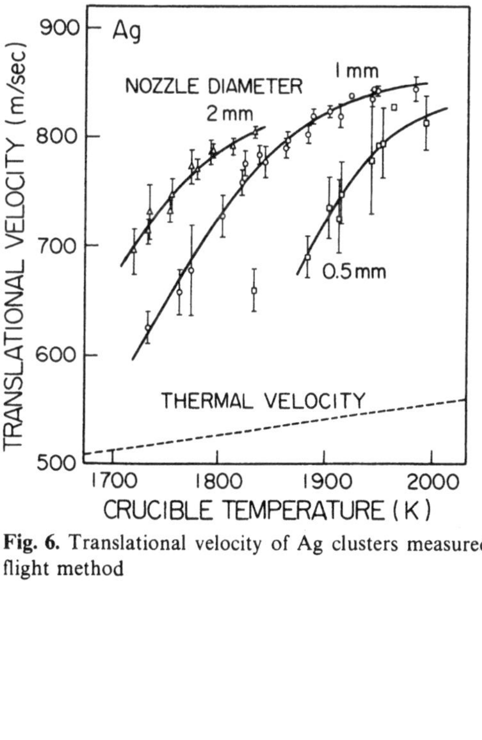

Fig. 6. Translational velocity of Ag clusters measured by time-of-flight method

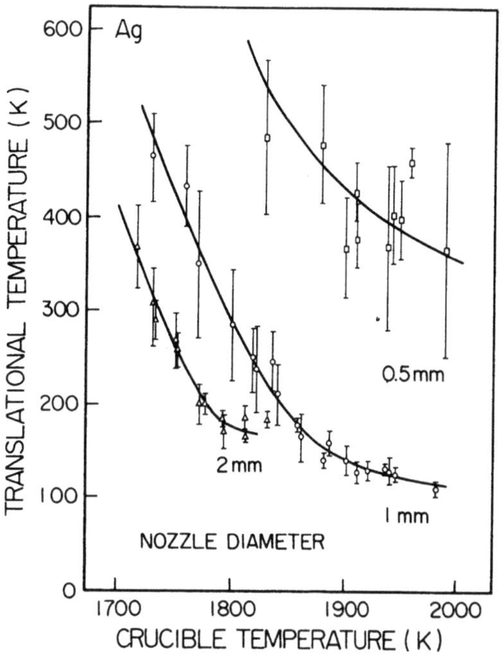

Fig. 7. Translational temperature of Ag clusters measured by time-of-flight method

The most common method for experimentally evaluating cluster size is energy analysis- either by the retarding field method or by the electrostatic energy analyser. Since the kinetic energy of a cluster is proportional to its mass and square velocity, it is possible to estimate the cluster size by measuring the energy. Figure 8 shows Ag cluster size distribution measured by the retarding field method. The size ranges broadly from several hundreds to two thousand atoms/cluster. Beam intensity and cluster size were increased by increasing the crucible temperature.

The cluster size has also been evaluated by using an electrostatic energy analyser [9]. The beam had energy in the range of 80–170 eV, agreeing well with the results from the retarding field method. Cluster size distribution obtained by the retarding field method was compared with that from the TOF method. Those data show a similar size distribution, which means that the cluster is ionized in singly charged state. Transmission electron microscope (TEM) observation have also been made [10]. The clusters were collected on carbon films which were cooled down to liquid nitrogen temperature. The deposition rate and collection time were chosen so as to avoid coalescence of the clusters on the sub-

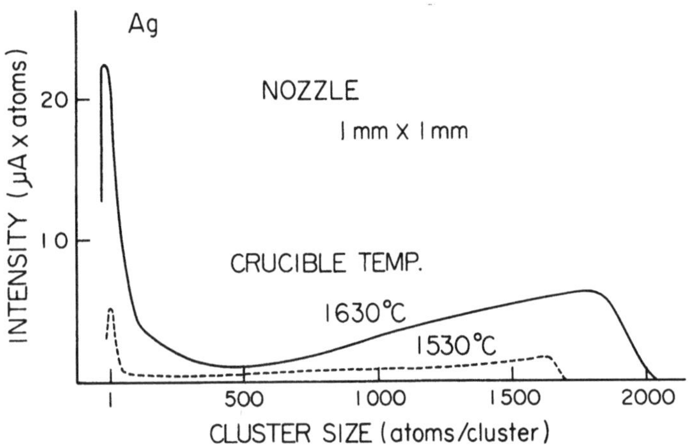

Fig. 8. Size distribution of Ag clusters measured by the retarding field method

strate surface. The diameter of the cluster was between 2–5 nm. The result of the TEM observation suggested that the cluster size ranged in the order of a thousand atoms.

The structure of metal clusters has been examined by electron diffraction [15]. The diffraction pattern of an Sb cluster beam, for example, shows that the clusters have a different structure from that of the microcrystallite and have no long-range ordering in their atomic arrangement. These special structures of the vapourized-metal cluster beam enable a unique film formation process.

These experiments have clearly shown that large size metal clusters are effectively formed. Considering the simulation results, one reason why clusters are formed is that they nucleate with high nucleation rates and start to grow in the upstream region of the nozzle.

V. Conclusions

Formation processes of large size vapourized-metal clusters have been investigated theoretically and experimentally. Detailed calculation of the nucleation rates indicated that, despite the large surface tension of metals, most metal vapours condense as easily as gases. The simulation has also shown that large size vapourized-metal clusters are effectively formed near the throat region in the nozzle. This theoretical treatment led to the postulation that metal clusters are easy to form. But one should not expect this simulation to be quantitatively exact, because the definition of surface tension is considerably ambiguous and the nucleation and growth models used in this simulation are rather simple. In spite of this, the results have been useful in understanding the cluster formation mechanism, and can direct cluster formation experiments qualitatively. Many different types of experiments have verified that cluster formation is thus possible. These theoretical and experimental results should be useful in further development in this field.

References

1. Takagi, T., Yamada, I., Kunori, M., Kobiyama, S.: Proceedings of the 2nd International Conference on Ion Sources. pp. 790–795. Vienna: Österreichische Studiengesellschaft für Atomenergie GmbH 1972
2. Takagi, T.: Z. Phys. D – Atoms, Molecules and Clusters **3**, 271 (1986)
3. Becker, E.W., Bier, K., Henkes, W.: Z. Phys. **146**, 333 (1956)
4. Stein, G.D.: In: Proceedings of the International Ion Engineering Conference – ISIAT '83 & IPAT '83 – Kyoto. Takagi, T. (ed.), pp. 1165–1176B. Tokyo: Inst. Elect. Eng. Japan 1983
5. Yamada, I.: ibid. pp. 1177–1192A
6. Yang, S.-N., Lu, T.-M.: J. Appl. Phys. **58**, 541 (1985)
7. Yamada, I., Takagi, T.: In: Proceedings of the Xth International Symposium on Molecular Beams. pp. VII-B1-B5. Cannes: DRET & CEA, France 1985
8. Theeten, J.B., Madar, R., Mircea-Roussel, A., Rocher, A., Laurence, G.: J. Cryst. Growth **37**, 317 (1977)
9. Yamada, I., Takagi, T.: Thin Solid Films **80**, 105 (1981)
10. Yamada, I., Takaoka, H., Inokawa, H., Usui, H., Cheng. S.C., Takagi, T.: Thin Solid Films **92**, 137 (1982)
11. Frankel, J.: Kinetic theory of liquids. New York: Dover Publication 1955
12. Merritt, G.E., Weatherston, R.C.: AIAA J. **5**, 721 (1967)
13. Thomson, W.: Proc. R. Soc. Edinburgh **7**, 63 (1870)
14. Yamada, I., Takagi, T., Younger, P.R., Blake, J.: SPIE Advanced Applications of Ion Implantation. Vol. 530, pp. 75–83. Los Angeles: Intern. Soc. Opt. Eng. 1985
15. Yamada, I., Stein, G.D., Usui, H., Takagi, T.: In: Proceedings of the 6th Symposium on Ion Sources and Ion-Assisted Technology. – ISIAT '82 –, Tokyo. Takagi, T. (ed.), pp. 47–52. Kyoto: Research Group of Ion Engineering 1982

I. Yamada
H. Usui
T. Takagi
Ion Beam Engineering
Experimental Laboratory
Kyoto University
Sakyo, Kyoto 606
Japan

On the Phase of Metal Clusters

J. Gspann
Kernforschungszentrum und Universität Karlsruhe,
Institut für Kernverfahrenstechnik, Karlsruhe, Federal Republic of Germany

Received April 10, 1986

Metal clusters of about 1,000 atoms formed in condensing flows of pure vapors are predicted to remain liquid in high vacuum since evaporation cooling essentially terminates before solidification is achieved. Radiation cooling does not contribute considerably in times comparable to the usual cluster flight times.

PACS: 3640; 6125M; 6470D

Introduction

The internal state of free metal clusters formed in partly condensing nozzle flows of pure vapors is unknown. For clusters of the rare gases except helium, as well as for other van der Waals bound clusters, electron diffraction studies have shown that all of these clusters are *solid*, though often not in the bulk crystal phase, and their internal temperature is about half the respective bulk melting temperature [1]. A simple correlation between the internal cluster temperature and the respective latent heat of sublimation has been found to predict the experimental data remarkably well [2]. Since the basic assumption underlying this correlation is solely that evaporation cooling determines the internal temperature of a free cluster in high vacuum, the same correlation should apply to metal clusters, too.

The following discussion will show that on this basis all metal clusters from pure vapor expansions are predicted to be *liquid*, in rather spectacular distinction to the van der Waals clusters. Radiation cooling will be estimated as negligible in comparison with evaporation cooling. Experimental results exist only for the semimetal antimony [3] which turns out to be a borderline case, with either liquid or amorphous solid clusters.

Internal Cluster Temperatures

The internal temperatures of clusters can be obtained from electron diffraction measurements in which an electron beam intersecting the cluster beam gives rise to Debye-Scherrer diffraction rings [1]. The ring diameters give the lattice parameters which allow to determine the lattice temperature by comparison with bulk lattice data if cluster size effects can be accounted for by extrapolation to infinitely large sizes. Obviously, this method of evaluation is only possible for clusters solidifying in the bulk crystal lattice. For the rare gas clusters, this requires clusters of about 1,000 or more atoms, since smaller clusters show non-crystalline icosahedral structures.

The experimentally observed values of the internal temperatures of the larger clusters turn out to be practically independent of the conditions of cluster generation, e.g. nozzle temperature or inlet pressure. Evidently, they represent a kind of final temperature characteristic for the respective cluster substance which can be estimated as follows [2]:

The time of residence τ of an atom on a surface of temperature T is given by the well-known relation [4]

$$\tau = \tau_0 \exp(u_0/kT) \qquad (1)$$

where τ_0 is the period of vibration of the atom on the surface, u_0 its heat of sublimation or vaporization, and k the Boltzmann constant.

Evaporation cooling of the free clusters in high vacuum terminates, at least in the time scale of 10^{-3} s which is the typical duration of flight from the source to the analyzing electron beam, as soon as the residence time τ of the surface atoms becomes comparable or longer than the flight duration, say 10^{-2} s. Using the typical 10^{-12} s for the period of vibration

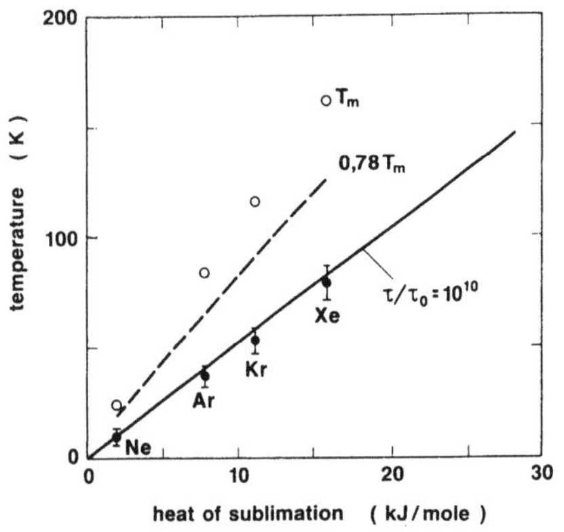

Fig. 1. Internal temperature of clusters from pure vapor expansions as a function of the heat of sublimation (solid line) according to Eq. (2) together with measured data (black dots) from [1], bulk melting temperatures (empty circles), and reduced melting temperatures of 1,000 atom clusters (dashed curve) for rare gases

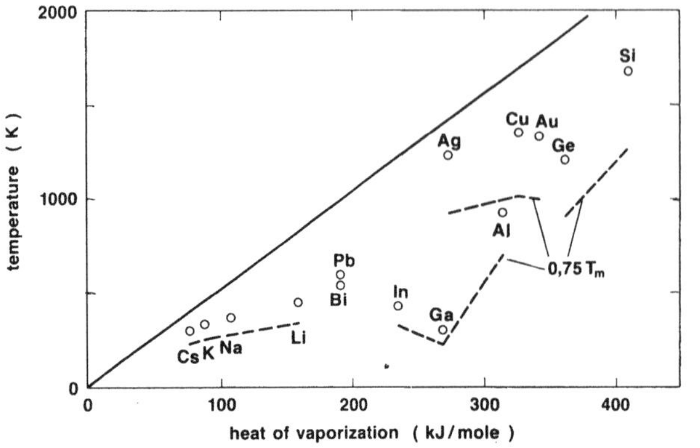

Fig. 2. Predicted internal temperature of clusters from pure vapor expansions as a function of the heat of vaporization (solid line) according to Eq. (2) together with bulk melting temperatures (circles) and reduced melting temperatures of 1,000 atom clusters (dashed lines) for several metals

τ_0 one gets for the internal cluster temperature

$$T = (10k \ln 10)^{-1} u_0, \tag{2}$$

Figure 1 shows that this relation fits nicely the internal temperatures of the rare gases except helium as reported by the Orsay group [1] if they are plotted versus the respective bulk data for the heat of sublimation [5].

Internal Phase

As already mentioned, van der Waals clusters always have been found to be solid. This is illustrated in Fig. 1 by the bulk melting points (empty dots) as well as by the reduced melting temperatures of 1,000 atom clusters (dashed line) which are well above the internal cluster temperatures. The reduced melting temperatures have been estimated to be 78% of the respective bulk values according to an interpolation formula by Ross and Andres [6].

Figure 2 shows the same plot for some representative metals. Obviously, the internal cluster temperatures (solid line) predicted by (2) are always higher that even the bulk melting temperatures, the more so for the depressed melting temperatures of 1,000 atom clusters (dashed lines). For the metals, the latter has been chosen to be 75% of the bulk melting temperature, according to the experimental results of Buffat and Borel for small gold particles [7]. Taking into account this melting temperature depression due to the finite cluster size, all the cluster species in Fig. 2 must be liquid.

This, of course, pertains only to clusters from pure vapor expansions. If condensation takes place in a nearly stagnant gas (gas evaporation technique) or in co-expansion with a carrier gas (seeded beams), collisional cooling may lead to solidification or to condensation into the solid phase from the very beginning. Experiments with antimony revealed crystalline clusters already at 1 Torr argon admixture [3].

Radiation Cooling

The high internal temperatures predicted by (2) for metal clusters from pure vapor expansions could be thought to lead to appreciable radiation cooling, considering the T^4 dependence of the Stefan-Boltzmann radiation law. However, the spectral emissive power of the black body given by Planck's law has to be reduced according to the spectral absorption efficiency factor E_λ which is the ratio of the spectral absorption cross section to the actual cluster cross section and contains the ratio of the cluster diameter D to the radiation wave length λ:

$$E_\lambda = 12\,\pi \frac{D}{\lambda} \frac{\varepsilon_2}{(\varepsilon_1+2)^2+\varepsilon_2^2}. \tag{3}$$

Therein, ε_1 and ε_2 denote the real and the imaginary part of the dielectric constant of the clusters.

Following Kreibig [8], the values of ε_1, ε_2 are calculated for silver clusters of 1,000 atoms, using a diameter D of 32 Å. For a cluster temperature of 1,500 K, the integral over the wavelength range from 0.5–7.5 μm of the black body spectral emissive power times E_λ as given by (3) turns out to be a factor 2.67 $\times 10^{-4}$ smaller than the corresponding black body radiation. The energy lost by radiation from the cluster surface πD^2 during 10^{-3} s flight duration is found to be 2.5×10^{-20} J which is more than an order of magnitude less than the latent heat of vaporization of a single atom which is about 4.5×10^{-19} J at 1,500 K. Thus, the small size of the cluster in comparison with the wavelengths of infrared radiation prevents them from cooling by radiation.

Experimental Data

Electron diffraction results for metal clusters from pure vapor expansions are only available for the semimetal antimony [3]. The pattern observed displays the diffuse halos characteristic of a liquid. However, the authors of that work prefer to consider these antimony clusters as being amorphous solid, mainly because in the radial distribution function obtained by Fourier transformation a tiny feature shows up, a knee at the second peak, which is often observed with amorphous films. Since antimony is known to form a glassy state rather readily such a phase cannot easily be excluded for Sb clusters.

The prediction of (2) for the internal phase is inconclusive as well: at the melting point the heat of vaporization drops enormously from about 190 to about 110 kJ/mole. Using the arithmetic mean value in (2), an internal cluster temperature of 770 K is obtained which is just about halfway in between the bulk melting temperature of 900 K and the 25% depressed value estimated for 1,000 atom clusters. Hence, antimony is a borderline case under various aspects. We are presently preparing studies with clusters of the alkali metals which are known not to form amorphous phases.

Conclusions

Based upon arguments on evaporation cooling, free metal clusters formed in pure vapor expansions are predicted to remain liquid in high vacuum. Radiation cooling does not contribute in times of the order of milliseconds, comparable to the usual flight duration of the clusters. Experimental phase determination is still needed.

The notion of metal clusters from pure vapor expansions being hot droplets could have implications for the understanding of the success of film formation by cluster deposition, e.g. the ICB process [9]: the cluster fluidity might explain the observed enhanced lateral mobility on the substrate while the heat carried by the cluster provides for effective but locally restricted annealing.

References

1. Farges, J., Feraudy, M.F. de, Raoult, B., Torchet, G.: Surf. Sci. **106**, 95 (1981)
2. Gspann, J.: Physics of electronic and atomic collisions. Datz, S. (ed.), pp. 79–96. Amsterdam, New York, Oxford: North-Holland 1982
3. Yamada, I., Stein, G.D., Usui, H., Takagi, T.: Proceedings 6th Symposium on Ion Sources and Ion Assisted Technology, Tokyo 1982, 47
4. Frenkel, J.: Kinetic theory of liquids. Chap. I. London, Oxford: Univ. Press 1946
5. Crawford, R.K.: Rare gas Solids. Klein, M.L., Venables, J.A. (eds.), Chap. 11. New York: Academic Press 1977
6. Ross, J., Andres, R.P.: Surf. Sci. **106**, 11 (1981)
7. Buffat, Ph., Borel, J.-P.: Phys. Rev. **A13**, 2287 (1976)
8. Kreibig, U.: J. Phys. **F4**, 999 (1974)
9. Takagi, T.: Z. Phys. D – Atoms, Molecules and Clusters **3**, 271 (1986)

J. Gspann
Institut für Kernverfahrenstechnik
Universität Karlsruhe
Postfach 3640
D-7500 Karlsruhe
Federal Republic of Germany

General Principles Governing Structures of Small Clusters

J. Koutecký
Freie Universität Berlin, Institut für Physikalische und Theoretische Chemie, Berlin, Germany

P. Fantucci
Dipartimento di Chimica Inorganica e Metallorganica, Centro CNR, Università di Milano, Italy

Received April 29, 1986

General rules which govern electronic and geometric structures of small clusters are formulated, and their validity is documented with the results of the MRD – CI investigations for Li_n, $BeLi_k$, Be_l ($n=2$–14, $k=2$–6, $l=2$–13) as well as on IIa and IVa tetramers. The MRD – CI results are compared with investigations performed with other methods.

PACS: 7300

1. Introduction

Theoretical investigations of the electronic structure of small clusters employing quantum chemistry methods make possible to elucidate the relevant properties of these very interesting and somehow intriguing entities. The understanding of the nature of bonds present in small clusters containing metal or semiconductor atoms should reveal the rules about cluster stability as a function of cluster size and cluster geometry. Two extreme points of view can be taken: either, the average binding energy of the clusters depends primarily on the electronic structure and the actual geometry of the framework of atomic nuclei is of secondary importance, or the shape of the framework of atomic nuclei is responsible for the cluster stability.

The first point of view offers the basis of the simple jellium theory successfully applied to metals [1]. The second point of view has been employed in the pair-potential theory of microclusters [2]. Due to the MO degeneracy, the jellium droplet model of clusters ("superatom model") leads necessarily to energetically favored closed-shell electronic structures for certain numbers of electrons in analogy with the behaviour of noble gas atoms [1]. The possible connection with the "magic numbers" for the nuclearity of clusters with especially high abundancies found experimentally is quite straightforward. On the other hand, it is evident that in the purely geometrical considerations about compactness of clusters the "closed-shell" of atoms plays a somehow similar role as the closed-shells of electrons ("super-atom model") of clusters.

Let us assume, for example, that an atom added to a "closed-shell" cluster with n atoms and an average number of nearest neighbours $\bar{N}(n) > 2k$, likes to have (k) atoms as nearest neighbours (e.g. $k=3, 4$). The increase of the cluster nuclearity from n to $n+1$ is connected with the drop of the average coordination number:

$$\bar{N}(n+1) - \bar{N}(n) = [2k - \bar{N}(n)]/(n+1) < 0.$$

The quantum chemical methods can in principle consider both the above mentioned aspects important for the estimate of the cluster stability and for the prediction of the most probable cluster forms.

In general, it is important to realize the evident fact that the abundancies observed experimentally in the preparation and identification of clusters result from very complicated processes during which also the ionization and possible dissociation or association can be of some importance. Consequently, the stability of the cluster ions, the vertical and adiabatic ionization potentials should be investigated before the explanation of experimental findings with theoretical results can be seriously attempted.

In this contribution some general rules on cluster stability are formulated. These laws are based on quantum chemical investigations of clusters Li_n, Be_n ($n=3$–14), Si_m, Ge_m ($m=3$–7), mixed clusters $BeLi_k$

($k=2$–8) as well as B_4 and C_4 tetramers. Quite extensive and systematic investigations of Ia and IIa clusters allow for some useful generalizations. The reported results have been obtained with the Hartree-Fock SCF method followed by the multireference diexcited configuration interaction procedure (MRDCI) [3]. In the case of Li_n [4] and $BeLi_k$ [5] clusters the SCF optimum geometry search has been carried out. The relevance of the electronic correlation for the theoretical study of such relatively unstable entities should not be underestimated. The comparison with the results obtained from other theoretical methods (Local Spin Density approximation, Hartree-Slater and X_α methods [6] Generalized Valence Bond approach [7] and jellium droplet model [1]) is very instructive because due to the agreements and disagreements among different methods, the reliability of the theoretical predictions can be evaluated.

Beside the shape and stability of the clusters, the spin multiplicity of the ground state is an other important characteristic cluster property. The energy differences among the low lying states of different multiplicity are for some clusters very small which can be considered as a sign of instability and high reactivity of these cluster shapes. Another interesting property is the distribution of electrons in a cluster. The differences between the charges between the "surface" atoms and "bulk" atoms in clusters may allow for some important conclusions about the applicability of the cluster model for chemisorption sites as well as of cluster model for impurities in metals.

2. General Rules on Cluster Stability

Electronic structure and geometry of stable small alkali metal clusters is determined by three factors (cf. [4]). Since they do not work always in the same direction, they must be considered altogether and simultaneously.

i) Compactness of the cluster geometry is a favorable factor because it allows for a high effective coordination number. Consequently, small clusters exhibiting five-fold symmetry cf. [8] and clusters with large numbers of tetrahedra as "building elements" are stable. Crystals with closed packed structures are naturally favored because fcc and hcp lattices exhibit such maximal number of tetrahedral subunits which does not conflict with the translational symmetry.

ii) The properties of characteristic relevant one-electron functions (molecular orbitals in one electron theories and natural orbitals in many electron theories) are of dominant importance for cluster stability and cluster geometry. The nodal properties, degeneracies and occupancies by available valence electrons,

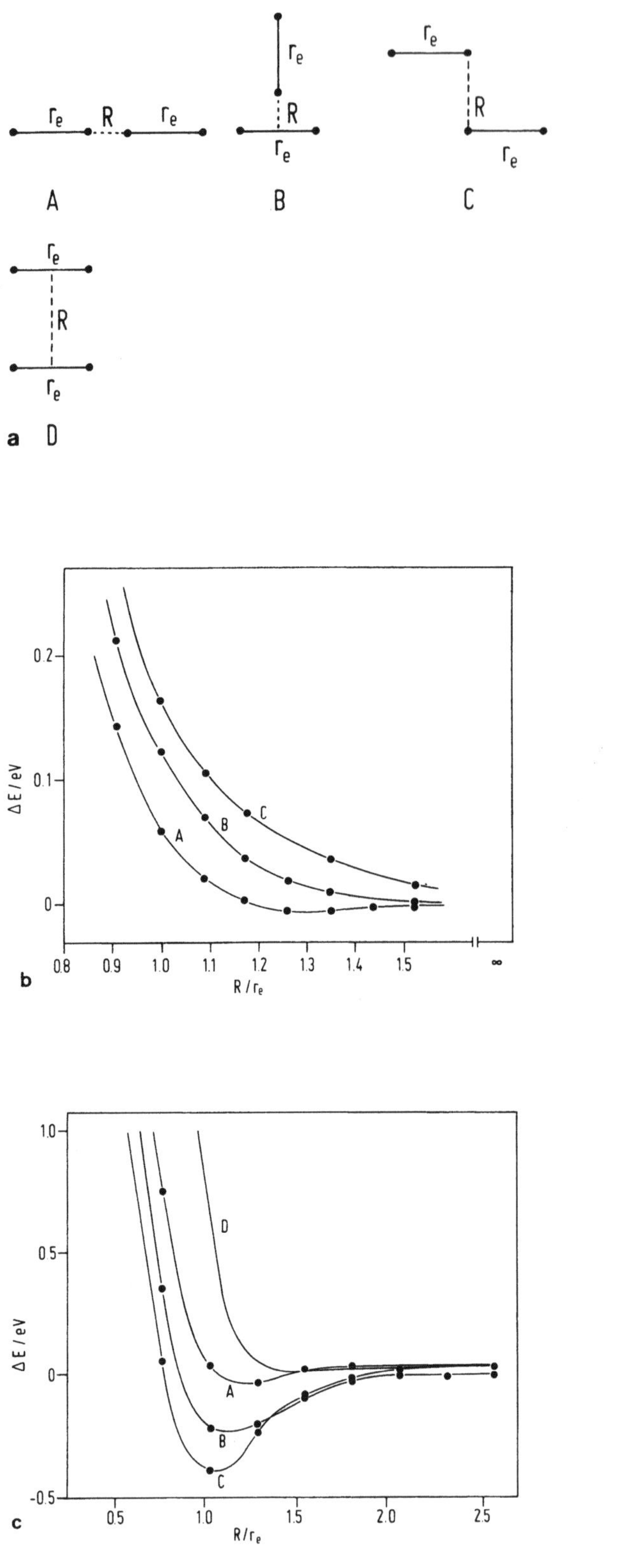

Fig. 1a–c. Energy change (ΔE) of the system Li_2+Li_2 as a function of the mutual distance R. The bond length in Li_2 moieties is kept constant: $r_e=3.02$ Å. **a** Approaches *A*, *B*, *C* and *D* of two Li_2. **b** The single reference diexcited CI (threshold 1 μ hartree) with the (10*s*) uncontracted AO basis set. The extrapolated values are plotted. **c** The MRD−CI procedure with an AO basis set (6*s* 1*p*/2*s* 1*p*) has been employed

allow for the qualitative predictions about clusters which are good candidates to be stable. Especially partial occupation of the degenerate or nearly degenerate one electron functions leads to the deformation of highly symmetrical (and consequently highly compact) cluster geometries toward less compact structures. The Jahn-Teller, pseudo-Jahn-Teller effect and related phenomena make sometimes the geometries which should be optimal according to the first rule concerning the cluster compactness, unfavorable. Large degeneracies are unfavorable for clusters with small number of electrons. Consequently, very small Ia and IIa planar clusters are very often quite stable.

iii) The consideration of higher coordination numbers of Ia and IIa atoms is necessary to explain the very existence of Ia and IIa clusters (as well as Ia and IIa metals). The *p*-AO's which are without importance for the description of the electronic structure of atoms play an important role for the stability of clusters. The *p*-type polarization functions determine also the direction and pathway of the geometry relaxation. The analogy with the Pauling's metal orbitals is evident. The importance of the inclusion of the *p*-type polarization function in the AO basis is illustrated by the consideration of various approaches of two Li_2 moieties (see Fig. 1). If the *p* polarization functions are not added to the relatively large basis set which yields an energy for Li atom near to the Hartree-Fock limit, the MRD–CI treatment gives rise to the repulsive curves (Fig.1b) for all the considered approaches of two Li_2 molecules (Fig. 1a). On the contrary the inclusion of the *p* polarization functions even in a modest AO basis set leads to the pronounced binding (Fig. 1c). The minima on the potential curve A and B are in reality only saddle points on the global energy surface and geometry optimization starting from these saddle points leads to minimum for the rhombic geometry of Li_4 tetramers [9].

Similarly, the planar rhombus is a very stable geometry of the Na_4.

3. Optimal Cluster Geometries and Their Stabilities

The geometries of the Li clusters have been optimized in the framework of the Hartree-Fock procedure starting from the very plausible cluster topologies which have been selected partlŷ according to the chemical intuition and partly according to the rules about difference sequences in the crystal growth theory [4]. The quantity named "average binding energy" obtained by subtracting the energy of the isolated atom from the energy E_n of the cluster divided by the number of atoms in the cluster is used as a measure of the cluster stability:

$$BE/n = -E_n/n + E_1 = (n\,E_1 - E_n)/n.$$

Notice, the connection of the difference of the average binding energies with the dissociation energy of Li_n into $Li_n + Li_1$:

$$n\,BE/n - (n-1)\,BE/(n-1) = E_n - E_1 - E_{n-1}.$$

The average binding energy obtained with the multireference diexcited CI method (MRD–CI) [3] for the singlet states of clusters in their optimal geometries as a function of the number *n* of atomic centers in Li clusters is diplayed in Fig. 2. Starting from the SCF optimal geometry the MRDCI partial geometry optimization has been carried out keeping the SCF cluster topology unchanged and varying only some characteristic bond lengths. The preliminary results for larger Li clusters are indicated by triangles. The optimal geometries are labelled by the numbers corresponding to the topologies shown in Fig. 3, in which also other topologies which have been consid-

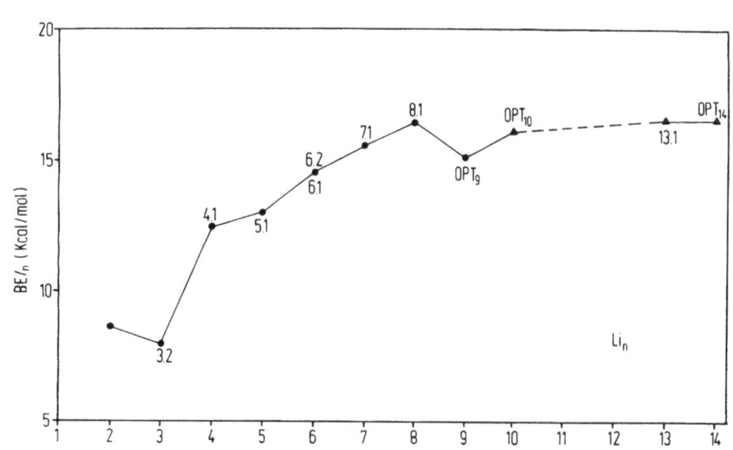

Fig. 2. Dependence of the MRD–CI binding energy per atom *BE/n* (kcal/mol) for the singlet states of Li clusters on the number *n* of atoms in clusters. The symbols ▲ label preliminary results

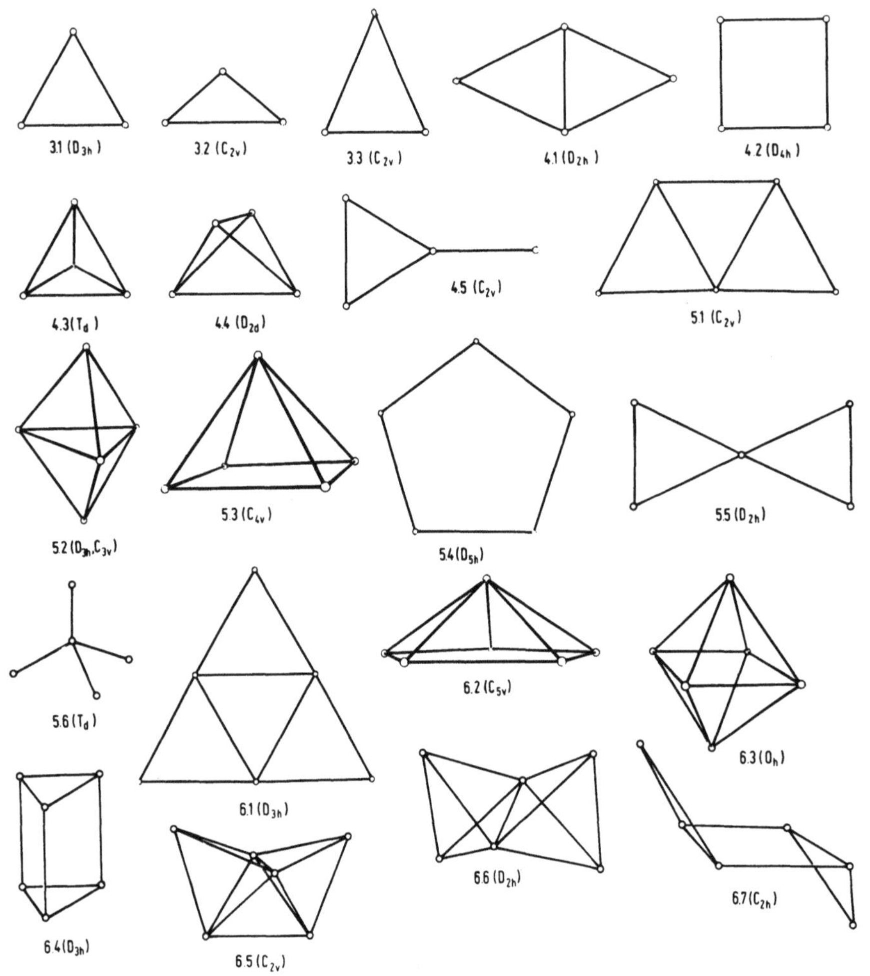

Fig. 3. Topologies of the clusters the description of the symmetry group in bracket

ered as starting points of the SCF geometry search are listed. The overall increase of cluster stability with increase in cluster size is clearly noticeable. A clear drop of the BE/n quantity for $n=9$ can be considered as symptom of a "magic number" $n=8$ parallel to the closed-shell structures proposed by Knight [1]. The moderate increase of BE/n for $n>9$ can be due to either to the fact that the performed geometry optimization leads to local minima only or to the size-consistency error of the CI treatment applied on relatively large systems. The large increase of BE/n for $n>9$ calculated with the SCF procedure (Fig. 4) seems to support the second possibility. The BE/n for cluster arrangements which can be considered as sections of the fcc crystal lattice are shown in the Fig. 4 as well.

Two different kinds of stable cluster geometries are possible to distinguish: optimal Li tetramers, pentamers and hexamers are deformed planar sections of (111) fcc lattice plane. Larger clusters ($n>6$) are composed from condensed deformed tetrahedral or triginal pyramids. The Li_6 with geometry of the pentagonal pyramid (6.2) has comparable energy as planar Li_6 (6.1). The pentagonal bipyramid Li_7, and mainly the Li_8 with the ideal T_d symmetry are much more stable than the best fcc cluster shapes with the same number of lithium atoms. Evidently, for still larger alkali metal clusters the topology requires the inclusion of some octahedral subunits representing the transition to the fcc crystal lattice which is built from both tetrahedral and octahedral subunits. The general trend of the increasing binding energy with the increasing nuclearity can be understood as due to the increasing compactness which can be measured by the average coordination number. It is necessary to keep in mind that the notion of the nearest neighbours for complicated cluster geometries has a limited meaning. The filling of the spherical shells of atoms in a cluster structure can generally lead to the oscillations in the binding energy per atom as a function of n. This phenomenon can be of course expected mainly for medium size clusters with relatively high average coordination number. For very small clusters the electronic properties as well as the shape of the framework of the atomic nuclei are important. This is illustrated by the stability of small planar cluster

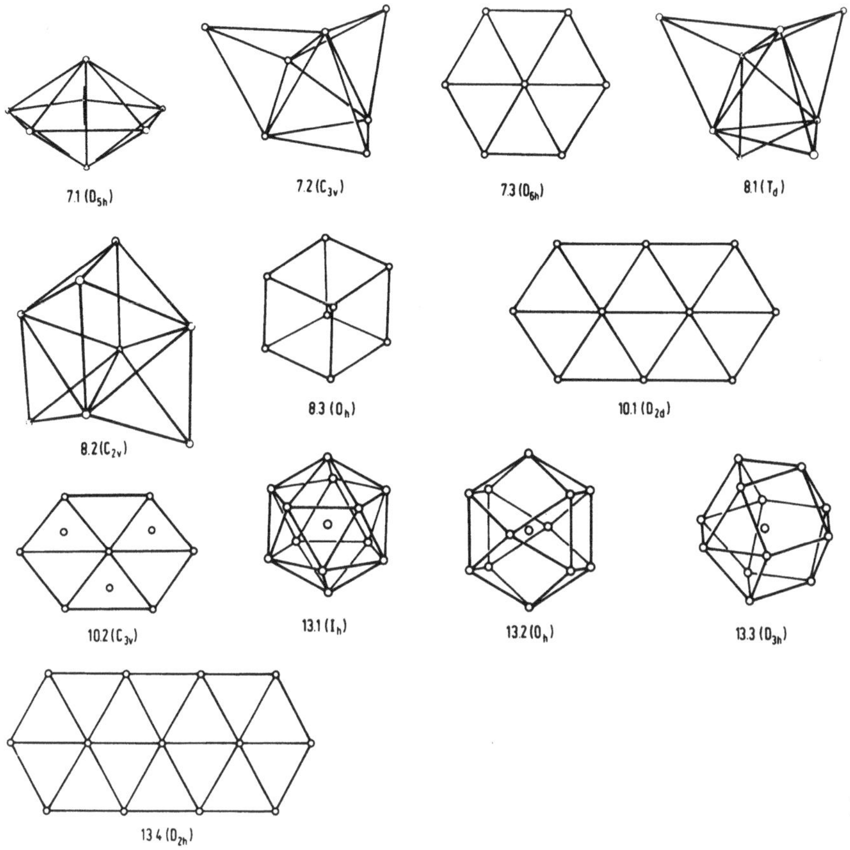

Fig. 3.1

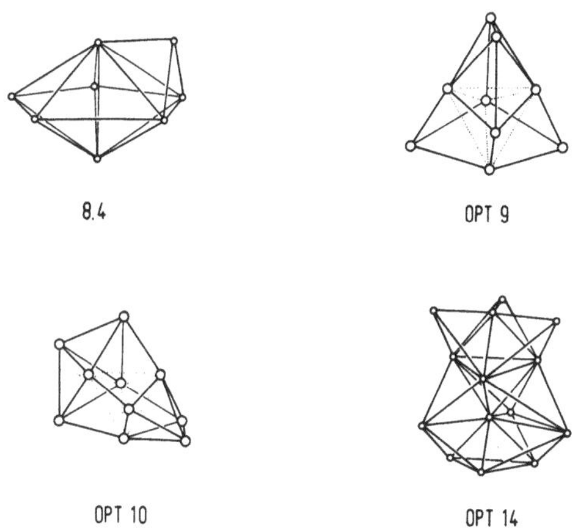

Fig. 3.2

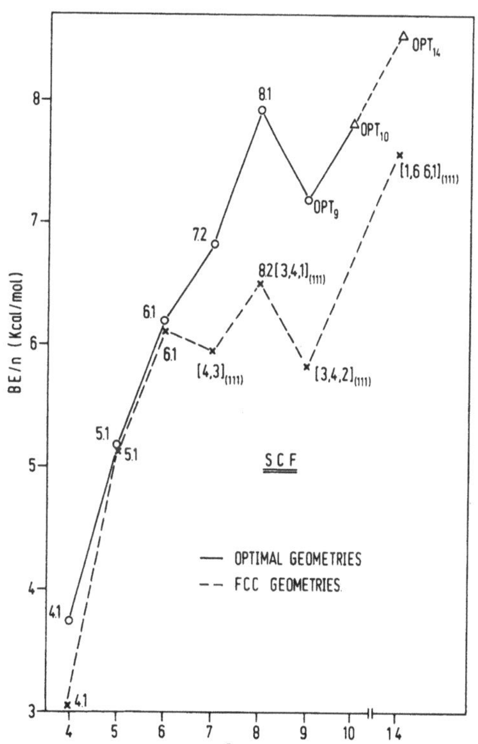

Fig. 4. BE/n (kcal/mol) for the SCF singlet states of Li clusters. The BE/n of Li clusters in optimal fcc geometry are displayed

geometries where the Jahn-Teller or pseudo-Jahn-Teller effect gives rise to a geometrical deformation which exhibits lower number of degeneracies or near-degeneracies of characteristic one-electron functions.

The high stability of a full electronic shell causes usually a high ionization potential. The vertical and adiabatic ionization potentials show a decreasing trend with the increasing nuclearity. The minima in ionization potentials for lithium clusters with odd number of atoms are evident from Fig. 5. The adia-

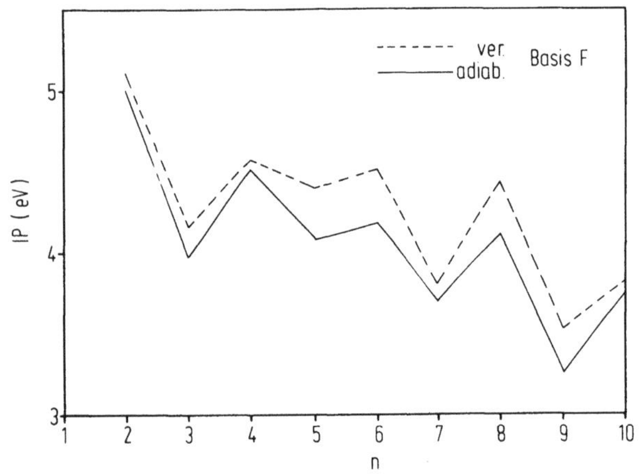

Fig. 5. The vertical and adiabatic ionization potentials (eV) as functions of the number of Li clusters. MRD–CI procedure with the basis (6s 1p/2s 1p)

batic ionization potential for Li_8 (T_d) is indeed quite high mainly if it is compared with small IP values for Li_7 and Li_9, but it does not exhibit any appreciable deviation from the decreasing trend of the IPs for Li_n with even n.

4. Cluster Geometries and Stabilities of Be_n [10–12]

No systematic search for optimal cluster geometries has been published until now. Nevertheless, it seems that the binding energy per atom is in general an increasing function of the number of the Be atoms in clusters which are sections of the hcp lattice. For Be_7, Be_{10} and Be_{13} very stable planar structures 7.3 (D_{6h}), 10.2 (C_{3v}) and 13.4 (D_{2d}) have been found, respectively. The average binding energy for Be_5 (5.2, D_{3h}) is higher than for Be_4 (4.3, T_d), which demonstrates that the number 8 ($8=2\times 1+2\times 3$) is not exceptionally favorable for the stability of the Be clusters. It is very well known that a strong hybridization among s and p atomic orbitals represents the main reason for the stability of the Be_n clusters with $n \geq 4$. Consequently the simple counting of valence electrons is not a sufficient measure of cluster stability.

5. Mixed Be – Li_k Clusters [5]

The consideration of the stability and other properties of the mixed Be – Li_k clusters allows to draw interesting conclusions about the general validity of the "super-atom" model beyond the simple case of alkali metal clusters. The SCF geometry optimization carried out until now shows can increasing stability of these clusters with an increasing even number k of lithium atoms ($2<k<6$). The optimal geometry exhibits high symmetries: $D_{\infty h}$, D_{4h}, and O_h for Li_2 – Be, Li_4 – Be and Li_6 – Be, respectively. Average binding per atom defined as

$$BE/k = (E_{\text{cluster}} - E_{\text{Be}} - k\,E_{\text{Li}})/(k+1)$$

where E_{cluster} is obtained from MRDCI method for the SCF optimized geometries, is equal to 6.6, 12.3 and 17.8 kcal/mol for Li_2 – Be, Li_4 – Be and Li_6 – Be, respectively. The HOMO or the second valence natural orbital according to the increasing occupation number is singly, doubly and triply degenerated, respectively. The number of available valence electrons in Li_k – Be clusters is sufficient to completely fill the degenerated highest molecular orbitals. Li_4 – Be prefers the less compact planar D_{4h} structure to the more compact one with T_d symmetry. The T_d symmetry group would allow for the triply degeneracy if the electronic occupancy would not cause the distortion of the cluster. An incomplete occupancy of the triply degenerated t orbitals causes Jahn-Teller or pseudo Jahn-Teller distortion. Further deformation towards rhombus structure (similar to the case of Li_4) is not favorable because the doubly degenerated HOMO in Li_4 – Be is fully occupied.

6. IIIa and IVa Clusters

The analysis of the electronic structure of IIIa and IVa tetramers based on the concepts of the $s-p$ hybridization leads to the prediction of their shape [13]. Since the tendency for hybridization is smaller in the second and third row in comparison with the first one, the reasons for the most stable forms built from second and third row atoms are also quite different from the reasons for the stability of B_4 and C_4. In general, the IVa clusters have not shapes corresponding to the sections of the diamond crystal lattice since the relatively small effective coordination number does not allow for the compensation of the promotion energy with a sufficient increasing number of bonds due to the sp^3 hybridization. The planar rhombic structure of Si_4 and Ge_4 are more stable than other more compact geometries of tetramers due to the pseudo Jahn-Teller effect which can be even qualitatively well understood when the symmetry orbitals built from the p – type AO's are considered. It is necessary of course to take into account the difficulties which can arise when the quantum chemical methods are applied on the systems with large number of electrons. Consequently, the careful investigation of the IIIa and IVa clusters it at the very beginning.

7. Comparison with Some other Theoretical Work on Ia Clusters

The effective core potential spin density scheme combined with the Hellman-Feynman forces calculations leads to the prediction of optimal geometries of Na clusters [6] which are in general agreement with the Hartree-Fock optimal Li cluster shapes [4, 8, 10]. Especially the rhombic form (4.1) for Na_4 tetramer, the planar geometry (5.1) of Na_5, the planar form (6.1, D_{3h}) and the pentagonal pyramid (6.2, C_{5v}) for Na_6 as well as the pentagonal bipyramid (7.1, D_{5h}) for Na_7 have stabilities in agreement with the predictions of the Hartree-Fock geometry search for Li_n clusters. Different results are obtained for Li_8 (8.1) and for Na_8 (8.2, C_{2v}).

The Generalized Valence Bond method yields also a high stability of planar Li_4 and Li_6 clusters and predicts that the clusters with optimal number of tetrahedral subunits should be very favorable [7].

The jellium droplet model for clusters agrees in predicting the large stability for Na_2 and Na_8, but it does not explain a relatively large abundancies for Na_7 and Na_{19} clusters [1].

8. Conclusions

1) The topology of the framework of the nuclei in a cluster influences strongly the electronic structure of clusters. The deviations from the spherical symmetry leads necessarily to interesting specific cluster properties.

2) Three factors must be considered when the optimal cluster geometries should be predicted.

i) the compactness of a cluster, which can be measured by the average coordination number, in general contributes favorably to the cluster stability.

ii) the nodal properties, degeneracies and occupancies of the relevant one-electron functions have a decisive influence on stability of given geometries of small clusters. Partial occupancy of valence molecular orbitals can cause destabilization of highly symmetric compact clusters shapes. The planar forms of very small stable clusters can result due to a lower dimension of the irreducible representations of the point groups to which the planar geometries belong.

iii) the polarization functions in the AO basis set play an important role in determining and for understanding of the cluster stability.

3) Local properties of the energy hypersurfaces are qualitatively well described in the one-electron approximation. The quantum chemical methods frequently yield several minima on the surfaces of the clusters. The inclusion of the electron correlation can easily inverse the energy sequence of the cluster structure corresponding to different SCF energy minima.

4) The hybridization of *s*- and *p*-type AO's is of large importance for the electronic properties of IIIa and IVa as well as of IIa clusters.

5) Theoretically predicted ionization potentials of Ia clusters show decreasing values with increasing nuclearity in agreement with the well known fact that the work functions of metal crystals have smaller values than the ionization potential of the corresponding atom.

6) In general, quantum chemical methods provide an interesting insight in the nature of interactions within clusters. Since the rules governing the cluster structures are very general the concurrent theoretical predictions of the clusters properties obtained from different quantum mechanical methods can be considered reliable.

This work was supported by the Deutsche Forschungsgemeinschaft Sonderforschungsbereich 6 "Structure and Dynamics of Interfaces" and by the Italian CNR. The authors would like to thank their collaborators H.-O. Beckmann, V. Bonačić-Koutecký, I. Boustani, G.-H. Jeung, G. Pacchioni, W. Pewestorf and D. Plavšić.

References

1. Knight, W.D., Clemenger, K., De Heer, W.A., Saunders, W.A., Chou, M.Y., Cohen, M.L.: Phys. Rev. Lett. **52**, 2141 (1984)
2. Hoare, M.R.: Adv. Chem. Phys. **40**, 49 (1979)
   Hoare, M.R., Pal, P.: Adv. Phys. **20**, 161 (1971)
3. Buenker, R.J., Peyerimhoff, S.D.: Theor. Chim. Acta **35**, 33 (1974)
   Buenker, R.J., Peyerimhoff, S.D., Butscher, W.: Mol. Phys. **35**, 771 (1978)
4. Koutecký, J., Boustani, I., Pewestorf, W., Fantucci, P., Bonačić-Koutecký, V.: (to be published)
5. Pewestorf, W., Koutecký, J.: (to be published)
6. Martins, J.L., Buttet, J., Car, R.: Phys. Rev. B **31**, 1804 (1985)
7. McAdon, M.H., Goddard, W.A. III: Phys. Rev. Lett. **55**, 2563 (1985)
   McAdon, M.H., Goddard, W.A. III: J. Non-Cryst. Solids **75**, 149 (1985)
8. Fantucci, P., Koutecký, J., Pacchioni, G.: J. Chem. Phys., **80**, 325 (1984)
   Pacchioni, G., Koutecký, J.: J. Chem. Phys. **81**, 3588 (1984)
9. Beckmann, H.-O., Koutecký, J., Botschwina, P., Meyer, W.: Chem. Phys. Lett. **67**, 119 (1979)
   Beckmann, H.-O., Koutecký, J., Bonačić-Koutecký, V.: J. Chem. Phys. **73**, 5182 (1980)
10. Pacchioni, G., Plavšić, D., Koutecký, J.: Ber. Bunsenges. Phys. Chem. **87**, 503 (1983)
    Pacchioni, G., Koutecký, J.: Ber. Bunsenges. Phys. Chem. **88**, 242 (1984)
    Pacchioni, G., Koutecký, J.: J. Chem. Phys. **71**, 181 (1982)
11. Pacchioni, G., Pewestorf, W., Koutecký, J.: J. Chem. Phys. **83**, 201 (1984)
12. Bauschlicher, C.W. Jr., Petterson, L.G.M.: J. Chem. Phys. **84**, 2226 (1986)
13. Koutecký, J., Pacchioni, G., Jeung, G.-H., Hass, E.-C.: Surf. Sci. **156**, 650 (1985)

J. Koutecký
Institut für Physikalische und Theoretische Chemie
Freie Universität Berlin
Takustrasse 3
D-1000 Berlin 33
Germany

P. Fantucci
Dipartimento di Chimica Inorganica e Metallorganica
Centro CNR
Università di Milano
Via Venezian, 21
I-20133 Milano
Italy

Geometrical Structure of Metal Clusters

J. Buttet
Institut de Physique Expérimentale, Ecole Polytechnique Fédérale de Lausanne, Lausanne, Switzerland

Received April 24, 1986

Extended Abstract

The geometrical structure of clusters as the size decreases is strongly influenced by the increasing importance of the surface energy. However for very small clusters electronic effects, such as the lowering of energy by Jahn-Teller distortions, have to be taken into account. In the first part we thus give some experimental and theoretical results for large clusters (diameter >20 Å) where a wealth of experimental information has been obtained by electron microscopy and electron diffraction. In the second part we describe theoretical results obtained for very small alkali metal clusters and compare them with the available experimental data.

It is well known that for some metals, e.g. platinum and aluminium, most of the aggregates are single crystals which retain the bulk structure, while for other metals, e.g. gold and silver, the surface induces changes in the structure which lead to the multiply twinned particles (MTP) described by Ino [1, 2]. The present day electron microscopes allow to study in detail the MTP's and their defects using high resolution lattice imaging, dark field and microdiffraction techniques. In a comparative electron diffraction study [3, 4] the lattice parameter changes and the integrated width of the (220) diffraction line have been observed as a function of size for gold and platinum particles of diameters comprised between 30 Å and 200Å. For both gold and platinum the lattice parameter a varies linearly as a function of the inverse diameter D, in agreement with a simple drop model [5]. The average surface stress $<\gamma>$, deduced from the drop model and the variation of a, is approximately equal to the surface tension σ of the bulk metal in the case of platinum. It is larger than σ for gold particles, which is in agreement with the elastic deformation of the tetrahedra forming the MTP [6]. The width Δ of the integrated diffraction peak follows the usual "Scherrer straight line" $\Delta = 1.3/D$ for platinum clusters. For gold clusters the different behavior reflects their multiply twinned structure.

Theoretical studies of the geometrical structure and electronic properties of particles having more than 100 atoms have only been carried using semiempirical methods. The change of structure when the size decreases from a cubooctahedral geometry to an icosahedral geometry has been studied for Ni and Pt clusters [7, 8]. A transition from a bcc to a fcc structure has been found for transition metal clusters using a tight-binding approximation [9]. Recently the cohesive energy of $3d$ transition metals has been calculated [10, 11] for different bcc and fcc clusters, using empirical N-body potentials proposed by Finnis and Sinclair [12]. No transition from bcc to fcc was found down to the smallest sizes. Using the same effective potential and starting from a dodecahedra, the forces were calculated on each atom and the relaxed geometry was studied for different size clusters (15–1,695 atoms).

With the cluster sources now available aggregates of essentially all elements, comprising between 3 and 100 atoms, can be formed in a beam. However the experimental information available on their structural geometry is very small and the comparison between the growing number of ab-initio calculations [13] and the experimental data is difficult. In the case of alkali aggregates a complete study of the Born-Oppenheimer surface of Li_3, K_3, Na_3 has been reported [14] using an approach based on the density functional theory. For Na, K and Li the equilibrium geometries as well as the dynamic behavior (passage from one equilibrium configuration to another through pseudorotation) are in agreement with the electron spin resonance measurements [15, 16] of alkali trimers in a rare gas matrix. The vibrational frequencies of the normal modes of sodium trimers in its ground state (obtuse isosceles triangle) have been obtained in analyzing the hot bands observed during a systematic study of the excitation spectrum of Na_3 [17, 18]. The measured values (49, 87 and 139 cm^{-1}) are in agreement with the theoretical results (58, 94, 142 cm^{-1}) deduced from an analytic representation [19] of the potential energy surfaces.

The structural and electronic properties of the neutral and ionized sodium clusters with $n \leq 8$ and $n=13$ have been investigated in a selfconsistent pseudopotential local spin-density calculation [20]. The equilibrium geometries have been obtained in starting from randomly generated cluster geometries and letting them relax under the action of the forces on the atoms [21]. The clusters with up to five atoms have closepacked planar equilibrium geometries, the six-atom cluster is quasiplanar with a planar isomer lying only 0.04 eV higher in energy, real three dimensional structures only occur when the number of atoms is greater than or equal to seven. Similar geometrical structures have been obtained for lithium [13]. The results are in good agreement with the measured ionization potentials [22–24]. For a seven atoms cluster, which is predicted to be a pentagonal bipyramid, the spin densities deduced from the hyperfine structure of electron spin resonance data [25] are in excellent agreement with the theoretical prediction. A simple model which takes into account the delocalized nature of the valence electrons and the Jahn-Teller effect explains the main features of the equilibrium geometries.

References

1. Ino, S.: J. Phys. Soc. Jpn. **27**, 941 (1969)
2. See also more recent thermodynamical treatments which take into account inhomogeneous strains, e.g.: Marks, L.D.: Surf. Sci. **150**, 358 (1985)
3. Solliard, C., Flüeli, M.: Surf. Sci. **156**, 487 (1985)
4. Solliard, C.: PhD Thesis, EPFL (1983); (to be published)
5. For a justification of the drop model in the case of a solid aggregate, see: Borel, J.P., Châtelain, A.: Surf. Sci. **156**, 572 (1985)
6. The surface stress coefficient γ is related to the surface tension σ by the relation $\gamma=\sigma+A\,\mathrm{d}\sigma/\mathrm{d}A$ where $A\,\mathrm{d}\sigma/\mathrm{d}A$ corresponds to the elastic deformation of the surface A
7. Gordon, M.B., Cyrot-Lackmann, F., Desjonquères, M.C.: Surf. Sci. **80**, 159 (1979)
8. Khanna, S.N., Bucher, J.P., Buttet, J.: Surf. Sci. **127**, 165 (1983)
9. Tomanek, D., Mukherjee, S., Bennemann, K.H.: Phys. Rev. B **28**, 665 (1983)
10. Marville, L.: Diplom work, EPFL (1985)
11. Marville, L., Andreoni, W.: (to be published)
12. Finnis, M.W., Sinclair, J.E.: Philos. Mag. A **50**, 46 (1984)
13. For a recent review, see: Koutecky, J., Fantucci, P.: Chem. Rev. (1986)
14. Martins, J.L., Car, R., Buttet, J.: J. Chem. Phys. **78**, 5646 (1983)
15. Thompson, G.A., Lindsay, D.M.: J. Chem. Phys. **74**, 959 (1981)
16. Garland, D.A., Lindsay, D.M.: J. Chem. Phys. **78**, 2813 (1983)
17. Delacrétaz, G.; PhD Thesis, EPFL (1985)
18. Broyer, M., Delacrétaz, G., Labastie, P., Whetten, R.L., Wolf, J.P., Wöste, L.: Z. Phys. D – Atoms, Molecules and Clusters **3**, 131 (1986)
19. Thompson, T.C., Izmirlian, G., Lemon, S.J., Truhlav, D.G., Mead, C.A.: J. Chem. Phys. **82**, 5597 (1985)
20. Martins, J.L., Buttet, J., Car, R.: Phys. Rev. B **31**, 1804 (1985)
21. Martins, J.L., Car, R.: J. Chem. Phys. **80**, 1525 (1984)
22. Herrmann, A., Schumacher, E., Wöste, L.: J. Chem. Phys. **68**, 2327 (1978)

23. Peterson, K.I., Dao, P.D., Farley, R.W., Castleman, A.W.: J. Chem. Phys. **80**, 1780 (1984)
24. Wöste, L.: Private communication
25. Thompson, G.A., Tischler, F., Lindsay, D.M.: J. Chem. Phys. **78** 5946 (1983)

J. Buttet
Ecole Polytechnique Fédérale
de Lausanne
Institut de Physique Expérimentale
PHB-Ecublens
CH-1015 Lausanne
Switzerland

Electronic Structure and Bonding in Clusters: Theoretical Studies

K. Hermann and H.J. Hass

Institut für Theoretische Physik, Technische Universität Clausthal
and Sonderforschungsbereich 126, Clausthal/Göttingen, Federal Republic of Germany

P.S. Bagus

IBM Almaden Research Center San Jose, California, USA

Received April 7, 1986

The electronic structures of small Al_n, $n=5$, 9, 13, clusters with bulk geometry are studied using the ab initio Hartree-Fock-LCAO method. The cluster ground states have always multiplicity higher than the lowest possible value. However, the energy difference between ground and lowest low spin state decreases with increasing cluster size. The energy range of the Al_n cluster valence levels is comparable with the width of the occupied part of the $3sp$ band in bulk Al. The different binding mechanisms that arise when a CO molecule interacts with Al_n clusters in different coordination sites are analyzed in detail with the constrained space orbital variation (CSOV) method. Electrostatic and polarization contributions to the interaction are found to be important. Among charge transfer (donation) contributions π electron transfer from Al_n to CO corresponding to π backbonding is energetically more important than σ electron transfer from CO to Al_n characterizing the σ bond.

PACS: 36.40.+d; 31.20.Ej; 71.45.Nt

1. Introduction

One of the features which makes studies of clusters extremely attractive is that these systems can provide a link between very small atomic or molecular systems and the extended solid. However, one expects and knows from experiment that a number of properties of small metal clusters can be quite different from those of respective bulk metal systems. For example, the spatial symmetry of clusters can differ substantially from that of the bulk crystalline structure. Fivefold symmetry axes postulated for free clusters [1] cannot exist in the ideal solid. Further, electronic excitations and ionization of clusters as observed by photoemission spectroscopy [2, 3], should be different to respective processes in the solid. Also the total spin symmetry of clusters does not necessarily reflect the bulk situation. Cluster states of high spin multiplicity can be energetically favoured whereas the solid system is not expected to have a net spin as will be discussed below. Finally, charging effects can occur in clusters as a consequence of variations in the local environment of the different cluster atoms. On the other hand, one should expect that local electronic properties of solids are rather similar to those of respective clusters which makes clusters very interesting model systems to study local phenomena in solid systems. This approach has been widely used with great success for solid surfaces and adsorbate systems [4].

In the surface cluster approach one assumes that the interaction between an adsorbed molecule or atom and a solid substrate is confined to a few substrate atoms near the surface. Then one can describe the interaction approximately by that in a fictitious cluster (surface cluster) which contains only the adsorbate and a few substrate atoms near the surface.

The electronic coupling of this cluster with the rest of the substrate (embedding) would have to be accounted for in a second step. Alternatively, one could try to increase the cluster by adding substrate atoms until the respective properties become independent of cluster size (cluster convergence). Since in this approach one tries to simulate a surface-like situation by a cluster one very often uses cluster geometries and electronic states that are deduced from bulk and surface situations. Therefore, one cannot expect these systems to reflect the true geometries and states of free metal clusters. However, it is believed that a number of qualitative conclusions that one can draw from these fictitious clusters are true for real cluster and surface systems.

The electronic states of the isolated clusters can be obtained from standard quantum chemical methods. However, these methods set an upper limit to the size of clusters that can be treated theoretically. Based on ab initio methods and the use of pseudopotentials for core electrons, one can study clusters of 50–100 atoms with computers that are presently available. These methods can give information about the physical parameters of the system. The local minima of the total energy E_{tot} of the cluster as a function of its geometry define possible equilibrium geometries. A comparison of the total energy at a given equilibrium geometry with that where parts of the cluster are separated from each other yields respective binding energies. Local ionicities can be obtained from an inspection of the electronic charge distribution. Further, the binding between atoms in the cluster can be characterized qualitatively with one-electron function or population analyses or using more sophisticated methods like the recently developed constrained space orbital variation (CSOV) analysis [5, 6] which will be discussed later. Vibrational properties of the clusters can be calculated and respective excitation energies determined e.g. using the harmonic approximation [7]. Finally, the various ionization potentials of the clusters can be obtained with different approximations [8]. In the surface cluster approach one identifies the physical quantities calculated for the isolated cluster with those of the surface system. The validity of this approach has to be tested for each system separately.

The general principles discussed so far will be illustrated in the following by results of recent model calculations on Al_n and Al_nCO clusters used to study the variation of the electronic properties with cluster size and to characterize the details of the interaction between a CO molecule and metal clusters. In Sect. 2, some of the technical details are briefly outlined while Sect. 3 presents a discussion of the numerical results. Finally, the conclusions are summarized in Sect. 4.

2. Technical Details

Figure 1 gives a schematic representation of the Al_n and Al_nCO clusters used in the present study. The smallest system is the Al atom. The geometries of the larger Al_n clusters, $n=4$, 5, 9, 13, are chosen to represent sections of bulk Al near the (100) surface and are kept fixed in the calculations. In the Al_nCO clusters, $n=1$, 5, 9, 13, the CO molecule is assumed to approach in a singly coordinated on top site. In addition, CO interacting with the metal in a two-fold coordinated bridge site is considered in Al_4CO. In the calculations the C–O and Al–CO distances of the clusters are varied to obtain equilibrium geometries.

The electronic states of the clusters are calculated with the ab initio Hartree-Fock-LCAO method [9] using flexible all-electron basis sets of contracted Gaussian-type orbitals for the central Al atoms and for C, O in the Al_nCO clusters. The Ne core of the peripheral Al centers is described by appropriate pseudopotentials [10]. In order to identify the cluster ground state a large number of cluster wavefunctions with different spatial and spin symmetry are considered.

The binding between a CO molecule and the metal clusters can be characterized qualitatively by population analyses or analyses of respective electron distributions. However, such analyses can fail particularly

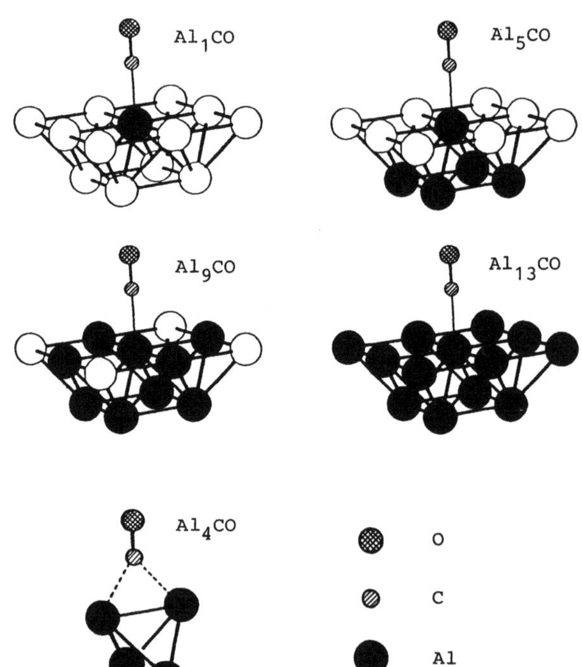

Fig. 1. Schematic representation of the Al_n and Al_nCO clusters, $n=1$, 4, 5, 9, 13, used in the present study

in larger clusters and when large basis sets with considerable overlap are used. Therefore, in this study the $CO-Al_n$ interaction is analyzed with the constrained space orbital variation (CSOV) method [5, 6]. This method has been developed recently to determine the energetic importance of different contributions to the interaction between a molecule and a cluster. Its basic ideas will be mentioned briefly in the following.

In the CSOV method one defines a CSOV step as the self consistent solution of the N-electron equations for the system MolClu consisting of two components, Mol (molecule) and Clu (cluster), where one imposes two additional boundary conditions. First, the charge density of one of the components (Mol or Clu) is fixed at a given density. Second, the variational space of the complementary component (Clu or Mol) is constrained on the basis of physical principles. To use this for the analysis of the interaction between Mol and Clu one calculates a sequence of CSOV steps with different boundary conditions.

The starting point is the superposition of the charge densities of the two separate components, Mol and Clu, at the equilibrium geometry of the combined system determined in a previous calculation. A comparison of the total energy of this situation with that of the two separate components yields an interaction energy which represents the purely electrostatic effect of superposition. In the first CSOV step one keeps the charge density of Mol fixed in its free state and solves the N-electron equations for Clu in the variational space of the cluster basis functions. This describes physically the polarization of the cluster in the presence of the frozen molecule. If the variational space for the Clu functions is increased by including virtual basis functions of Mol then possible charge transfer from the cluster to the molecule can be described in addition. This charge transfer can be split into e.g. different symmetry contributions if the virtual functions are selected appropriately. In the third CSOV step one freezes the charge distribution of Clu from the previous step and solves the N-electron equations for Mol in the variational space of Mol. This accounts for polarization of the molecule in the presence of the frozen cluster. Finally, in the fourth step virtual functions of Clu are included in the variational space of Mol which allows for charge transfer from the molecule to the cluster.

For a complete description of the interaction between molecule and cluster one would have to repeat the above CSOV sequence until self-consistency is achieved. However, in the systems we have studied so far the electronic structure at the end of the first sequence was always close to the self-consistent solution so that further sequences were not necessary. The main virtue of the analysis lies in the fact that one can calculate the total energy of the combined system MolClu for each CSOV step separately. Therefore, one can determine the energy contributions of different physical effects like polarization and charge transfer to the total binding energy and make quantitative statements about the relative importance of these effects for the molecule-cluster interaction.

3. Results and Discussion

3.A. Al_n *Clusters*

Table 1a lists the ground state configurations and calculated atomization energies E_{at} of the Al_n clusters, $n=5, 9, 13$. Here, E_{at} is defined as the total binding energy of the cluster with respect to separated atoms divided by the number of cluster atoms. First, we note that the clusters are all stable with respect to separated atoms and the atomization energy increases with cluster size. However, the binding of 0.3 eV per atom seems to be rather weak compared to what one would expect for free clusters. One reason for this is the fact that we chose cluster geometries corresponding to sections of bulk Al which will certainly not reflect the true geometry of the free clusters. Another reason lies in the method that we use to calculate the electronic states and total energies. In the Hartree-Fock method electron correlation is neglected which in general underestimates binding energies.

The ground states of the clusters are always found to be states of high spin multiplicity, quartet states as indicated in Table 1a. This is in contrast to the Al bulk and surface situation where low spin states are expected to be energetically favoured. In order

Table 1. Configurations and calculated atomization energies E_{at} of the Al_n clusters, $n=5, 9, 13$. The results are given for the cluster ground and surface states defined in the text

a) Ground states

n	Configuration	E_{at}(eV/atom)
5	$(6a_1^1 1b_1^2 1b_2^2 3e^2, {}^4X)$	0.26
9	$(7a_1^1 1b_1^2 2b_2^2 5e^2, {}^4X)$	0.33
13	$(8a_1^2 3b_1^1 2b_2^2 1a_2^2 6e^2, {}^4X)$	0.39

b) Surface states

n	Configuration	E_{at}(eV/atom)
5	$(6a_1^1 1b_1^2 1b_2^2 3e^2, {}^2X)$	0.06
9	$(7a_1^2 1b_1^2 2b_2^2 5e^1, {}^2X)$	0.28
13	$(8a_1^2 2b_1^2 2b_2^2 1a_2^2 6e^3, {}^2X)$	0.37

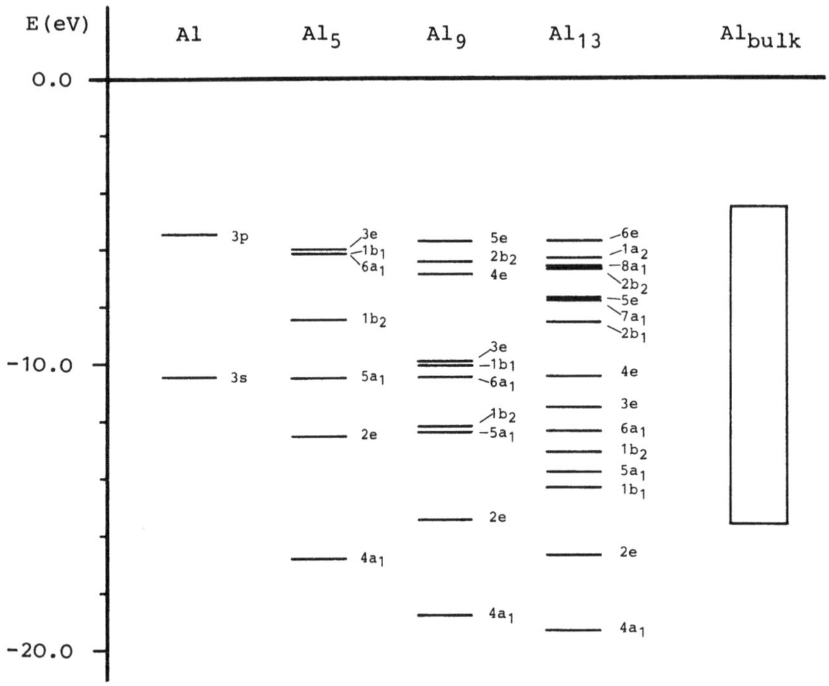

Fig. 2. Valence level diagrams of the Al_n clusters, $n=1, 5, 9, 13$. The levels are labeled according to the irreducible representations of the C_{4v} symmetry group describing the cluster symmetry. The width of the occupied part of the $3sp$ band in bulk Al obtained from band structure methods [11] is shown for comparison

to simulate a surface-like situation in our clusters we considered in addition to the cluster ground states the energetically lowest doublet (low spin) states. The results for these states named "surface" states are listed in Table 1b. The atomization energies of the surface states are somewhat below those of the ground states but the differences turn out to be quite small.

The valence level diagrams of the Al clusters are shown in Fig. 2. Here the levels refer to Hartree-Fock one-electron energies of the occupied $3sp$-derived cluster orbitals of Al_n, $n=1, 5, 9, 13$. Obviously, the energy range of these levels does not vary substantially with cluster size. One finds a width of about 14 eV which is rather similar to the width of the occupied part of the $3sp$ band in bulk Al. Band calculations [11] yield 12 eV as indicated on the right of Fig. 2. However, this comparison should not be taken too quantitatively since Hartree-Fock one-electron energies differ in their meaning from one-electron energies of band theories based on the local density approach [12]. Cluster orbitals within the valence region can be rather different in their character. For example, in Al_{13} the $5a_1$ orbital of the lower part of the valence region lies inside the cluster volume as can be seen from the contour plot of Fig. 3a. In contrast, the energetically highest orbital, 6e, is localized dominantly on edge atoms of the cluster and extends well beyond the cluster volume. Orbitals of this type are expected to become important when small molecules approach the cluster and covalent bonds are formed.

Müller and Bagus [13] have studied large Cu_n clusters with up to 50 atoms with crystalline structures analogous to the ones used in the present study. Their Hartree-Fock valence level diagrams show essentially the same behavior as we find. The energy range of the Cu sp-derived valence levels does not vary substantially with cluster size. One finds a width of about 9 eV compared to 8 eV for the occupied part of the 4sp band in bulk Cu [11].

3.B. Al_nCO *Clusters*

Table 2 contains numerical results of the Al_nCO clusters. The cluster configurations, CO – Al binding energies and equilibrium distances are listed for those clusters considered for CO on top site stabilization. While Table 2a shows the cluster ground state results Table 2b refers to surface states defined as the energetically lowest low spin states. The last column of Table 2b gives also the excitation energies E_{ex} between ground and surface states.

Obviously, CO does not bind to a single Al atom. For a CO – Al separation where the molecule stabilizes on the larger clusters there is a repulsion of 0.14 eV. The origin of this repulsion has been discussed in a previous paper [14] and will become evident later. In the ground state of Al_5CO CO is only

a) Al_{13} $5a_1$ orbital

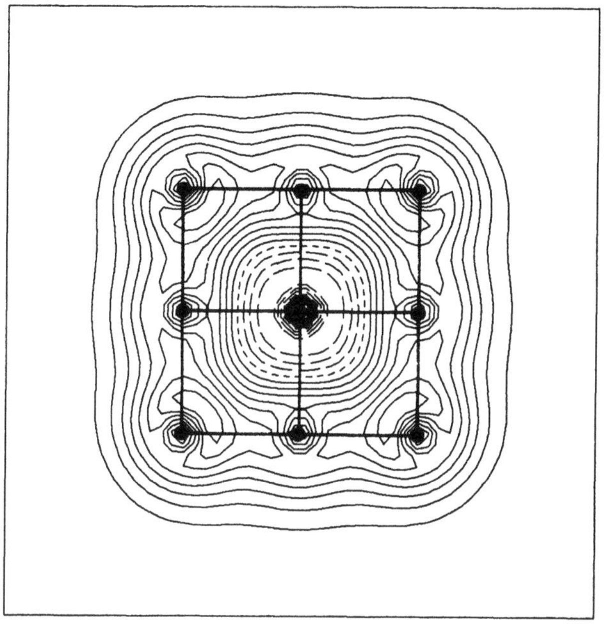

b) Al_{13} 6e orbital

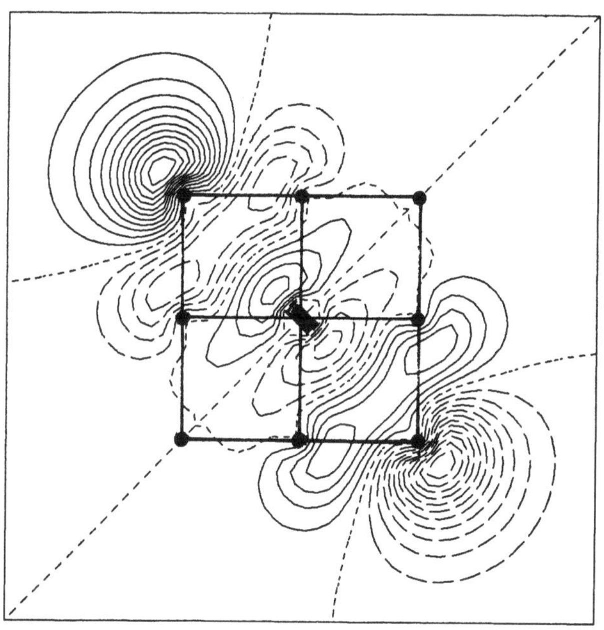

Fig. 3. Contour plots of the $5a_1$ (Fig. 3a) and 6e orbitals (Fig. 3b) of Al_{13} for the planar section containing 9 Al centers indicated in the plots. The contour values vary between −0.055 a.u. and +0.055 a.u. in equidistant steps of 0.005 a.u. Dashed (full) lines refer to negative (positive) values

weakly bound by 0.13 eV whereas the binding is stronger on the larger clusters, 0.90 eV for Al_9CO and 0.80 eV for $Al_{13}CO$. These results might suggest a convergence of the binding energy with cluster size.

Table 2. Results for CO on top site binding in Al_nCO, $n=1$, 5, 9, 13. Here E_b^{CO} denotes the CO−Al binding energy and d_{min} is the C−Al equilibrium distance. The results are given for the cluster ground and surface states defined in the text. Table 2b also lists the excitation energies E_{ex} between ground and surface states

a) Ground states

n	Configuration	E_b^{CO} (eV)	d_{min} (bohr)
1	$(9a_1^2 3e^1, {}^2X)$	−0.14	3.93
5	$(11a_1^1 1b_1^2 1b_2^2 4e^2, {}^4X)$	0.13	3.85
9	$(12a_1^1 1b_1^2 2b_2^2 6e^2, {}^4X)$	0.90	3.83
13	$(13a_1^2 3b_1^1 2b_2^2 1a_2^2 7e^2, {}^4X)$	0.80	3.90

b) Surface states

n	Configuration	E_b^{CO} (eV)	d_{min} (bohr)	E_{ex} (eV)
1	$(9a_1^2 3e^1, {}^2X)$	−0.14	3.93	0.0
5	$(11a_1^1 1b_1^2 1b_2^2 4e^2, {}^2X)$	0.56	3.77	0.56
9	$(11a_1^2 1b_1^2 2b_2^2 6e^3, {}^2X)$	0.85	3.82	0.94
13	$(13a_1^2 2b_1^2 2b_2^2 1a_2^2 7e^3, {}^2X)$	0.73	3.91	0.37

However, the number and size of the clusters considered here does not make such a statement very meaningful. A binding energy of about 0.2 eV has been observed for the CO/Al(111) adsorbate system [19].

A comparison of the ground state binding energies with those of the surface states shows major differences only for Al_5CO where CO is much more strongly bound in the surface (0.56 eV) than in the ground state (0.13 eV). For the larger clusters the differences are rather small; the variations lie between 0.05 and 0.1 eV. This is interesting in view of the fact that numerous cluster model studies for surface systems have been published [4] where the authors did not consider cluster ground states but concentrated on low spin surface states in the calculations. The present data suggest that for large enough clusters the differences between ground and surface states can be neglected.

The naive interpretation of the CO−Al binding is based on a molecular orbital scheme. In this picture, one would describe the CO−Al bond formation in analogy to the traditional metal carbonyl binding scheme [15]. One assumes that of the occupied CO valence orbitals 4σ, 5σ, and 1π only the 5σ orbital is involved in the interaction with the metal. This orbital which is localized on the C side of CO admixes metal $3sp$ valence contributions to form the σ bond. Further, metal valence functions of π character can admix contributions of CO 2π orbitals (which are unoccupied in the free molecule) leading to the π back-bonding effect. The σ bond results in a charge transfer

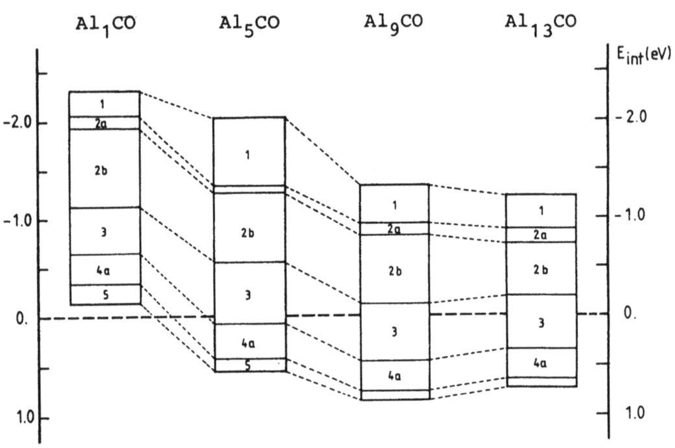

Fig. 4. CSOV results for CO on top site binding in Al_nCO, $n=1$, 5, 9, 13. For a definition of the interaction energies E_{int} see text

from CO to the metal which is compensated by the π backbonding effect leading to an overall neutral CO-metal coupling. This binding picture can be tested with the CSOV analysis.

The results of the CSOV analysis for the Al_nCO clusters are shown in Fig. 4. Here the interaction energies E_{int} of the different CSOV steps are given as block diagrams for each cluster where the analysis refers to the respective CO – Al equilibrium geometry of the surface state. The interaction energy E_{int} is defined as

$$E_{int} = E_{tot}(Al_n) + E_{tot}(CO) - E_{tot}(Al_nCO;\ \text{CSOV step})$$

where $E_{tot}(Al_nCO;$ CSOV step) is the total energy of Al_nCO at the respective CSOV step and $E_{tot}(Al_n)$, $E_{tot}(CO)$ are the total energies of the separate components. Negative E_{int} values indicate repulsion between CO and Al_n and positive values mean attraction.

The electrostatic interaction of the charge superposition of the separate components represented by the top of each block in Fig. 4 yields a repulsive energy of 1.3–2.3 eV. This repulsion is reduced by 0.3–0.5 eV due to polarization of the Al_n clusters in the first CSOV step. An analysis of the charge redistribution in this first step shows that the polarization is dominated by $\sigma(a_1)$ symmetry orbitals of Al_n which are described as Al $3s$-type valence functions hybridizing with $3p$ functions to move away from the CO molecule. This effect has been interpreted in a number of CO-metal systems as σ repulsion [14]. Obviously, $3sp$ hybridization costs more energy in the Al atom than in the larger clusters which explains why the energy gain due to polarization is smallest for the Al atom. This effect together with the strongest repulsion of the charge superposition is the origin of the overall repulsive interaction between CO and the Al atom. Electron transfer from Al_n to CO included in the second CSOV step increases the interaction energy by 0.7–0.9 eV. This contribution can be split into electron transfer of $\sigma(a_1)$ symmetry, 2a, and $\pi(e)$ symmetry, 2b. It is clear from the diagrams of Fig. 4 that the dominant contribution originates from π electron transfer from Al_n to CO. The interaction energy is further increased by 0.5–0.7 eV in the third CSOV step which accounts for polarization of the CO subunit resulting in an overall attractive CO-Al interaction in the larger clusters. The fourth CSOV step which describes charge transfer from CO to Al_n increases the interaction energy by 0.3 eV roughly independent of cluster size. Obviously, this charge transfer is dominated by σ electrons, step 4a. The fourth CSOV step is rather close to the fully self-consistent result without any constraints defining the bottom of each block. The difference is always below 0.1 eV. Thus, the first CSOV sequence is sufficient to describe the interaction contributions almost quantitatively.

The important results of this analysis are first that electrostatic and polarization contributions which cannot be assigned to specific electron functions become important for the CO-metal interaction. Second, among the charge transfer (donation) contributions π electron transfer from Al_n to CO, which corresponds to a mixing of metal valence with CO 2π functions or π backbonding in the MO picture, is energetically more important than σ electron transfer from CO to Al described as a mixing CO 5σ with metal valence functions characterizing the σ bond. The importance of polarization contributions for the CO-metal interaction has not been recognized so far and the second result concerning donation contributions does not agree with the Blyholder [15] scheme of metal carbonyl binding mentioned above. This scheme has been widely applied in the interpretation of photoemission spectra. However, we have shown recently [16, 17] that the CO-metal binding picture suggested by the CSOV analysis is consistent with results from photoemission spectroscopy.

The details of the CO-metal binding can be quite different for different binding sites of the cluster. This will be illustrated by comparing results of the CO on top site binding discussed above with CO binding in a 2-fold bridge site. For the bridge site an Al_4CO cluster (see Fig. 1) with tetrahedral symmetry of the metal subunit is considered which seems to be sufficient for a qualitative comparison. In addition to the various binding contributions frequencies for C – O and CO – Al stretching vibrations are determined using the harmonic approximation. Table 3 compares the binding and vibrational results for the on top site Al_5CO and the bridge site Al_4CO clusters. The

Table 3. Comparison of the binding and vibrational results for Al_5CO (on top site) and Al_4CO_1 (bridge site). The calculations refer to surface states, $(11a_1^1\ 1b_1^2\ 1b_2^2\ 4e^2,\ ^2X)$ of Al_5CO and $(12a_1^2 7b_1^2 3b_2^2 1a_1^2,\ ^1X)$ of Al_4CO. Table 3a lists Co–al binding energies E_b^{CO}, equilibrium distances d_{min} (C–O) and d_{min}(C–Al), and frequencies of the C–O and CO–Al stretching vibrations. Values obtained for free CO are also shown for comparison. Table 3b contains results of the CSOV analysis. For a definition of E_{int}, given in eV, see text

a) Binding Results

Cluster		Al_5CO (on top)	Al_4CO (bridge)	CO
E_b^{CO}	(eV)	0.56	0.03	–
d_{min}(C–O)	(bohr)	2.14	2.22	2.124
d_{min}(C–Al)	(bohr)	3.76	3.18	–
$\hbar\omega$(C–O)	(meV)	261	210	280
$\hbar\omega$(CO–Al)	(meV)	35	23	–

b) CSOV Analysis

	E_{int} (on top)	E_{int} (bridge)
0. Charge superposition	−2.03	−4.24
1. Polarization Al_n	−1.33	−3.31
2a. $\sigma(a_1)$ charge transfer Al_n to CO	−1.25	−3.02
2b. total charge transfer Al_n to CO	−0.54	−1.39
3. Polarization CO	+0.08	−0.53
4a. $\sigma(a_1)$ charge transfer CO to Al_n	0.43	−0.22
4b. total charge transfer CO to Al_n	0.44	−0.19
5. unconstrained SCF	0.56	+0.03

equilibrium distance d_{min}(C–Al) of CO in the bridge site is considerably smaller than in the on top site and the bridge site binding energy E_b^{CO} is much samller than that of the on top site. Naively one would expect from symmetry arguments that CO stabilizing in a bridge site leads to larger π electron transfer from the metal cluster to CO compared to the on top site. This is confirmed by the results from CSOV analyses for the two clusters in their respective equilibrium geometries (cp. Table 3b). The difference of the interaction energies E_{int} between CSOV steps 2a and 2b which accounts for π(e) electron transfer from Al_n to CO is more than twice as large in the bridge (1.63 eV) than in the on top site (0.71 eV). This π electron transfer leads to a partial occupation of the CO 2π orbitals which are anti-bonding in CO so that the C–O bond is weakened and as a consequence the C–O stretching frequency is lowered. Based on the CSOV results, this effect should be larger in the bridge than in the on top site which is consistent with the calculated C–O stretching frequencies of Table 3a. However, the increased π electron transfer in the bridge site does not result in an overall stronger CO-metal binding. In contrast, the binding energy in the bridge site is rather weak (0.03 eV) compared to the on top site (0.56 eV). This is mainly due to difference in the electrostatic repulsion when the CO and Al_n charge densities are superimposed. In the bridge site this repulsion (4.24 eV) is much larger than in the on top site (2.03 eV) since CO stabilizes closer to the cluster and therefore penetrates more into the cluster charge distribution. These results show clearly the importance of electrostatic contributions to the CO-metal binding when different binding sites are considered.

A similar behavior was found in a study of the interaction of NH_3 with an Al_{10} cluster [18]. Here the NH_3–Al interaction in the on top site leads to stabilization with a binding energy of 0.8 eV whereas the interaction curve of the 3-fold hollow site is always repulsive. This difference is a consequence of much larger electrostatic repulsion and smaller polarization contributions in the 3-fold hollow compared to the on top site [18].

4. Conclusions

In summary, the theoretical methods applied in this study can give useful information about the electronic structure of small metal clusters. The ground states of Al_n, $n=5$, 9, 13, clusters with bulk geometry are always found to be of high spin multiplicity. However, the energy difference between ground and lowest low spin state ("surface" state) of Al_n decreases with increasing cluster size. It is found that the energy range of the Al_n valence levels is rather similar to the width of the occupied part of the $3sp$ band in bulk Al. However, the character of the cluster orbitals is quite different to that of the one-electron functions in the bulk.

The different bonding mechanisms that arise when metal clusters interact with molecules can be characterized in detail with the constrained space orbital variation (CSOV) method [5, 6]. For the interaction of Al_n clusters with CO this analysis indicates that electrostatic and polarization contributions which cannot be assigned to specific electron functions become important. Further, among the charge transfer (donation) contributions π electron transfer from Al_n to CO, corresponding to π backbonding in the MO picture, is energetically more important than σ electron transfer from CO to Al_n characterizing the σ bond. A comparison of CO binding to Al_n in different coordination sites gives further evidence for the importance of electrostatic contributions to the CO-metal interaction.

References

1. Farges, J., Feraudy, M.F. de, Raoult, B., Torchet, G.: Surf. Sci. **156**, 370 (1985)
2. Baetzold, R.C.: Surf. Sci. **106**, 243 (1981)
3. Schmidt-Ott, A., Federer, B.: Surf. Sci. **106**, 538 (1981)
4. See e.g. Müller, J.E.: Proceedings ECOSS-8, Jülich 1986; Surf. Sci. (to appear)
5. Bagus, P.S., Hermann, K., Bauschlicher, C.W.: J. Chem. Phys. **80**, 4378 (1984)
6. Bagus, P.S., Hermann, K., Bauschlicher, C.W.: J. Chem. Phys. **81**, 1966 (1984)
7. Hermann, K., Bagus, P.S., Bauschlicher, C.W.: Phys. Rev. **B30**, 7313 (1984)
8. Bagus, P.S., Hermann, K., Seel, M.: J. Vac. Sci. Technol. **18**, 435 (1982)
9. For these calculations the Hartree-Fock-LCAO cluster program system implemented at the Technical University Clausthal was used
10. The basis sets and pseudopotentials used here were identical to those given in Ref. 5
11. Moruzzi, V.L., Janak, J.F., Williams, A.R.: Calculated Electronic Properties of Metals. New York: Pergamon Press: 1978
12. See e.g. Sham, L.J., Schlüter, M.: Phys. Rev. **B32**, 3883 (1985)
13. Müller, W., Bagus, P.S.: Private communication
14. Bagus, P.S., Nelin, C.J., Bauschlicher, C.W.: Phys. Rev. **B28**, 5423 (1983)
15. Blyholder, G.: J. Phys. Chem. **68**, 2772 (1964)
16. Bagus, P.S., Hermann, K.: Appl. Surf. Sci. **22/23**, 444 (1985)
17. Bagus, P.S., Hermann, K.: Phys. Rev. **B33**, 2987 (1986)
18. Hermann, K., Bagus, P.S., Bauschlicher, C.W.: Phys. Rev. **B31**, 6371 (1985)
19. Chiang, T.C., Kaindl, G., Eastman, D.E.: Solid State Commun. **36**, 25 (1980)

K. Hermann
H.J. Hass
Institut für Theoretische Physik
Technische Universität Clausthal
und Sonderforschungsbereich 126
Leibnizstrasse 10
D-3392 Clausthal-Zellerfeld
Federal Republic of Germany

P.S. Bagus
IBM Almaden Research Center
San Jose, CA 95120-6099
USA

Note Added in Proof

It is important to emphasize that the order in which the constrained variations are performed in the CSOV analysis does not have a large effect. Further, our present CSOV approach is restricted to cases where the covalent bonding is dative.

Metallic Ions and Clusters: Formation, Energetics, and Reaction

A.W. Castleman, Jr. and R.G. Keesee
Department of Chemistry, The Pennsylvania State University,
University Park, Pennsylvania

Received May 6, 1986

Several aspects of the properties of metal clusters and metals in cluster ions are discussed. In particular, results from our laboratory on the photoionization and reactions of sodium clusters, and the thermochemistry and formation of clusters containing metallic ions are presented.

PACS: 35.20; 33.80; 82.30

I. Introduction

Interest in the field of clusters in general, and metal clusters in particular, is growing at an ever increasing rate. Major impetus is derived from the belief that the results of such research will serve to elucidate the behavior of condensed matter and surfaces at the microscopic molecular level. Furthermore, it is expected that research on metal clusters will elucidate the physical basis of catalysis and will contribute to an understanding of the influence of topology and density on metallic conductivity. Finally, metal atoms, ions, and particles are present in the effluent of many combustion and smelting processes and also result from meteoritic ablation into the atmosphere, but their reactions and influence on various chemical and physical processes are not well known.

A problem of long-standing interest is the influence of the degree of aggregation of the metallic system on the ionization potential. In this regard, comparison of predictions from theories such as the Jellium model, classical electrostatics, or quantum calculations with experimental results are important. In the case of bulk metals, it is well known that the adsorption of molecules leads to an alteration (usually a reduction) in the work function. A problem of some interest is the degree to which cluster systems which have either been reacted with molecules or contain adsorbed species reflect such trends.

Another stimulus for research on metal clusters with emphasis on chemical reactions has been derived from recent findings [1–4] which show that certain cluster sizes are more reactive than others. Of particular interest are trends reported by Kaldor and coworkers [5] which suggest a correlation between the ionization potential of a metal and its reactivity. In view of the difficulty in selecting specific neutral metal clusters for investigation, and hence in deriving accurate rate coefficients, this interest is also promoting studies of ionized systems where specific cluster sizes can be mass selected and subjected to subsequent examination. This paper summarizes some of our findings in these areas and sets them in general perspective.

II. Photoionization of Sodium Clusters

Systems comprised of alkali metals are particularly interesting for study due to their simple one-electron nature, and the ease with which they can be compared with theoretical calculations. Photoionization spectra for Na_x, with x ranging from 1 to 8, have been obtained in our laboratory using a molecular beam coupled with a tunable UV light source and quadrupole mass spectrometer detection system. A schematic of the molecular beam apparatus is shown in Fig. 1, while the details of the sodium oven and dual expansion nozzle are depicted in Fig. 2. In order to produce clusters, sodium metal is vaporized at temperatures (ca. 750 K which correspond to a sodium vapor pressure between one and two torr) and is expanded with argon carrier gas at pressures ranging between 70 and 100 torr. Sodium clusters are formed by expansion

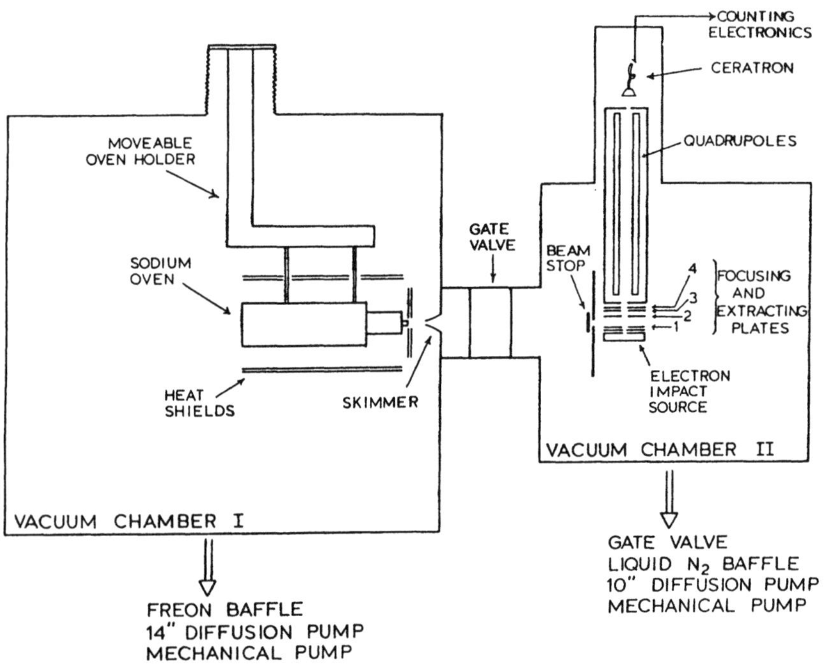

Fig. 1. Schematic of molecular beam apparatus

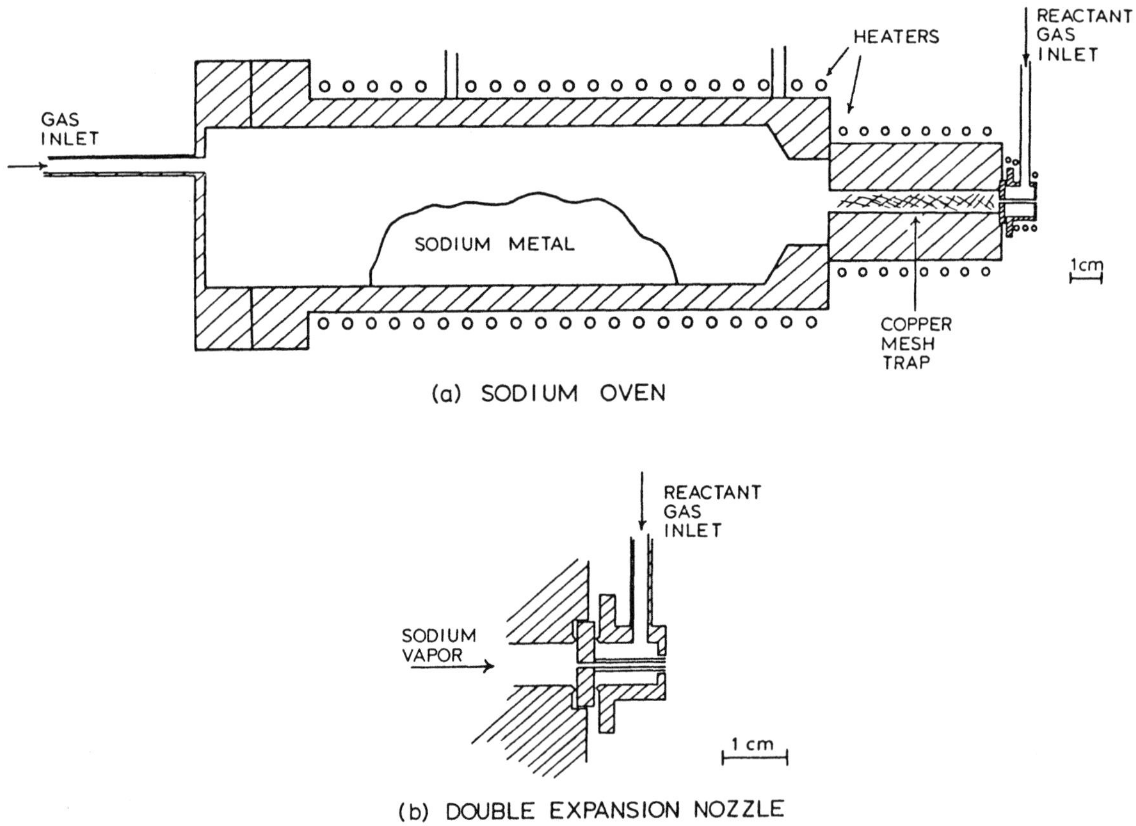

Fig. 2. a Detailed drawing of a sodium oven; **b** enlarged view of double expansion nozzle

through a 300 µm diameter tubular nozzle having a length of 14 mm. Once formed, the clusters pass through a skimmer into a detection chamber where they are photoionized and mass analyzed. The ionization region is located about 30 cm down from the skimmer. Photoionization is accomplished by use of a 500 watt xenon arc lamp equipped with a 68 mm diameter, 4 element $f/0.7$ UV condensing lens. The desired wavelength is selected with a 0.25 m, $f/3.0$ monochromator with a grating blazed at 240 nm.

Table 1. Appearance potentials of Na_x

X	Present work [6] (± 0.04)	Schumacher et al. [7] (± 0.05)
1	5.14	5.14
2	4.91	4.934 (0.011)
3	3.98	3,97
4	4.28	4.27
5	3.95	4.05
6	3.97	4.12
7	4.06	4.04
8	$\simeq 3.9$	4.1

The linear dispersion is 33 Å/mm and the slit width is 1 mm. The light intensity on the monochromator is continually monitored using a thermopile detector and the collected spectra are corrected for variations in power and the monochromator slit function.

Data from our laboratory [6] agree extremely well with measurements reported by Schumacher and coworkers [7] for all clusters with the exception of $x=5$, 6, and 8 (see Table 1). This is particularly evident in view of the somewhat different source and source conditions employed in the experiments. Even the measured value for Na_5 is essentially within experimental error, but Na_6 and Na_8 are unexpectedly different. Recent measurements for Na_8 reported at this conference by Buttet [8] are in good agreement with our findings.

A closer inspection of our data for Na_6 (see Fig. 3) reveals two breaks in the photoionization efficiency curve. One appears at 3.97 eV, which is the assigned appearance potential from our measurements [6] and another at 4.12 eV, which is identical to the appearance potential assigned by Schumacher and coworkers [7]. In fact, it is worthy of note that the data from the two laboratories are essentially identical and it is only the assigned appearance potential which is different. At one time this was thought to be due to the different method in choosing the appearance potential, correcting for light intensity and allowing for the nature of the monochromator slit function, but it is now evident from the data in Fig. 3 that there is a distinct two-region onset in the intensity data. At the present time, the reasons for this are unclear. It is important to note that the data in Fig. 3 have been corrected for lamp intensity variations and the monochromator spectral properties.

There are several possible reasons for two onsets in the appearance potential. In addition to possible influences due to Franck-Condon factors, or hot bands, neither of which are expected to relate to the problem at hand [6], questions exist concerning the possible existence of isomers or the possibility that the results can be influenced by fragmentation of larger clusters. In terms of the last question, it is instructive to examine the photoionization efficiency curve for Na_7 shown in Fig. 4. A local maximum is seen in the intensity followed by an abrupt drop in signal intensity at the region of 4.12 eV which coincidentally corresponds to the second increase in the Na_6 spectrum. Nevertheless, based on the relative intensities, it is difficult to say that there is a sufficiently large decrease in the number of Na_7^+ ions to account for the increased intensity of Na_6^+.

We have calculated the bond energies for atoms bound to metal cluster ions using our appearance potential measurements and three sets of calculations [9–11] for the bonding of sodium atoms to neutral clusters. The derived values for the energy to remove a sodium atom from Na_x^+ are shown in Table 2. Based on the calculations of Martins et al. [9] and Lindsay [10], one would conclude that the atoms are sufficiently strongly bound that fragmentation of Na_7^+ is unlikely. On the other hand, using the results of Flad et al. [11] indicates that this ion is nearly

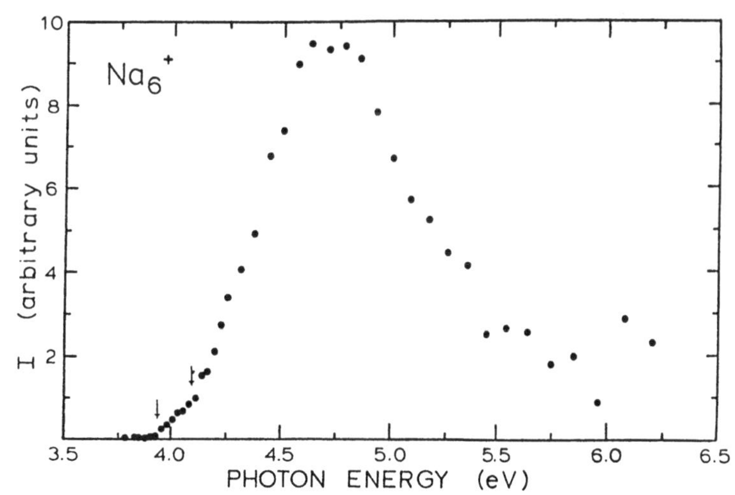

Fig. 3. Photoionization signal of Na_6^+ versus photon energy. Note two breaks in the curve indicated by arrows: the first appears at 3.97 eV while the second is at 4.12 eV

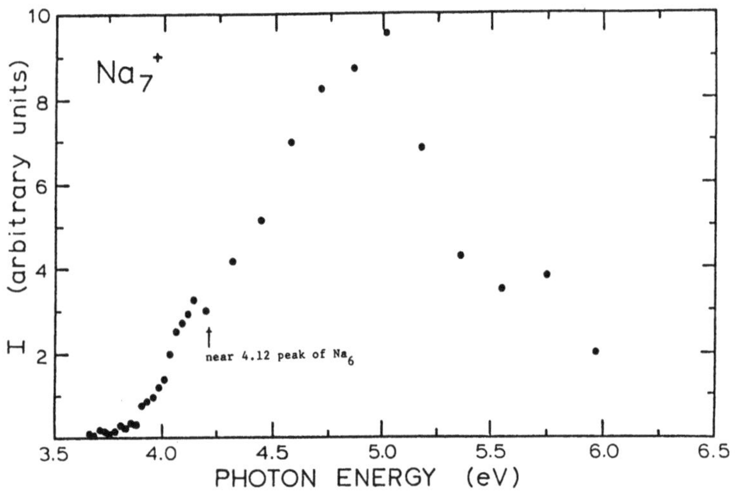

Fig. 4. Photoionization signal of Na_7^+ versus photon energy

Table 2. Bond energies in ionized sodium clusters (values expressed in eV)

x	$Na_x^{+\,a}$	$Na_x^{+\,b}$	$Na_x^{+\,c}$
2	1.13	0.99	1.06
3	1.32	1.31	1.19 1.16
4	0.84	0.52	0.2 0.17
5	1.09	0.92	0.69 0.63
6	1.16	0.97	0.44 0.34
7	1.27	0.57	0.04 0
8	1.30	1.32	1.48 1.35

Bonding of metal cluster ions determined from present ionization potential measurements and the neutral bond energies from:

[a] Reference 9
[b] Reference 10
[c] Reference 11

unbound and could fragment easily into Na_6^+. Hence the question of fragmentation remains open.

Another intriguing possibility is that the data provide evidence of isomers. Martins et al. [9] suggest that Na_6 should have two isomers separated by only 0.06 eV, while the experimental data would lead to a difference in appearance potential of 0.15 eV if the difference for the two steps in the ionization efficiency curve is attributed to the two isomers. This intriguing questions awaits further study.

Knight and coworkers [12] found that a simple Jellium model in terms of shell closings could describe trends in the relative abundances observed in the mass spectra of cluster distributions for sodium and potassium. However, the trends in ionization potentials appear to be at variance with predictions from the Jellium model. Several other levels of theoretical calculations [9, 10] have been made which predict relatively large variations (compared to experimental observations) in ionization potential between odd- and even-numbered clusters. Interestingly, the experimental data correlate reasonably well with the classical electrostatic expression relating the work function of a system of radius r to the bulk work function, W_∞ (see Fig. 5).

III. Reactions of Sodium Clusters with Halogens and Oxygen Containing Molecules

Using the reaction source shown in Fig. 2, subchlorides and suboxides of sodium containing clusters were generated through reactions with Cl_2, HCl, N_2O, and O_2. While reaction of Na_2 with Cl_2 and N_2O to form the respective subchloride and oxide species are exothermic, reactions with HCl and O_2 are endothermic and only become exothermic for the trimer and larger clusters (see Table 3). Hence a four centered reaction is likely to account for the observed reaction products in these experiments [13]. Photoionization experiments [14] lead to an appearance potential of 4.15 eV for Na_2Cl, in good agreement with the theoretically predicted range of adiabatic values (see Table 4). Despite large equilibrium geometry differences the adiabatic and vertical ionization potentials are expected to be similar due to the rather shallow potential surfaces which have been calculated for the ion and neutral [15]. This has been rationalized by the fact that the electron with very low ionization potential in Na_2Cl must be very diffuse and is not expected to exert a very strong influence on bonding.

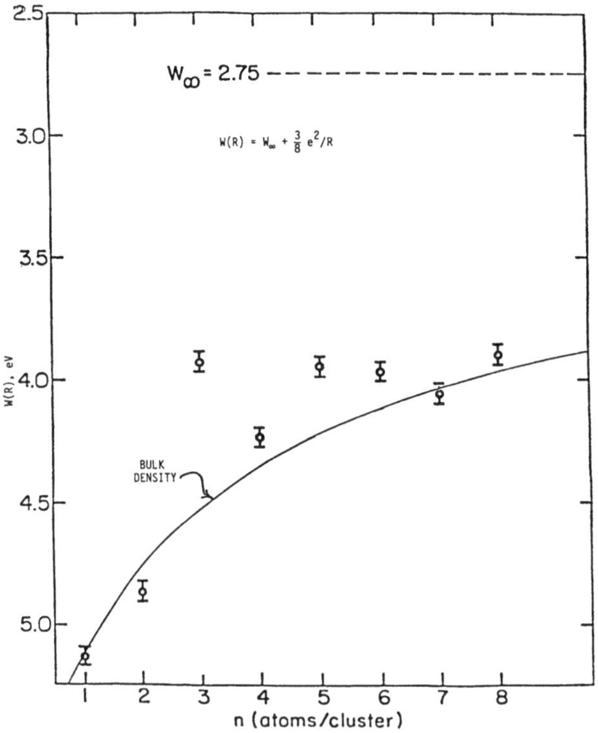

Fig. 5. Plot of work function vs. number of atoms in a cluster. The line is derived from classical electrodynamics, $W(R) = W_\infty + 3e^2/8R$. The conversion between number and equivalent spherical size was made using the bulk density of sodium

Table 3. Thermochemistry of metal cluster reactions[a]

Reaction	ΔH (eV)
$Na_2 + Cl_2 \rightarrow Na_2Cl + Cl$	−1.78
$Na_2 + HCl \rightarrow Na_2Cl + H$	+0.22
$Na_3 + HCl \rightarrow Na_2Cl + NaH$	−1.52
$Na_2Cl \rightarrow Na + NaCl$	+0.87
$Na_2 + N_2O \rightarrow Na_2O + N_2$	−2.60
$Na_2 + O_2 \rightarrow Na_2O + O$	+0.87
$Na_3 + O_2 \rightarrow Na_2O + NaO$	−1.52
$Na_2O \rightarrow Na + NaO$	+2.30

[a] Data taken from table in Ref. 6

Table 4. Appearance potential for the photoionization of Na_2Cl

$Na_2Cl + h\nu \rightarrow Na_2Cl^+ + e$		
Experimental result (Ref. 14)		−4.15 eV
Theory	Adiabatic (Ref. 15)	−3.97 eV −4.17 eV
	Vertical (Ref. 16)	−4.33 eV −4.43 eV

Table 5. Comparison of bond energies of sodium suboxide and subchloride species

Suboxides	Bond energy (eV)	Subchlorides	Bond energy (eV)
$Na_2^+ + O$	−4.08	$Na_2^+ + Cl$	−5.01
$Na_2 + O$	−4.28	$Na_2 + Cl$	−4.29
$Na_2^+ + O^-$	−7.68	$Na_2^+ + Cl^-$	−5.54

Table 6. Difference in Bond Energy[a] for O Atoms Bound to M_2 and M_2^+

	$\Delta \equiv \Delta E_{M_x^+ - O} - \Delta E_{M_x - O}$	
X	Na Δ, (eV)	K Δ, (eV)
1	−1.36	–
2	−0.15	−0.65
3	0.08	−0.35
4	0.33	−0.02

[a] Results calculated from data given in Ref. 19

From tabulated thermochemical data and the appearance potentials measured in our laboratory [14], bond energies have been derived for Na_2^+ with X and X^- and for Na_2 with X, where $X = O$ or Cl (see Table 5). The bond energies for O bound to Na_2^+ and Na_2 differ by only 0.2 eV. By contrast, the bond energy for Na_2^+ with Cl lies between that of Na_2 with Cl and Na_2^+ with Cl^-. The similarity in the bond energy of O to Na_2^+ and Na_2 and the fact that the ionization potential of Na_2O is close to that of the metal is suggestive that the electron removed upon photoionization comes largely from a nonbinding orbital perturbed by the Na(3s). In the case of the chlorine containing species, the ionization potential of Na_2Cl (4.15 eV) lies between the electron affinity of chlorine (3.617 eV; [17]) and the ionization potential of sodium (5.13 eV; [18]). This result suggests that the highest occupied level may be spatially localized around Cl and influenced by the electron affinity of Cl.

It is also interesting to compare in Table 6 the difference in the bond energy for O atoms bound to Na_x and Na_x^+, where measurements are available from our studies for x ranging up to 4. The bond energy in NaO is much greater than that in NaO^+. The difference between the neutral and ionic species in atomic oxygen bond energies becomes diminishingly small for $x = 3$ and reverses for $x = 4$, i.e., the ion more strongly binds atomic oxygen (see Table 6). This trend is in accord with the observation that the presence of adsorbed impurities on polycrystalline so-

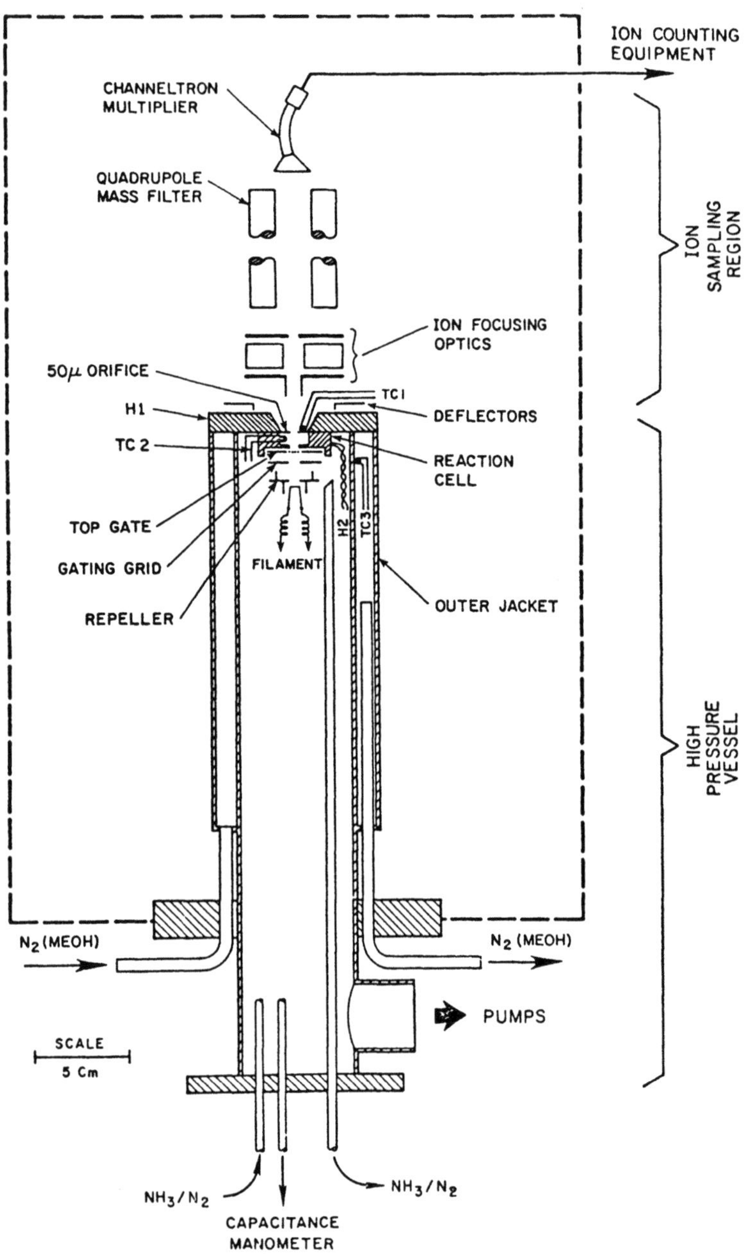

Fig. 6. Schematic of the ion clustering apparatus. Internal supports are omitted for clarity. The broken line in this figure delineates a region of high vacuum. For further discussion refer to the text

dium surfaces lowers the bulk work function. The potassium system, while not displaying a reversal in stability between neutral and ion, does show a diminishing trend with cluster size. It is tempting to speculate that there would be a reversal for a larger cluster (see results in Table 6).

IV. Thermochemical Considerations of Ligand Bonding to Metal Ions

For a number of years we have employed a high pressure reaction system which enables the measurement of ion-molecule bond energies [20]. The apparatus used in these studies is shown in Fig. 7. The advent of high pressure ion source mass spectrometry has been particularly valuable in quantitatively determining the stability of ion clusters. In this technique, ions effuse from a high pressure source (typically a few torr) through a small aperture into a mass filter where the distribution of ion clusters is determined. Ionization may be initiated by various methods including radioactive sources, heated filaments, and electric discharges. The pressure of the ion source is maintained sufficiently high such that ions reside in a region of well-defined temperature for a time adequate to ensure the attainment of equilibria among the various ion cluster species of interest, but

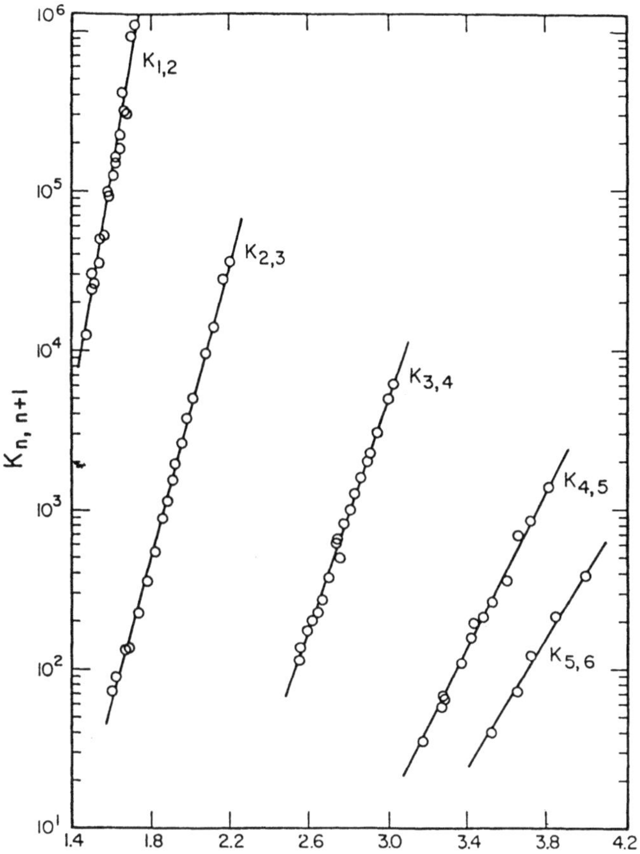

Fig. 7. van't Hoff plot, logarithm of the equilibrium constant vs. reciprocal absolute temperature, for $Na^+(NH_3)_n + NH_3 \rightleftharpoons Na^+(NH_3)_{n+1}$

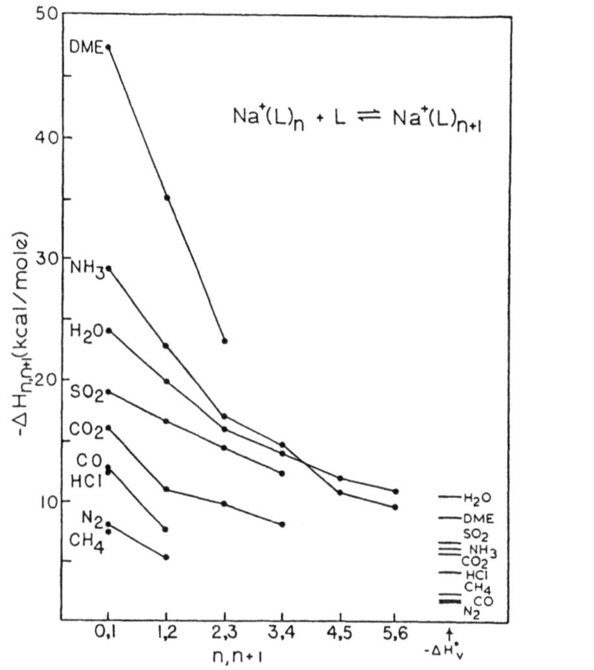

Fig. 8. $-\Delta H^0_{n,n+1}$ versus clustering reaction step, $n, n+1$, for DME, NH_3, H_2O, SO_2, CO_2, CO, HCl, N_2, and CH_4 onto Na^+

the pressure must be low enough to avoid additional clustering via adiabatic expansion as the gas exits the sampling orifice.

Cluster formation can be represented by the general association reaction:

$$I\cdot(n-1)\,L+L+M=I\cdot nL+M. \quad (1)$$

Here, I designates a positive or negative ion, L the clustering neutral (ligand), and M the third body necessary for collisional stabilization of the complex. Taking the standard state to be 1 atm, and making the usual assumptions [21] concerning ideal gas behavior and the proportionality of the chemical activity of an ion cluster to its measured intensity, the equilibrium constant $K_{n-1,n}$ for the *nth* clustering step is given by

$$\ln K_{n-1,n}=\ln\frac{C_n}{C_{n-1}\,P_L} = -\frac{\Delta G^0_{n-1,n}}{RT} = -\frac{\Delta H^0_{n-1,n}}{RT}+\frac{\Delta S^0_{n-1,n}}{R} \quad (2)$$

Here, C_{n-1} and C_n represent the respective measured ion intensities, P_L the pressure (in atm) of the clustering species L, $\Delta G^0_{n-1,n}$, $\Delta H^0_{n-1,n}$, and $\Delta S^0_{n-1,n}$ the standard Gibbs free energy, enthalpy, and entropy changes, respectively, R the gas-law constant, and T absolute temperature. By measuring the equilibrium constant $K_{n-1,n}$ as a function of temperature, the enthalpy and entropy change for each sequential association reaction can be obtained from the slope and intercept of the van't Hoff plot ($\ln K_{n-1,n}$ vs. $1/T$). Thermodynamic information also can be obtained by studying switching or exchange reactions of the form

$$I\cdot nL+L'=I\cdot(n-1)\,L\cdot L'+L. \quad (3)$$

The thermodynamic quantities for the association of L' onto $I\cdot(n-1)\,L$ are the sum of those for reaction (1) plus reaction (3).

A typical van't Hoff plot is shown in Fig. 8. A wide variety of different metal ions, bound to neutrals with various dipole moments, polarizabilities, and quadrupole moments have been selected for investigation in our laboratory. The data in Fig. 9 demonstrates the general bonding behavior as a function of cluster size for a variety of ligands including dimethoxyethane (DME), NH_3, H_2O, SO_2, CO_2, CO, HCl, N_2, and CH_4 bound to Na^+. The trend for the various ligands shows the general importance of the dipole moment but with closer inspection the importance of polarizability and quadrupole moment

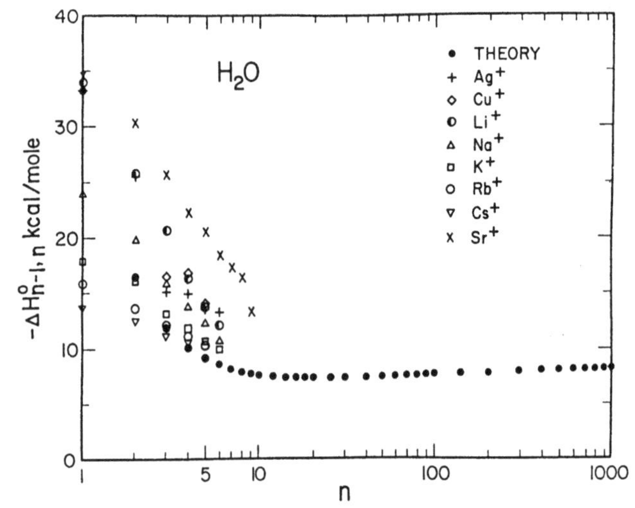

Fig. 9. Gas-phase hydration enthalpies. The theoretical points are derived from the Thomson equation at 313°K

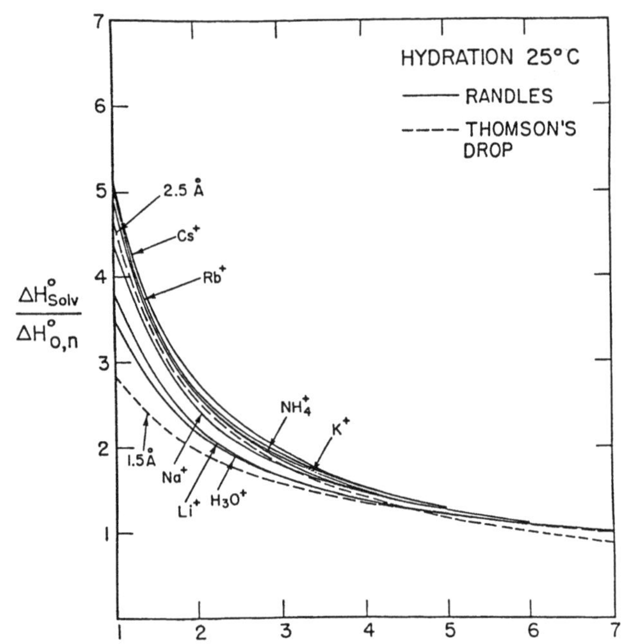

Fig. 10. Ratio of Randle's total enthalpy of hydration to the partial gas-phase enthalpy of hydration for positive ionic cluster size n

is evident. For example, consider the bonding of ammonia and water. The dipole of ammonia is smaller but its polarizability and quadrupole moment are greater and hence the bonding is enhanced. Conversely, both ammonia and SO_2 have similar dipole moments and similar polarizabilities, but the quadrupole moment of SO_2 is repulsive while that of ammonia is attractive and ammonia bonds much stronger. Similar trends can be seen for various ions such as Bi^+ [22] which have more complex electronic structures than the simple closed-shell alkali metal ions.

A comparison of the enthalpies of attachment of one ligand to various ions shows the importance of the electronic nature of the ion on bonding. See Fig. 9. The gas-phase clustering energetics relate to the solution (condensed phase) through a simple correlation as depicted in Fig. 10. The ratios of the heat of solvation of a gaseous ion into the condensed phase, ΔH^0_{solv}, to the sum of the stepwise enthalpies of clustering $\Delta H^0_{o,n}$ converge at larger cluster sizes ($n \geq 5$) [23–24]. This finding is related to the Born equation with correction for the influence of "surface tension" for ligands bound to an ion in the gas phase (the Thomson drop model).

V. Clustering Reactions to Metal Ions

In order to investigate the individual reaction kinetics for a metal ion (or metal cluster ion) with a specific ligand, a fast-flow reaction apparatus (Fig. 11) has been assembled in our laboratory and tested using a number of charge transfer reactions [25]. Figure 11 depicts a selected ion flow tube. Ions are produced in the ion source and ions of a particular mass-to-charge ratio are selected with a quadrupole mass filter and injected into the bulk flow. While most of the gas is pumped away, a small fraction is sampled through an orifice where the ions are mass identified and counted. Reactant gases can be added at any inlet position in the flow, so kinetic rates can be determined by monitoring the reactant ion signal and varying the position of reactant injection, by varying the flow rate of reactant into the flow tube, or by varying the bulk blow velocity.

Recently, we investigated the clustering of a variety of ligands including NH_3, CH_4, CH_3F, CH_3Cl, CH_3Br, and CO to a variety of ions including Pb^+, Al^+, Li^+, Ag^+, Cu^+, and the dimer ion Al_2^+. In the low pressure limit, the association reactions for the attachment of a ligand to an ion (or cluster ion) is related to the collision rate k_1 between the ion and ligand, the stabilization rate k_2 which with unit efficiency is equal to the collision rate for the cluster with a third body M, and the lifetime τ of the collision complex against unimolecular dissociation. Using ADO theory [26] for polar ligands and Langevin theory [27] for collisions with systems having no permanent dipole moment, we have calculated the forward rates, k_1 and k_2, and deduced lifetimes of the collision complexes from the experimental measurements of the overall forward rate ($k_1 k_2 \tau$). In the case of ammonia, these values are found to range from 21 picoseconds for Pb^+ to 1.9 ns for Cu^+. The life-

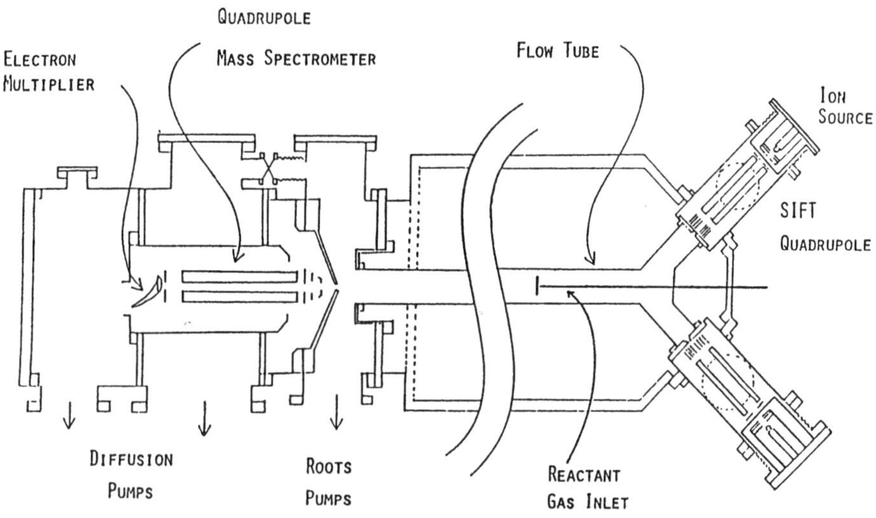

Fig. 11. A schematic of the Selected Ion Flow Tube (SIFT). The break in the flow tube indicates a 90° rotation of the ion source for a more detailed view of the apparatus

Table 7. Bonding of NH_3 of M^+

M^+	ΔE (eV)	Collison Complex Lifetime, τ (picosecond)
Pb^+	1.52[a]	21
Li^+	1.68[b]	140
Ag^+	>1.9[c]	410
Cu^+	>2.0[c]	1,900

[a] Ref. 28
[b] Ref. 29
[c] Ref. 30

times follow the trend in the ion-molecule bond energy as seen from the data in Table 7. For the case of CH_3Cl, the complexes have lifetimes from 140 picoseconds (Pb^+) to 11 ns (Cu^+). By contrast, the methane systems have extremely short lifetimes. The trends in lifetimes are in rough accord with expectations based on the angular momentum coupling model [31].

Of particular significance are comparisons of reactions for the aluminum dimer ion Al_2^+ compared to the atomic ion Al^+. The dimer ion is found to react from 10 to 50 times faster with CO, CH_3Br, and NH_3, which indicates corresponding increases in lifetimes of the collison complex. Such variations in lifetimes and reaction rates are of innate interest in themselves, but are expected to reflect on the trends observed for the reactivity of metal clusters [1–4]. Such work is currently in progress in our laboratory.

Support by the National Science Foundation, Grant No. ATM-82-04010, the Department of Energy, Grant No. DE-ACO2-82-ER60055, and the Army Research Office, Grant No. DAAG29-85-K-0215, is gratefully acknowledged. The authors are indebted to various colleagues whose work is referenced, and to current group members Sterling Sigsworth, Robert Leuchtner, Robert Farley, and Hideyuki Funasaka who are engaged in studies of the lifetimes and reactivity of metal ions and cluster ions and to Professor K. Weil (on leave from the University of Darmstadt) for stimulating discussions and suggestions concerning this area of research.

References

1. Whetten, R.L., Cox, D.M., Trevor, J., Kaldor, A.: J. Phys. Chem. **89**, 566 (1985)
2. Geusic, M.E., Morse, M.D., Smalley, R.E., J. Chem. Phys., **82**, 590 (1985); Morse, M.D., Geusic, M.E., Heath, J.R., Smalley, R.E.: J. Chem. Phys. **83**, 2293 (1985)
3. Liu, K., Parks, E.K., Richtsmeier, S.C., Pobo, L.G., Riley, S.J.: J. Chem. Phys. **83**, 2882 (1985)
4. Castleman, A.W., Jr., Keesee, R.G.: Clusters: properties and formation. Annu. Rev. Phys. Chem. (Invited review article, submitted for publication)
5. Whetten, R.L., Cox, D.M., Trevor, D.J., Kaldor, A.: Phys. Rev. Lett. **54**, 1494 (1985)
6. Peterson, K.I., Dao, P.D., Farley, R.W., Castleman, A.W., Jr.: J. Chem. Phys. **80**, 1780 (1984)
7. Herrmann, A., Leutwyler, S., Schumacher, E., Wöste, L.: Helv. Chim. Acta **61**, 453 (1978)
8. Buttet, J.: Z. Phys. D – Atoms, Molecules and Clusters **3**, 155 (1986)
9. Martins, J.L., Buttet, J., Car, R.: Phys. Rev. B **31**, 1804 (1985)
10. Calculations of Lindsay, D.: Personal communication 1986
11. Flad, J., Stoll, H., Preuss, H.: J. Chem. Phys. **71**, 3042 (1979)
12. Knight, W.D., Clemenger, K., Heer, W.A. de, Saunders, W.A., Chou, M.Y., Cohen, M.L.: Phys. Rev. Lett. **52**, 2141 (1984); Cleland, A.N., Cohen, M.L.: Solid State Commun. **55**, 35 (1985); Chou, M.Y., Cleland, A.N., Cohen, M.L.: Solid State Commun. **42**, 645 (1984)
13. Crumley, W.H., Gole, J.L.: J. Chem. Phys. **76**, 6439 (1982)
14. Peterson, K.I., Dao, P.D., Castleman, A.W., Jr.: J. Chem. Phys. **79**, 777 (1983)
15. Struve, W.S.: Mol. Phys. **25**, 777 (1973)
16. Lin, S.M., Wharton, J.G., Grice, R.: Mol. Phys. **26**, 317 (1973)
17. Miller, T.H.: Electron Affinities CRC Handbook of Chemistry and Physics (in press)
18. Moore, C.E.: Atomic Energy Levels, NBS Circ. 467 (1949)
19. Dao, P.D., Peterson, K.I., Castleman, A.W., Jr.: J. Chem. Phys., **80**, 563 (1984)

20. Castleman, A.W., Jr., Holland, P.M., Lindsay, D.M., Peterson, K.I.: J. Am. Chem. Soc., **100**, 6039 (1978)
21. Castleman, A.W., Jr., Keesee, R.G.: Swarms of ions and electrons in gases. Lindinger, W., Mark, T.D., Howorka, F. (eds.), pp. 167–193. Berlin, Heidelberg, New York: Springer-Verlag 1984
22. Castleman, A.W., Jr.: Chem. Phys. Lett. **53**, 560 (1978)
23. Lee, N., Keesee, R.G., Castleman, A.W., Jr.: J. Colloid Interface Sci. **75**, 555 (1980)
24. Castleman, A.W., Jr., Keesee, R.G.: Clusters: bridging the gas and condensed phases. Accts. Chem. Res. (Invited review, submitted for publication)
25. Shul, R.J., Upschulte, B.L., Passarella R, Keesee, R.G., Castleman, A.W., Jr.: Thermal energy charge transfer reactions of Ar^+ and Ar_2^+. J. Phys. Chem. (submitted for publication)
26. Su, T., Bowers, M.T.: Int. J. Mass Spectrosc. Ion Phys. **12**, 347 (1973)
27. McDaniel, E.W., Cermak, V., Dalgarno, A., Ferguson, E.E., Friedman, L.: *Ion-Molecule Reactions* Barnett, C.F., Cobble, D.M. (eds.) p. 16. New York: Wiley-Interscience 1970
28. Estimated from data in Reference 22
29. Keesee, R.G., Castleman, A.W., Jr.: Thermochemical data on gas-phase ion-molecule association and clustering reactions. J. Phys. Chem. Ref. Data (in press)
30. Estimated based on data from: Holland, P.M., Castleman, A.W., Jr.: J. Chem. Phys. **76**, 4195 (1982)
31. Schelling, F.J., Castleman, A.W., Jr.: Chem. Phys. Lett. **111**, 47 (1984)

A.W. Castleman, Jr.
R.G. Keesee
Department of Chemistry
The Pennsylvania State University
University Park, PA 16802
USA

Experiments on Size-Selected Metal Cluster Ions in a Triple Quadrupole Arrangement

P. Fayet and L. Wöste
Institut de Physique Expérimentale, Ecole Polytechnique Fédérale de Lausanne, Switzerland

Received May 12, 1986; final version July 26, 1986

A high flux of positively and negatively charged metal cluster ions was produced in a sputtering arrangement, energy-analyzed and mass-filtered. The resulting monodispersed cluster ion beam was introduced into a quadrupole drift tube, where it interacted with a laser beam or reacted with an introduced gas. All inelastic scattering events were recorded with a subsequent quadrupole mass filter. The results exhibited a high sensitivity of positively and negatively charged silver clusters $Ag_n^{\pm}$ ($n \leq 16$) with respect to photofragmentation. Ion-molecule reactions of nickel clusters with carbon monoxide allowed to synthesize very interesting organometallic and carbonyl compounds, and the maximum number of ligands provided interesting structural indications.

PACS: 0775; 3640; 8240 D

1. Introduction

Molecular beam expansion combined with mass spectrometry is a well established technique to investigate optical and chemical properties of metal clusters in the gas phase. Experiments, however, are strongly handicapped by fragmentation phenomena, which commonly occur during reactive collisions, electronic transitions, neutralization or ionization processes. Therefore parent molecules and fragments cannot easily be distinguished.

For this reason we developed an apparatus, that allowed us to perform the above-mentioned experiments with metal clusters of a specific size and charge. As cluster source we chose a sputtering arrangement [1]. This allowed us to generate positively and negatively charged clusters, as well as neutrals from nearly all materials. The emerging cluster ions were energy-filtered, mass-separated and then introduced into an ion drift tube, where they were kept at low kinetic energy in a radiofrequency confinement. The confined ions were then irradiated with laser light or interacted with inert and reactive gases. All photodissociation events, collision-induced fragmentations or chemical reactions of the confined cluster ions were then analyzed with another mass spectrometer at the exit of the ion drift tube. The results showed very distinct properties of individual cluster sizes with respect to stability, structure and chemical reactivity.

2. Experimental Setup (Fig. 1)

For generating the primary ion beam we used a modified cold reflex discharge ion source (CORDIS) [2], which was developed by R. Keller for use in the heavy ion accelerator at GSI Darmstadt. The ion source is placed inside a differentially pumped vacuum chamber at a distance of 50 cm from the target, with an angle of incidence of 50°. Operating at an ion energy of 20 keV and a primary beam diameter of 10 mm, typical ion currents of 2 mA were measured. The target is mounted on a manipulator device for rapid exchange and optimum positioning. It is temperature controlled and can be kept at any desired bias voltage. A metal cylinder around the target reduces the contamination of the vacuum chamber with sputtering deposits. The chamber contains window flanges on both sides of the target in order to reionize sputtered neutrals by photoionization. The system is pumped with a 500 l/s turbomolecular pump.

Symmetrically to the primary ion beam a Knudsen beam source is also directed to the target. This allows to deposit small amounts of alkali metal, which enhances the formation of negative ions by

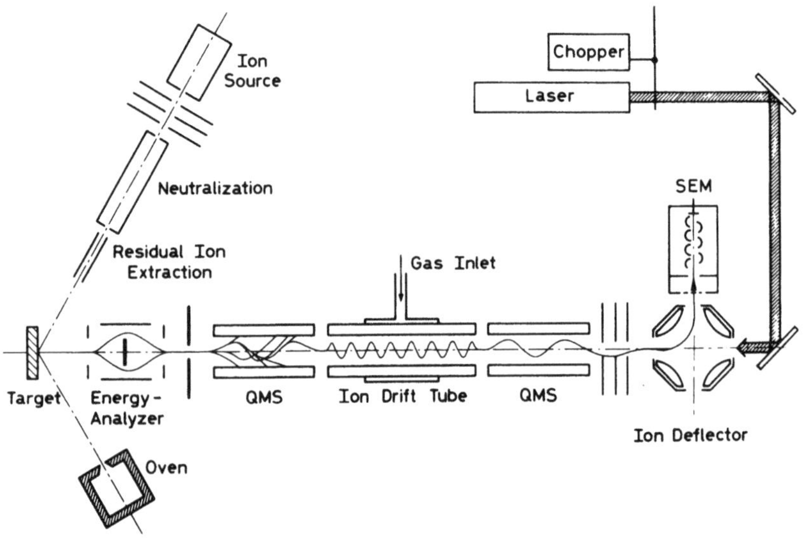

Fig. 1. Experimental setup

a factor of more than 100, due to an increased amount of negative surface charges. Hydrogen or oxygen, on the other hand, increase the amount of positive secondary ions by a factor of 50. The Knudsen beam source also served for the in-situ preparation of pure alkali targets.

The kinetic energy distribution of sputtered particles is typically between 0 and 10 eV. A cylindrical retarding field energy analyzer cuts out of this range an energy band of 4 eV. In addition to that it blocks all fast primary ions and sputtered neutrals. The energy filter is directly mounted to the entrance of the quadrupole mass filter at a distance of about 4 cm from the target. We used a commercial quadrupole rod system of 9.5 mm pole diameter and a length of 20 cm (Extranuclear Model 4-162-8), which reaches under normal operating conditions a maximum mass value of 1700.

The monodispersed cluster beam is directly guided from the quadrupole filter into the drift tube, which is 60 cm long and has an identical sectional view as the mass filter. The drift tube is operated with the same RF-frequency as the mass filter at a controlled phase shift. Since, however, no DC-voltage is applied, there is no mass discrimination, and all ions, photofragments and reaction products, are kept in confinement. The relative electric potential of the rod system allows to slow down the confined ions to almost thermal velocities, which corresponds to residence times of up to 10 ms. The drift tube is surrounded by a scattering chamber of 10 cm length, where reactant gas can be introduced. The complete drift tube assembly is placed inside another differentially pumped ultra-high vacuum chamber, pumped by a 500 l/s turbomolecular pump. The background pressure under normal operation conditions is $<10^{-8}$ mbar, which is sufficient to prevent undesired background collisions. It can be operated, however, at pressures up to 10^{-2} mbar, when reactant gas is introduced.

At the exit of the drift tube the confined ion beam enters another quadrupole mass filter (Extranuclear Model 4-162-8). It is operated at the same frequency as the first mass filter and the drift tube at a controlled phase shift, and it is placed inside a differentially pumped ultra-vacuum chamber, pumped by a 500 l/s turbomolecular pump. The same chamber also contains a 21-stage Cu – Be secondary electron multiplier for detecting the ions. It is mounted on an electrostatic ion deflector, in order to allow the laser beam to enter the triple quadrupole system.

3. Mass-Spectroscopic Results

Figure 2 shows a mass spectrum of negatively charged silver clusters. The first mass filter was tuned, and the second filter as well as the ion drift tube were kept in a non-discriminative mode on RF only. The intensity distribution of the Ag_n^- particles shows a distinct odd-even alternation. This is due to the preferred stability of compounds having paired electron configurations, as was explained by Joyes and Leleyter [3]. A similar behaviour is also found for positive mass spectra [4, 5]. The increasing width of the mass peaks with growing cluster size is due to the growing number N of isotopic combinations, given by

$$N(Ag_n) = n + 1$$

as silver has two natural isotopes of almost equal abundance. Figure 3 shows the isotopically-resolved mass peak of the six isotopic components of Ag_5^-.

Besides homonuclear silver aggregates the cluster distribution in Fig. 2 shows some mixed compounds

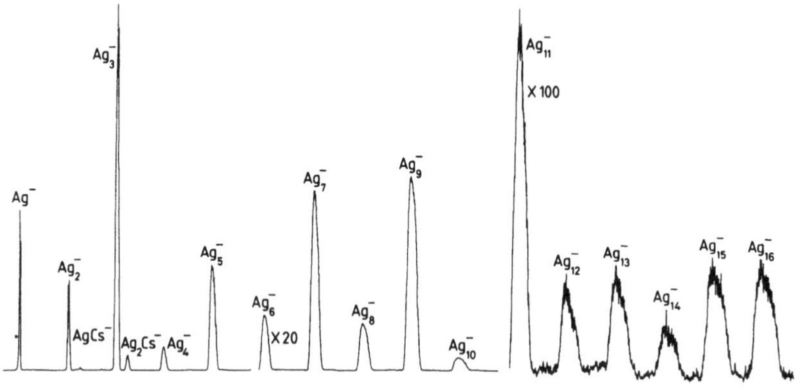

Fig. 2. Mass spectrum of negatively charged silver clusters

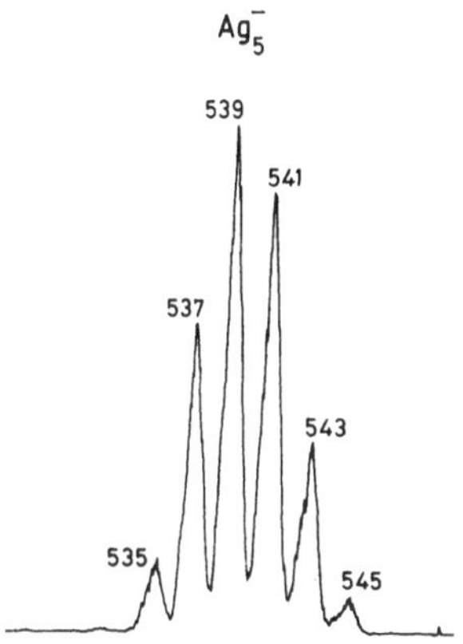

Fig. 3. Mass spectrum of the isotopic distribution of Ag_5^-

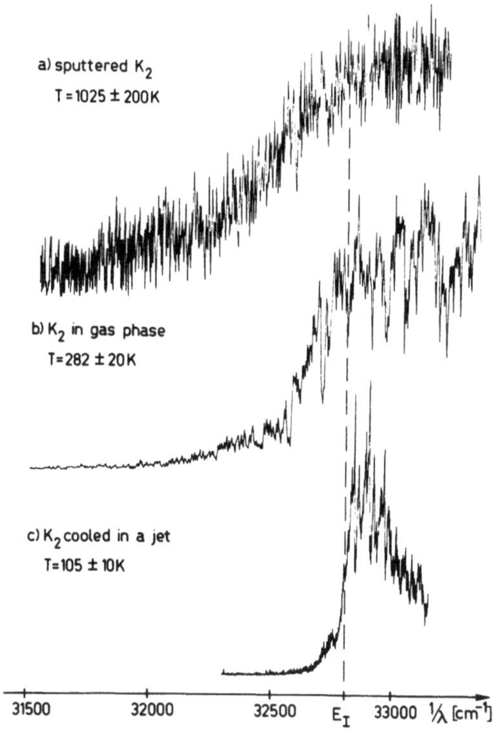

Fig. 4 a–c. One-photon ionization of **a**. sputtered potassium dimers, **b**. K_2 in the gas phase, **c**. jet-cooled K_2

like $AgCs^-$ and Ag_2Cs^- as well. This is due to the effusive beam of cesium, which was simultaneously evaporated on the silver target in order to enhance the negative ion current [6]. In this way the arrangement became capable to generate very significant currents of monodispersed cluster ions. With a 15 keV, 0.5 mA primary Ar^+-beam we measured for example, with a Faraday cup detector, an Ag_3^- particle flux of 10^{12} ions per second. This flux is sufficient for controlled surface deposition or matrix generation [7].

4. Internal Energy Distribution

The cluster formation process by means of sputtering is not yet very well understood. We therefore performed an experiment, which allowed us to determine the internal energy of sputtered compounds. The experiment was performed on neutral alkali dimers. We applied a tunable UV-laser beam in front of the alkali target, in order to reionize the sputtered neutral dimers. The target was prepared before this by vapor deposition with the Knudsen source. When the laser was tuned across the ionization potential curve, the Boltzmann tail, which is characteristic for the internal energy distribution of the electronic ground state, appeared. Figure 4a shows an example for sputtered K_2, whereas in Fig. 4b the corresponding results for K_2 in the gas phase is shown, and Fig. 4c shows the result for jet-cooled potassium dimers. The result clearly indicates, that the sputtered compounds are hot, a fact which has to be taken into consideration. Complementary experiments and a systematic interpretation, performed in collaboration with J.P. Wolf, are published elsewhere [8].

5. Photofragmentation and Photodetachment Experiments

Experiments on monodispersed cluster ions were performed as follows: The first quadrupole mass filter was set to a particular cluster size, for example Ag_7^+,

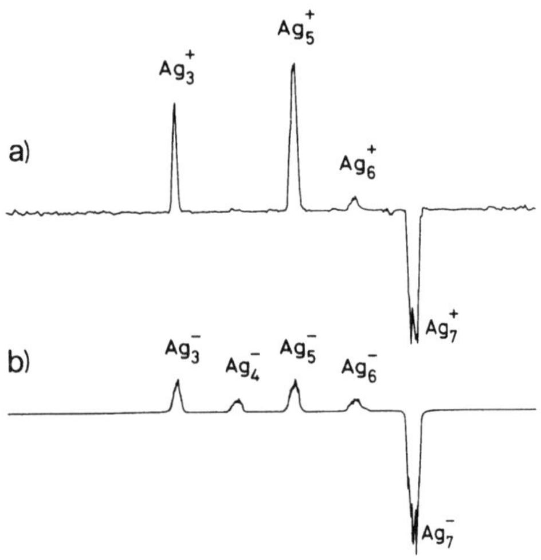

Fig. 5 a and b. Photofragmentation and photodestruction pattern of **a**. Ag_7^+ and **b**. Ag_7^-. The irradiated laser wavelength was 488 nm

which was then alone introduced into the drift tube. Consequently, a scan on the second mass filter shows only the mass peak of Ag_7^+. When, however, a laser beam was introduced into the drift tube, photofragmentation events occurred. Such a result is shown in Fig. 5a, where Ag_7^+ was irradiated with 1 W of the 488 nm line of an Ar^+-laser. The phase-sensitively recorded signal clearly exhibits the depletion (negative peak) of the Ag_7^+ parent ion, which predominantly produces Ag_5^+ and Ag_3^+ fragments. As we use a long photofragmentation interaction region, it could be possible that the photofragments themselves absorb another photon and dissociate again. For the laser power of 1 W, however, this process is still very unlikely and we only observe linear power dependencies for all fragment channels. The series was systematically measured for Ag_n^+ cluster sizes up to $n=16$. The results for the relative intensities of the different fragmentation channels are given in Table 1. Several interesting features arise from these data: The particles dissociate preferably with growing cluster size into a small neutral fragment and a large fragment ion. From energetic reasons this can be understood, because the ionization potentials of metal clusters decrease with growing cluster size. The formation of a smaller ion fragment requires therefore more energy. Cluster ions with an even number of atoms preferably loose an odd number of neutral atoms (Ag or Ag_3). This way they form a fragment ion with an odd number of atoms, which is more stable due to its total number of paired electrons. Cluster ions with an odd number of atoms, on the other hand, prefer to loose an even number of atoms (Ag_2). This allows them to maintain their preferred stability. There are certain odd-numbered particle sizes, like Ag_3^+, Ag_9^+

Table 1. Relative intensities of the Ag_n^+-photofragmentation patterns obtained at 488 nm. The $Ag_{12}^+ Ag_{11}^+$ and $Ag_{11}^+ Ag_{10}^+$ fragmentation channels could not be recorded because of spontaneous decay

Size of parent ions (columns); Size of fragment products (rows)

Ag_n^+	2	3	4	5	6	7	8	9	10	11	12	13	14	15	16
1	100	43.3	20.9	0.9	0	0	0	0	0	0	0	0	0	0	0
2		56.7	38.8	0.9	0	0	0	0	0	0	0	0	0	0	0
3			40.3	89.8	91.3	38.9	21.7	3.7	6.3	5.6	5.7	0	0	0	0
4				8.3	4.3	2.0	1.1	0	3.1	0.6	0	0	0	0	0
5					4.3	55.7	30.0	9.8	13.8	10.4	10.1	0	0	0	0
6						3.4	3.7	0	0	0.7	0.4	0	0	0	0
7							43.4	52.8	35.2	25.0	20.6	0	0	0	0
8								25.6	12.6	14.9	16.6	1.3	0	0	0
9									28.9	42.8	46.6	30.3	12.8	0	0
10										\	0	0	0	0	0
11											\	55.3	19.1	11.6	0
12												13.2	8.5	10.9	0
13													59.6	57.4	37.8
14														20.2	22.2
15															40

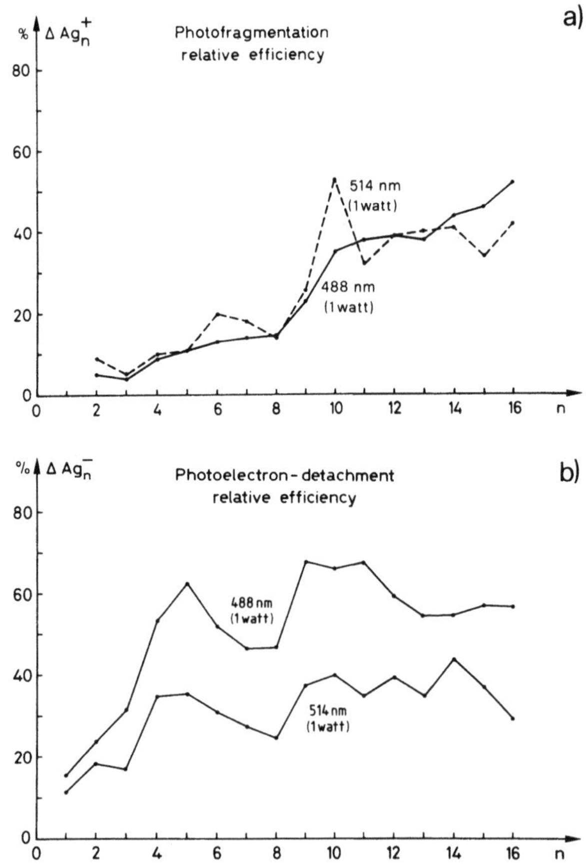

Fig. 6 a and b. Relative photofragmentation and photodestruction efficiency for **a**. positively and **b**. negatively charged silver cluster at 488 and 514 nm

and Ag_{13}^+, which seem to be more stable than others, because they appear more frequently. Certain cluster ions among the even-numbered particle sizes, like Ag_2^+, Ag_6^+ and Ag_{10}^+, on the other hand, seem to be very unstable, because they appear very rarely in the fragment ion pattern.

The relative photofragmentation efficiency, as it was obtained in the photofragmentation experiment, is given in Fig. 6a. The results were obtained for a laser power of 1 W at wavelengths of 488 and 514 nm and at an ion energy of approx. 10 eV. Both curves indicate for growing particle size, a similar increase in their fragmentation efficiency. The 514 nm curve, however, shows size effects more elaborately, and cluster sizes like Ag_2^+, Ag_6^+ and Ag_{10}^+ appear less stable than others. The average photofragmentation cross section, which is computed from these results, is of the order of 10^{-16} cm^2, which is surprisingly large for bound-free transitions.

When we irradiated negatively charged silver clusters, we expected to observe photoelectron detachments, whereby the mother ion should mainly disappear without the formation of photofragments. The result in Fig. 5b shows that this is not the case; the negatively charged silver clusters rather fall apart than loose their electron. Again we observe an odd-even alternation in the fragmentation pattern. Contrary to positively charged silver clusters, however, there is a significant difference between values obtained at 488 nm and at 514 nm (see Fig. 6b). The fact that the signal at 514 nm is about 30% lower, might indicate an onset of photodetachment processes. Experiments with a tunable dye laser will soon provide more information.

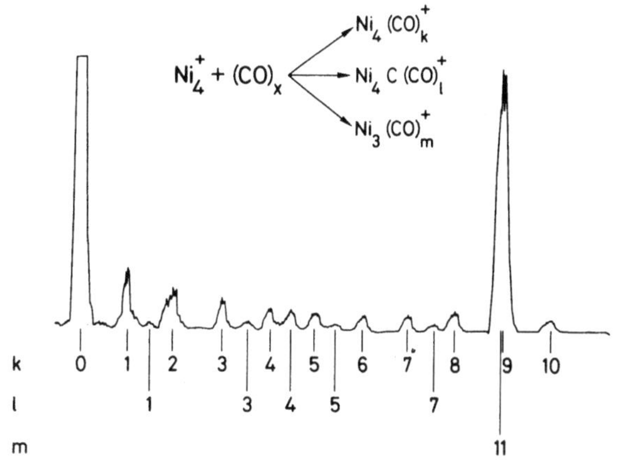

Fig. 7. Mass spectrum obtained from the reaction of Ni_4 with carbonmonoxide

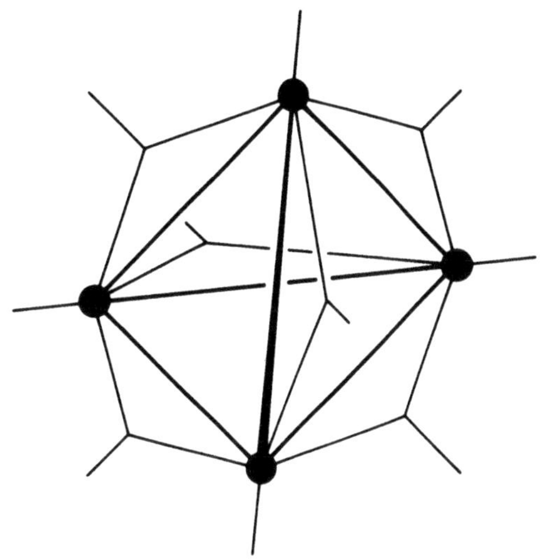

Fig. 8. Structure of $Ni_4(CO)_{10}$

6. Cluster-Molecule Reactions: The System $Ni_n^+ + (CO)_x$

When we introduced size-selected nickel clusters and carbon-monoxide into the drift tube, we were able to perform the controlled synthetization of nickel carbonyls ($Ni_n(CO)_k^+$) and nickel carbides ($Ni_nC(CO)_l^+$). A typical example, as it was obtained for Ni_4^+ at a kinetic energy of 2.0 eV and a CO-pressure of $2.4 \cdot 10^{-3}$ mbar is shown in the mass spectrum in Fig. 7. The result clearly exhibits a resonance on the mass of $Ni_4(CO)_9^+$ and a CO-ligand periodicity at lower mass peaks. Beyond $Ni_4(CO)_{10}^+$ no further species were observed. This indicates, that the Ni_4^+ compound with ten CO ligands is saturated.

The stoichiometry of these carbonyls can be related to recently developed bonding models for organometallic cages and clusters. The main idea is that transition metals have nine valence atomic orbitals used as acceptor or for containing unbonded electrons. This is the basis of the 18-electron rule followed by the majority of the organometallic complexes. Deviations from this rule appear to be due to the large s to p promotion energies found for the free atoms. This effect becomes more and more important when one moves to the heavier transition elements. Thus the bonding capacity of a given metal will depend upon the relative energy of its p orbitals.

Obviously, the situation is more complicated in transition metal clusters. The presence of the metal-metal interactions greatly destabilizes the energy levels of the p atomic orbitals and influences their aptitudes to bond ligands. Generally, one or more p orbitals will be empty and unbonded, the resulting organometallic complex is then electron deficient. Nevertheless, simple electron-counting rules exist to correlate molecular geometries of metal clusters with the total number of valence electrons.

These rules are mainly based on the scheme of delocalized molecular orbitals and were developed empirically by Wade and Mingos (polyhedral skeletal electron pair theory) [9, 10, 11].

A tetrahedral metal cluster will have maximum stability when the total number of cluster valence electrons is 60. This total is made up of the metal valence electrons augmented by those supplied by the ligands, two electrons per molecule in the case of carbon monoxide. The $Ni_4(CO)_{10}^+$ ion, produced by the reaction of excess CO with Ni_4^+ will possess the tetrahedral arrangement of metal atoms, shown in Fig. 8. This structure with four terminals and six bridging CO molecules admits a full T_d symmetry and formally assigns to each nickel atom an 18-electron configuration. The systematic interpretation of the results, which were obtained in collaboration with M.J. McGlinchey for cluster sizes up to Ni_{13}^+ are published elsewhere [12].

The authors wish to express their gratitude to J.P. Wolf and M.J. McGlinchey for stimulating discussions and valuable assistance. They also acknowledge the support and cooperation of their colleagues in the Institute of Experimental Physics.

References

1. Devienne, F.M., Roustan, J.C.: Org. Mass Spectrom **17**, 173 (1982)
2. Keller, R.: Symp. on Acc. Asp. of Heavy Ion Fusion, Darmstadt (1983)
3. Leleyter, M., Joyes, P.: J. Phys. (Paris) **C 2**, 11 (1976)
4. Begemann, W., Dreihöfer, S., Meiwes Broer, K.H., Lutz, H.O. Z. Phys. D – Atoms, Molecules and Clusters **3**, 183 (1986)
5. Fayet, P., Wöste, L.: Surf. Sci. **156**, 134 (1985)
6. Hortig, G., Müller, M.: Z. Phys. **221**, 119 (1969)
7. Fayet, P., Granzer, F., Hegenbart, G., Moisar, E., Pischel, B., Wöste, L.: Phys. Rev. Lett. **55**, 3002 (1985)
8. Fayet, P., Wolf, J.P., Wöste, L.: Phys. Rev. **B 33**, 6792 (1986)
9. Wade, K.: Adv. Inorg. Chem. Radiochem. **18**, 1 (1976); Wade, K.: Transition metal clusters Johnson, B.F.G. (ed.), pp. 193–263. New York: John Wiley & Sons, Pub. 1980
10. Mingos, D.M.P.: J. Chem. Soc. Dalton Trans. **133**, (1974)
11. Lauher, J.W.: J. Am. Chem. Soc. **100**, 5305 (1978)
12. Fayet, P., McGlinchey, M.J., Wöste, L.: J. Am. Chem. Soc. (submitted for publication)

P. Fayet
L. Wöste
Institut de Physique Expérimentale
Ecole Polytechnique Fédérale de Lausanne
CH-1015 Lausanne
Switzerland

Sputtered Metal Cluster Ions: Unimolecular Decomposition and Collision Induced Fragmentation ⋆

W. Begemann, S. Dreihöfer, K.H. Meiwes-Broer, and H.O. Lutz
Fakultät für Physik, Universität Bielefeld, Federal Republic of Germany

Received April 7, 1986; final version May 9, 1986

Cluster ions are produced by ion bombardment of thick metal targets and mass selected in a Wien filter. The unimolecular decomposition of Al_n^+, Cu_n^+, Mo_n^+, W_n^+, and Pb_n^+ is investigated under UHV conditions. The time evolution of the decay allows a glimpse into the cluster formation/fragmentation process. Highly excited metal cluster ions decompose mainly by evaporating single neutral atoms with rates reaching 100%. The collision induced fragmentation (CIF) of stable mass selected metal cluster ions in a low pressure Ar and O_2 gas target will be compared to the unimolecular decay.

PACS: 36.40.+d; 07.75.+h; 79.20.Nc

1. Introduction

While most experiments on small gas phase metal clusters have examined neutrals, working with charged ones offers several experimental advantages: It is possible to select a particular cluster size, the energy in collision experiments can be varied in a simple way. In this paper another significant attribute is used in addition, namely the possibility to set (in a wide range) arbitrary time windows between the production of highly excited sputtered cluster ions and their detection.

When investigating cluster properties most of our knowledge arises from mass spectrometric studies. To what extend does a mass spectrum reveal the "true" intensity distribution at the time of production or ionisation? Several investigations deal with the fragmentation and metastable decay of gas phase clusters after photo- or electron impact ionisation [1–6]. Strong decay of sputtered alkali-halide [7] and Ag_n [8] cluster ions have been reported, too. Such work shows that cluster ion intensities alone cannot provide an unambiguous identification of stable structures. In particular it is often not possible to separate spectral features due to the properties of the parent neutral clusters from those of daughter ions.

⋆ Dedicated to Professor Dr. N. Riehl on occasion of his 85th birthday

Experimental work on reactive and non-reactive collisions of clusters under low pressure conditions is scarce. Most of it has been performed on hydrogen and helium clusters [9, 10] where collision induced fragmentation (CIF) is dominant. The kinematical behaviour of argon clusters is studied in a collision experiment with helium [11]. CIF experiments in the 0...3 eV range of Mn_2^+ yield the metal dimer bound energy [12]. Sputtered molecular ions and small clusters dissociate in gas targets [13, 14]. Reactions of small metal cluster ions with ethylene have been reported previously [15].

The experiments described here investigate *a)* the spontaneous behaviour of sputtered metal cluster ions, i.e. the time evolution of their unimolecular decomposition, as well as *b)* their fragmentation induced in collisions with $1.5 \cdot 10^{-5}$ mbar Ar and O_2 gas targets.

2. Experimental

The experiments are performed in a four-stage differentially pumped UHV machine with pressures of $<10^{-8}$ mbar (base pressure) and $<10^{-7}$ mbar (during sputtering) inside the target chamber. No target contamination or collisional relaxation disturbs the unimolecular decay of the sputtered clusters. In par-

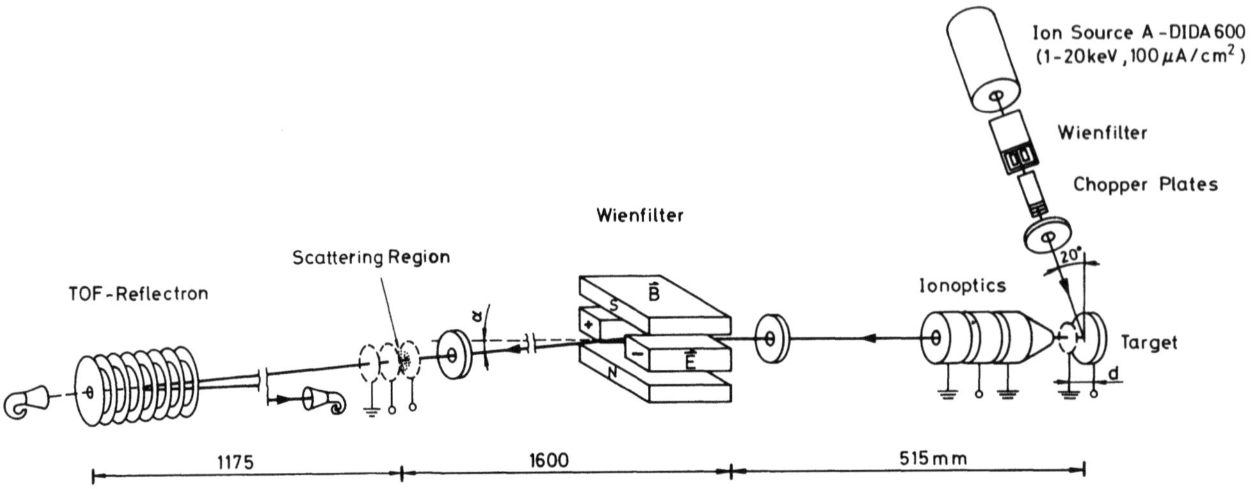

Fig. 1. Experimental setup

ticular the production conditions will not seriously be affected by scatter gas as the collision region is well separated from the target.

The apparatus consists of four subgroups (see Fig. 1): The primary ion source, the target device incl. acceleration and focussing optics, the mass separator, and the collision region with subsequent mass analyzer.

a) Primary Ion Source

We use a single discharge ion source (ATOMICA A/DIDA 600) which is designed for secondary ion mass spectrometry (SIMS). Rare gas ions are accelerated up to 22 keV and purified by means of a constant field Wien filter. The primary ions are focussed over a distance of 0.6 m where a spot diameter of 2 mm is obtained. This long distance allows simple positioning of the beam with two pairs of electrostatic plates. An additional pair of plates in conjunction with a 3 mm diaphragm (pressure stage) can serve as beam chopper.

b) Target Device and Ion Optics

The target holder supports up to eight electrically isolated thick target plates (15 mm diameter) which can be set and adjusted without the need of opening the machine. As the sputtering yield is not strongly affected by temperature changes of the target [16] no temperature control is installed. The primary beam hits the target 70° to its surface normal for highest sputtering yield. By applying a target voltage (300...3,000 V, typically 1,500 V) the secondary ions are accelerated either to a grounded net 1.7 mm or to the (grounded) entrance of an einzel lens 18.5 mm in front of the target. Over a distance of 1.9 m the clusters are slightly focussed into the scatter chamber where a beam diameter of about 10 mm is achievable, depending on the cluster mass (translational energy distribution) and on the mass selection conditions.

c) The Mass Separator

For mass separation we use a COLUTRON Wien filter model 600 B. Actually this instrument is a velocity filter which consists of a liquid-cooled 3000 Gauss magnet and an electric deflection field perpendicular to the magnetic field. In order to ensure a uniform electric field between the plates and to compensate for unwanted focussing effects a set of 18 guard rings is provided with adjustable potentials.

The Wien filter works in straight line configuration at a transmission of up to 50%; no retardation devices are required. In this experiment the drift tube is tilted ($\alpha = 2...5°$, see Fig. 1) with respect to the target-filter direction in order to suppress neutral sputtered particles. The theoretical resolution of about 400 is not achieved in this experiment because the sputtered cluster ions have broad energy distributions. The translational energy distribution of sputtered clusters varies with the mass and so does the Wien filter resolution. Typical resolution at 1.5 keV is $m/\Delta m = 20...30$ with a distance between Wien filter and entrance slit of the scatter chamber of 1.4 m.

d) Scattering Zone and Mass Analyzer

Scatter gas is injected into the starting region of a time-of-flight (TOF) reflectron by means of a tube

(0.5 mm diameter) with the flow controlled by a needle valve. The mass selected cluster ions, now pulsed, as well as nascent fragments pass the field-free drift region and are rapidly decelerated to about one third of their kinetic energy. Then they enter the reflecting field with low velocity where their differences in the time-of-flight, due to the energy spread obtained during the pulse production, is corrected. The reflected ion beam (including an angle of 3.4° with the incident beam) is detected by a tandem channel plate and recorded after time to pulse height conversion in a multichannel analyzer. With the second detector behind the reflector (and the reflector voltage set to zero) the instrument can be used as normal TOF analyzer.

The benefits of a TOF reflectron have been demonstrated before [17, 3]. Beside offering an improvement of the mass resolution it is useful for the study of metastable decompositions and collision or photofragmentation processes. Different modes of operation are possible; e.g., collisions can be investigated at variable energies when the scatter region is electrically floated. Here we report on measurements where the scattering zone is grounded so that the cluster ions hit without deceleration onto the gas target. The parent masses as well as CIF products enter the reflecting field. Their penetration depth is a function of their energy and so is the time of arrival at the detector.

The unimolecular decay is additionally studied in a continuous beam experiment. Therefore the TOF reflection is replaced by a simple repeller energy analyzer, mounted in front of a 17 stage CuBe electron multiplier EMI 9643/3B. Transmitted current measurements as function of the repeller voltage reveal the spontaneous behaviour of the cluster ions.

3. Results

3.1. Mass Spectra of Sputtered Metal Cluster Ions

In the first stage of the experiment the sputtering conditions are studied and optimized. The total sputtering yield is a smoothly increasing function of the kinetic energy of the primary ion beam (source voltage) and increases when the Ar^+ beam is changed to Kr^+ and to Xe^+. The ratio of the cluster intensity compared to that of dimer ions does not change with the source voltage. Therefore all the following measurements are performed with a 20...22 keV Xe^+ primary beam.

The cluster ion beam content is either studied by chopping the primary ion beam and performing a subsequent time-of-flight analysis, or by scanning the deflection voltage of the Wien filter and monitoring

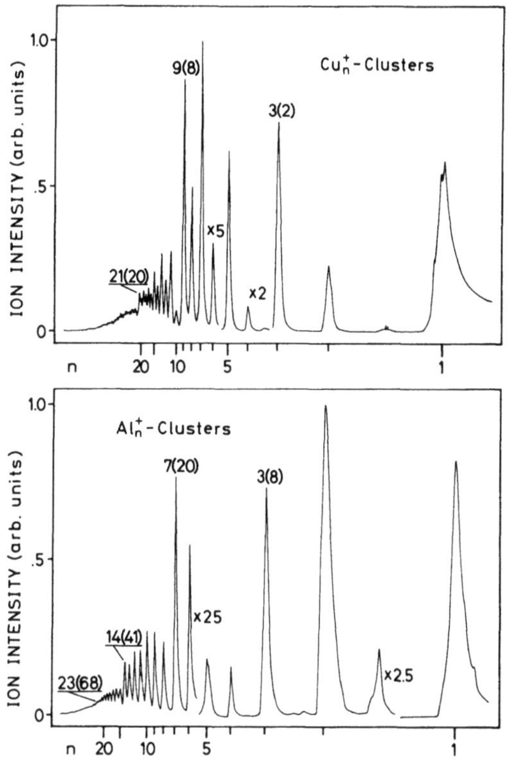

Fig. 2. Wien filter spectra of Cu_n^+ and Al_n^+, sputtered by 20 keV Xe^+. The figures in parentheses give the number of valence electrons of particulary stable cluster ions

the transmitted ion current. The second method yields spectra with poorer resolution but high intensity. Figure 2 shows such Wien filter spectra of Al_n^+ and Cu_n^+, Fig. 3 spectra of Mo_n^+, W_n^+ and Pb_n^+. The apparently low resolution of the Wien filter originates from the broad velocity distributions of sputtered ions. Continuous, mass identified beams of all cluster ions as shown in Figs. 2 and 3 and of a variety of other metals and silicon have been produced and can be subjected to different investigations.

3.2. Unimolecular Decay

For the study of the connection between measured intensity distributions and cluster stability the time evolution of the metastable decomposition is investigated: The ion beam is velocity separated by means of the Wien filter which is tuned to one (parent) cluster mass of interest. All particles whose velocities after acceleration differ from those of the selected clusters will be deflected out of the beam. Only the chosen ions and their positively charged fragments (of the same velocity) reach the detection chamber. Consequently the detected signal is an image of the number of selected ions which have survived the acceleration [18]. The acceleration time t_0, determined by the ac-

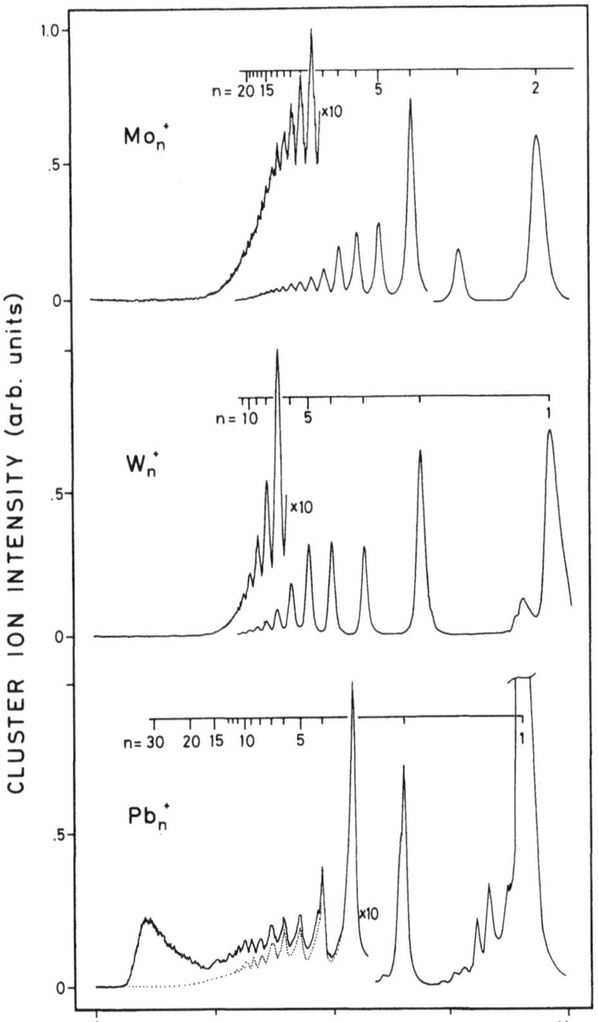

Fig. 3. Wien filter spectra of Mo_n^+, W_n^+, and Pb_n^+. In the dotted part of the lower spectrum the contribution of charged fragments is suppressed

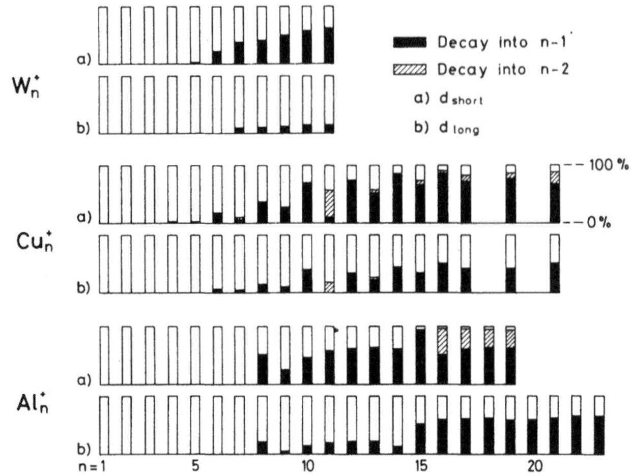

Fig. 4. Decay probabilities of sputtered Al_n^+, Cu_n^+ and W_n^+ cluster ions into positive fragments for the long (d_{long}) and short (d_{short}) acceleration distance. Values in percent, the uncertainty being ±2.5%

celeration voltage, the acceleration distance d (1.7 and 18.5 mm, respectively) and the selected mass, varies between $t_0=35$ ns (Al^+, $d=1.7$ mm) and $t_0=2.7$ μs ($Cu_{21}^+=18.5$ mm) and opens a time window. The time window closes at $t_0+\Delta t$ when the ions reach either the repeller grid of the energy analyzer (2.45 m away from the target, 20...200 μs after the production at $t=0$), or the end of the TOF reflectron (3.25 m).

Using the repeller grid energy analyzer the fragment ion content in the cluster beam at the end of the time window ($t_0+\Delta t$) is measured by scanning the repeller voltage. Intensity steps reveal uniquely the unimolecular decay of the separated clusters into positive fragment ions of defined mass. The results for W_n^+, Cu_n^+ and Al_n^+ are given in Fig. 4. The maximum height of the bars is 100%, corresponding to the decay (in percent) as it is observed for the *a*) short (d_{short}) and *b*) long (d_{long}) acceleration distance. In all cases small metal cluster ions with $n \leqq 3$ are practically stable within Δt. Strong decomposition starts at larger n and rises with increasing n. When t_0 is shifted towards the production event ($d_{long} \rightarrow d_{short}$) the decay probability rises. A variation of Δt by enlarging the total drift length by 1 m affects the outcome of the experiment only insignificantly.

The dominating charged decay product is in most cases the $(n-1)$ fragment (black bars in Fig. 4), $(n-2)$ decomposition (shaded bars) occurs in some cases in Cu_n^+ and Al_n^+. The metastable decay of W_n^+ rises smoothly with n. In Cu_n^+ a marked even-odd alternation is superimposed on a general increase of the decomposition rate. This regularity is disturbed by the high rate at $n=10$ and the exceptional $(n-2)$ fragmentation for $n=11$ into $n=9$. Al_n^+ decomposes irregularly with a rate reaching nearly 100% for large n (d_{short}).

3.3. Collision Induced Fragmentation (CIF)

All CIF experiments are performed with the long acceleration distance (d_{long}) so that relatively stable cluster ions are selected. The mass identified 1.8 keV clusters impinge without retardation onto the $1.5 \cdot 10^{-5}$ mbar gas target. Nascent fragments are separated and analyzed in the TOF reflectron. When the voltages of the reflector are chosen to optimize the resolution not all fragment masses will be recorded simultaneously. E.g., in the case of $Cu_9{}^+$ only the parent mass and fragments with $n=8$ and 7 are recorded with good resolution whereas fragments

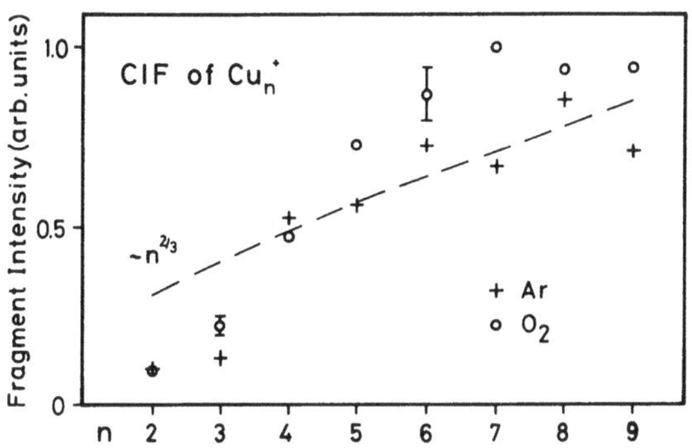

Fig. 5. Relative total fragment intensities I_{frag} for 1.8 keV Cu_n^+, colliding with $1.5 \cdot 10^{-5}$ mbar argon and oxygen. The dashed line gives the relative geometrical cross section for a drop with invariable compactness

with $n<7$ are hard reflected at the entrance to the field, thus producing an unresolved broad line. Next the reflecting voltage is lowered below the cluster energy E_{cl}. Now the parent mass is missing in the spectrum whereas lower fragments enter the reflecting field and are recorded as well resolved peaks. Iteration of this procedure unrolls the whole CIF pattern of Cu_9^+.

In this manner the CIF of Cu_n^+, $2 \leq n \leq 9$, has been measured in $1.5 \cdot 10^{-5}$ mbar argon and oxygen gas targets. For each n fragmentation is observed into Cu_i^+, $i<n$, where the fragment intensities oscillate in the same manner as the ion intensities in the Cu_n^+ mass spectrum, see Fig. 2. A complete presentation of the CIF results goes beyond the scope of this paper; instead, Fig. 5 accumulates the relative total fragment intensities (I_{frag}). These are the sums of the integrals of all fragment peaks which result from mass identified parent clusters (related to the parent cluster intensity) as a function of n, for Ar and O_2, respectively. Contributions of chemical reaction products are omitted in this analysis. I_{frag} shows an overall increase with n for the O_2 target, where the values for Ar oscillate, remaining generally below I_{frag} of O_2.

4. Discussion

The unimolecular decay probabilities of sputtered Al_n^+, Cu_n^+ and W_n^+ cluster ions (Fig. 4) turn out to be mirror images of the ion abundances in the mass spectra of Figs. 2 and 3. This finding should be elucidated with more care: The Wien filter spectra have been taken with a grounded repeller grid in front of the multiplier. All ions which have been produced on or just in front of the target (<0.1 mm) *and* survived the acceleration distance d will be recorded on the mass according to the Wien filter setting. The decay probabilities, on the other hand, are measured in a large window of width Δt which opens at t_0. It is strongly suggested that differences in cluster *ion* stabilities are responsible for both the shape of the mass spectra and the decay pattern.

A connection between the intensity of sputtered cluster ions and their stabilities has been pointed out before [19]. In this reference cluster stability is regarded as the dominant factor in the *production* process of sputtered cluster ions. Our experiments, however, indicate that the complete ion intensity distribution is the result of cascades of metastable decay processes of an unknown initial distribution. One striking example of a mass spectrum which is (for $n>15$) completely determined by the metastable decay is given in the lower part of Fig. 3 for Pb_n^+: The dotted curve has been taken with the repeller in front of the multiplier set to 50 eV below E_{cl}, i.e. all fragments of a parent n are repelled. Contributions of masses with $n>15$ are missing. The situation changes drastically when the fragments are also allowed to reach the detector (upper part of the Pb_n^+ mass spectrum), thus showing that parent clusters with $n>30$ have been produced, but have fragmented before reaching the detector.

The metastable decay rates of excited Cu_n^+ clusters have been calculated by Klots [20]. These calculations combine continuum based expressions of evaporation and general properties of unimolecular rate constants. Collisions and radiative losses are supposed to be negligible. The decay probabilities as function of n, calculated with this model, are in reasonable agreement with the experimental results, for d_{short} and d_{long}. These findings support the assumption that highly excited copper cluster ions evaporate single neutral atoms, cool down and gain stability.

We note that the size distribution of sputtered copper clusters (see Fig. 2) is similar to that which has been observed after photoionisation of cold Cu_n, produced in a laser vaporization source (Fig. 2 in Ref. [21]). Consequently the ion abundances in Ref. 21 are probably not an image of the neutral cluster ionisation potentials as has been stated there. In particular, in accordance with the low intensity of Cu_4^+ in the mass spectra, pseudopotential calculations of the dissociation energies of Cu_n^+ give an extremely low value for $n=4$ [22]. The decisive influence of the metastable decay and the stability of cluster decay products on the observed cluster ion mass spectrum are of rather general nature: The internal energies of the sputtered metal cluster ions (after cooling at $t<t_0$) can be estimated with the model of Klots and are in the order of 0.2 eV. Similar energies will easily be absorbed by neutral clusters during photo- or electron impact ionisation.

The decay probabilities of Cu_n^+ and Al_n^+ (Fig. 4) as well as the cluster ion abundances (Fig. 2) vary irregularly with n. The intense lines are explainable with particular stabilities of the cluster ions: In the case of copper strong even-odd variations are superimposed on intensity drops after $n=3$, 9, 21 (35, 41 and 59, not resolved in Fig. 2). Al_n^+ displays pronounced lines at $n=3$, 7, 14 and 23. Even-odd oscillations of Cu_n^+ have been attributed to the increased stability of clusters possessing spin-paired binding electrons. The prominent lines of Cu_n^+ and Al_n^+ are explainable by the shell model of Knight et al. [23] in which the valence electrons of the cluster atoms (see numbers in parentheses, Fig. 2) are assumed to fill the levels in a square well potential.

In the CIF experiment the collision energies exceed by far possible excitation energies of the cluster ions. Nevertheless the fragmentation pattern of each Cu_n^+ is governed by the fragment stability (for the small angle scattering observed here). Therefore also CIF is probably caused by evaporation of neutral atoms after the cluster absorbed energy in the collision. A similar behaviour has been observed for the fragmentation of 0.5 keV hydrogen cluster ions in a helium gas target [9].

I_{frag} of Cu_n^+ (proportional to a total fragmentation cross section) is in rough accordance with a relative geometrical cross section of a spherical drop of invariable compactness [9] ($\approx n^{2/3}$, see Fig. 5) for $n>3$. Nearer inspection shows deviations which are not surprising since the cluster stability and with it the tightness of the bonds is strongly varying with n.

We thank J. Simon for helpful discussions. Financial support by the Deutsche Forschungsgemeinschaft is gratefully acknowledged.

References

1. Stace, A.J., Shukla, A.K.: Chem. Phys. Lett. **85**, 157 (1982)
2. Stephan, K., Märk, T.D., Castleman, A.W., Jr.: J. Chem. Phys. **78**, 2953 (1983)
3. Castleman, A.W., Jr., Echt, O., Morgan, S., Dao, P.D., Stanley, R.J.: Ber. Bunsenges. Phys. Chem. **89**, 281 (1985)
4. Kummel, A.C., Haring, R.A., Haring, A., Vries, A.E. de.: Int. J. Mass Spectrom. Ion Proc. **61**, 97 (1984)
5. Birkhofer, H.P., Haberland, H., Winterer, M., Warsnop, D.R.: Ber. Bunsenges. Phys. Chem. **88**, 207 (1984)
6. Kreisle, D., Echt, D., Knapp, M., Recknagel, R.: Phys. Rev. A **33**, 768 (1986)
7. Ens, W., Beavis, R., Standing, K.G.: Phys. Rev. Lett. **50**, 27 (1983)
8. Katakuse, I., Ichihara, T., Fujita, Y., Matsuo T., Sakurai T., Matsuda, H.: Int. J. Mass Spectrom. Ion Proc. **67**, 229 (1985)
9. Van Lumig, A., Reuss, J., Ding, A., Weise, J., Rindtsich, A.: Mol. Phys. **38**, 337 (1979)
10. Gspann, J., Ries, R.: Surf. Sci. **156**, 195 (1985)
11. Buck, U., Meyer, H.: Ber. Bunsenges. Phys. Chem. **88**, 254 (1984)
12. Ervin, K., Loh, S.K., Aristor, N., Armentrout, P.B.: J. Phys. Chem. **87**, 3593 (1983)
13. Devienne, F.M., Roustan, J.-C.: Org. Mass Spectrosc. **17**, 173 (1982)
14. Fayet, P., Wöste, L.: Spectrosc. Int. J. **3**, 91 (1984)
15. Hanley, L., Anderson, S.L.: Chem. Phys. Lett. **122**, 410 (1986)
16. Oechsner, H.: Z. Phys. **261**, 37 (1975)
17a. Mamyrin, B.A., Karataev, V.I., Shmikk, D.V., Zagulin, V.A.: Sov. Phys. - JETP **37**, 45 (1973)
17b. Boesel, U., Neusser, H.J., Weinkauf, R., Schlag, E.W.: J. Phys. Chem. **86**, 4857 (1982)
18. Begemann, W., Meiwes-Broer, K.H., Lutz, H.O.: Phys. Rev. Lett. **56**, 2248 (1986)
19. Joyes, P., Sudraud, P.: Surf. Sci. **156**, 451 (1985)
20. Klots, C.E.: J. Chem. Phys. **83**, 5854 (1985); Klots, C.E.: Private communication
21. Powers, D.E., Hansen, S.G., Geusic, M.E., Michalopoulos, D.L., Smalley, R.E.: J. Chem. Phys. **76**, 2866 (1983)
22. Flad, J., Igel-Mann, G., Preuss, H., Stoll, H.: Chem. Phys. **90**, 257 (1984)
23. Knight, W.D., Clemenger, K., Heer, W.A. de, Saunders, W.A., Chou, M.Y., Cohen, M.L.: Phys. Rev. Lett. **52**, 2141 (1984); Knight, W.D.: Private communication

W. Begemann
S. Dreihöfer
K.H. Meiwes-Broer
H.O. Lutz
Fakultät für Physik
Universität Bielefeld
Postfach 8640
D-4800 Bielefeld 1
Federal Republic of Germany

A Penning Trap for Studying Cluster Ions

H.-J. Kluge
CERN, Geneva, Switzerland and Institut für Physik, Universität Mainz, Mainz, Federal Republic of Germany

H. Schnatz and L. Schweikhard
Institut für Physik, Universität Mainz, Mainz, Federal Republic of Germany

Received April 18, 1986

We propose to use a Penning trap for spectroscopy of stored cluster ions. A similar device has been built for the purpose of mass measurements of short-lived nuclei produced at the on-line isotope separator ISOLDE/CERN. A resolving power of 500,000 in a mass measurement of ^{39}K and an accuracy of 2×10^{-7} for the $^{85}Rb/^{39}K$ mass ratio were obtained. An efficiency for in-flight capture as high as 70% was achieved. The method provides very high sensitivity since typically only 10 to 100 ions are stored in the trap. We intend to perform laser spectroscopy on trapped Na clusters as a first application of the trap technique.

PACS: 36.40+d; 35.10.Bg; 29.20.−c

1. Introduction

Although enormous progress has been made in cluster spectroscopy during the past years, only very little is known for example, about, electronic structures of clusters, life times, and reaction, catalysis, or electron detachment processes. The reasons can be found in essentially two general drawbacks of all techniques applied so far to the study of clusters:

(i), the limited time of observation because atomic or ionic beams are used,

(ii), the difficulty of preparing a sample of one specific cluster size with sufficient purity and abundance.

Both problems can be solved by confining and accumulating the cluster ions of the desired size in an ion trap. The device best suited for this purpose is the Penning trap where a combination of static magnetic and electric fields is used to establish a three-dimensional trapping potential: In such a trap [1, 2] the cluster ions would be almost at rest in space, confined in a small volume, essentially free of any undesired perturbation, and at hand for almost infinite times limited only by the life time of the cluster itself. The high resolving power of mass measurements in a Penning trap would allow a precise identification of the cluster size and even enable studies of isotope effects if they exist at all.

In the past many experiments made use of the ion trap technique for the investigation of charged particles and led to a variety of high-precision experiments [3]. These experiments include laser spectroscopy on single ions [4, 5], accurate mass measurements [6, 7], ultra-high resolution microwave spectroscopy [8], and determination of the *g*-factors for the electron and positron [9], which represents the most accurate fundamental constant known today.

All charged particles investigated so far in an ion trap were created inside the potential well produced by the quadrupole field. However, in case of cluster ions it is desirable to produce these species by well established techniques [10] outside the trap, eventually to separate them in mass and to guide them with high efficiency into a trap.

Recently we reported on the first in-flight capture of ions in a Penning trap starting from a continuous ion beam [11]. Simultaneously and independently Alford et al. [12] were successful in capturing bunches of cluster ions in a similar device which differs from our set-up mainly with respect to the geometry and the detection scheme of the cyclotron resonance.

In the following, our method and our apparatus will be described. It was developed for precise mass measurements of short-lived isotopes. Emphasis will be put on the discussion of the geometry of the Penning trap, the harmonic motion of the charged particles in the trap, and the detection scheme which differ from those described in [12]. The performance of the existing device and finally its potential for a future application to cluster spectroscopy will be discussed.

2. Principle of a Penning Trap

Figure 1 shows the essential layout of a Penning trap. In the homogeneous magnetic field B directed along the z-axis particles with charge e and mass m perform a cyclotron motion with a frequency given by

$$\omega_c = (e/m)\,B. \tag{1}$$

In order to obtain a restoring force along the z-axis in addition to the confinement by the magnetic field in $x-y$ direction, a positive potential for positively charged ions is applied to the upper and lower electrodes (so-called endcaps, see Fig. 1) relative to the ring electrode. Due to the special (hyperbolical) shape of the electrodes the motion of a stored ion in the z-direction is decoupled from those in the $x-y$ directions. Furthermore all motions are harmonic. They can be separated into an oscillation ω_z along the z-direction and the modified cyclotron frequencies ω_+ and ω_-. These frequencies deviate from the pure cyclotron frequency ω_c due to the presence of the electrostatic field. The harmonic oscillations are described by the frequencies

$$\omega_z = \sqrt{(2eU/m r_0^2)}, \tag{2}$$

$$\omega_+ = \omega_c/2 + \sqrt{(\omega_c^2/4 - \omega_z^2/2)}, \tag{3}$$

$$\omega_- = \omega_c/2 - \sqrt{(\omega_c^2/4 - \omega_z^2/2)}. \tag{4}$$

The voltage U and the radius r_0 are explained in Fig. 1. In case of an ion with mass $m = 40$ amu, $B = 5.87$ T, $U = 8$ V and $r_0 = 0.8$ cm the frequencies are $\nu_c = 2{,}253$ kHz, $\nu_z = 124$ kHz, $\nu_+ = 2{,}250$ kHz, and the magnetron frequency $\nu_- = 3.4$ kHz, almost independent of the mass of the stored particle.

Note that ω_z as well as the modified cyclotron frequencies depend on the voltage U applied to the electrodes. These frequencies correspond to one-

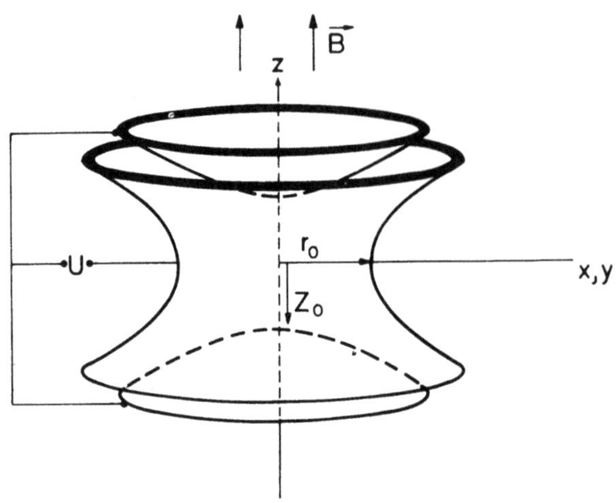

Fig. 1. Scheme of a Penning trap. The shapes of the electrodes are hyperbolical with rotational symmetry around the z axis which is also the direction of the magnetic field. The distance of the endcaps $2z_0$ is related to the minimum radius of the ring electrode by $r_0^2 = 2z_0^2$

quantum transitions in the harmonic oscillators. However, the sum frequency $\omega_+ + \omega_-$ equals ω_c. This frequency depends only on B (see (1)) and represents a two-quantum transition. Although the ω_c resonance has a small strength because it is a two-quantum transition, it has the advantage of being independent of the trapping voltage. Even more important, and in contrast to ω_+ this resonance is insensitive to space charge effects caused by a larger number of stored ions and is not very much influenced by imperfections of the hyperbolical shape of the electrodes caused, for example, by the necessary holes for capture and ejection of the ions.

3. Detection of the Cyclotron Resonance by Time of Flight Measurements

In resonance the trapped ions gain energy out of the applied radio frequency (RF) field. This is detected by a time-of-flight method [6]. The charged particles are ejected out of the trap by applying an electrical pulse to the ring electrode. For a short time the potential between the endcaps and the ring electrode is lowered so that the ions with highest energy can just escape the trap. These ions drift to a channel plate detector and their time of flight is determined. This procedure is repeated by continuously lowering the trapping potential as shown in Fig. 2 until the trap is empty. The mean time of flight as a function of the RF frequency shows a resonance which gives the mass of the stored ions (Fig. 3).

The basic principle of this detection technique is easily understood. In resonance the cyclotron orbit

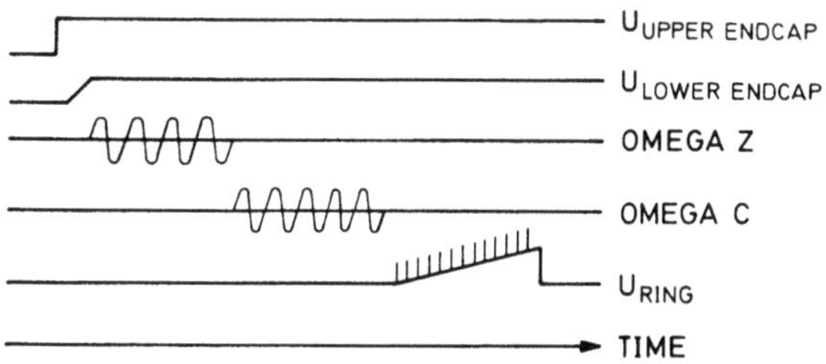

Fig. 2. Timing sequence for trapping charged particles, inducing the radio frequency, and for ejection of the ions. This sequence was used for mass measurements in a Penning trap using the time-of-flight technique

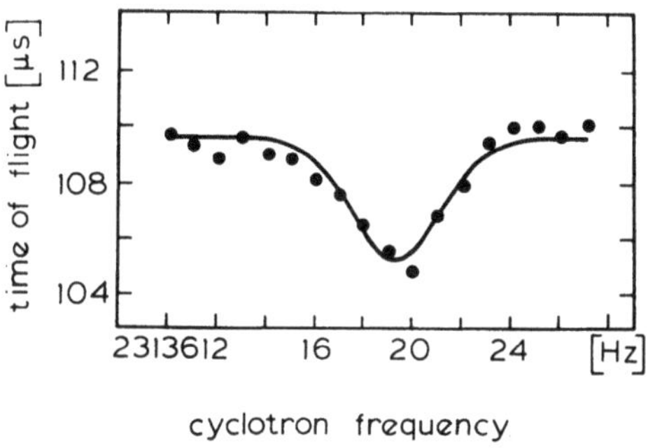

Fig. 3. Cyclotron resonance of potassium ions as obtained by the mean time of flight as a function of the frequency of the applied RF field. The resolution observed is 5×10^5

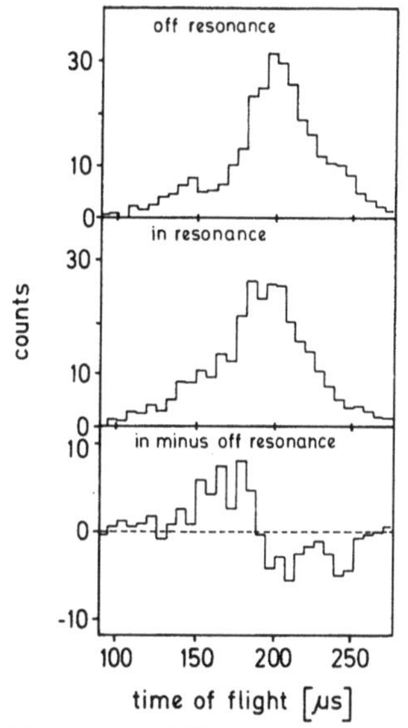

Fig. 4. Time-of-flight spectrum of N and N_2 ions. At $t=0$ the ions are ejected out of the trap and are then detected by the channel-plate detector. Top: Spectrum obtained with the radio frequency off resonance. Middle: Spectrum obtained in resonance at $\nu=\nu_c$. Bottom: Difference spectrum obtained by substracting the spectra in resonance and off resonance

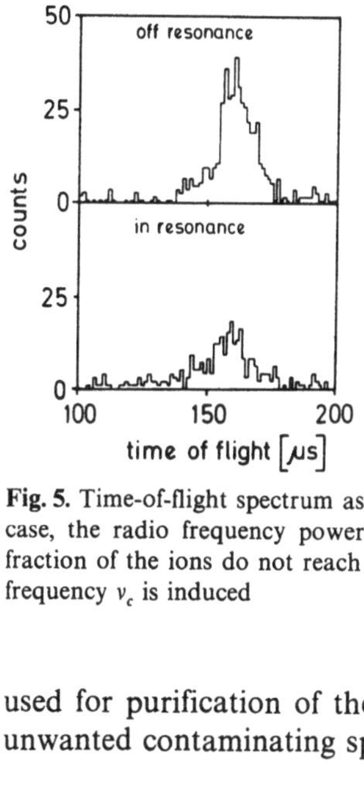

Fig. 5. Time-of-flight spectrum as in Fig. 4 but for He ions. In this case, the radio frequency power is much higher. Hence a large fraction of the ions do not reach the detector when the resonance frequency ν_c is induced

becomes larger. As a consequence the orbital magnetic moment μ_L increases, leading to a higher energy of the charged particle in the magnetic field. If the ions are ejected out of the trap into a region of lower magnetic field, the ions experience a force $\mu_L \delta B/\delta z$ in the direction of the magnetic field gradient. Conservation of energy demands that the radial energy $\mu_L B$ is transformed into longitudinal kinetic energy. This is just the reverse of the magnetic-bottle effect. Hence, the ions reach the detector faster in resonance than out of resonance.

Figure 4 shows the time-of-flight effect for the case of a mixture of N and N_2 ions. If the frequency $\nu_c(N_2)$ is applied, the mean time of flight is shifted to smaller values. This can be seen in the bottom part of Fig. 4 which shows the difference between the time-of-flight pattern in and off resonance.

The change in the mean time of flight has to be kept small if high resolution of the mass determination is required. Hence, the change in time of flight displayed in Fig. 4 only amounts to a few percent. If the strength of the RF field is increased, the ν_c resonance is power broadened. Finally at very high RF power, the ions gain such a high orbital energy that they can no longer escape through the hole of the ion trap. In this case, they strike the electrodes and are lost (Fig. 5). This effect can be used for purification of the sample, i.e. to get rid of unwanted contaminating species in the trap.

4. Experimental Setup

The experimental setup for the mass measurements is shown in Fig. 6. The apparatus consists of an alkali ion source, a Penning trap (trap 1) placed in the pole gap of an electromagnet, a transfer line with a number of electrostatic lenses and deflectors, a Penning trap (trap 2) placed in the stable and homo-

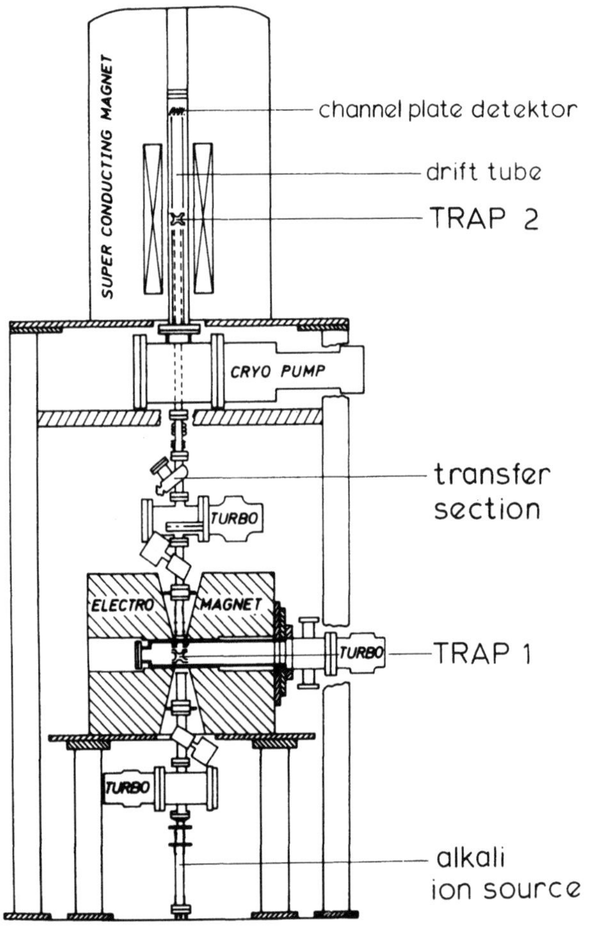

Fig. 6. Experimental set up for direct mass determination of externally created ions

geneous field of a superconducting magnet, a drift tube and finally a channel plate detector.

The first trap is essentially a bunching device to collect the continuous beam of the ion source. The second trap is used for high-precision measurements of the cyclotron frequency. Hence ultra-high vacuum, excellent homogeneity of the magnetic field and perfect geometry of the second trap are essential. Since the operating conditions of a bunching trap are incompatible with those of a precision trap, the layout of Fig. 6 was chosen with two completely separated traps connected only by a transfer tube. This configuration allows efficient differential pumping between traps 1 and 2 and the installation of ion optical devices for steering the ion beam.

Bunching Trap: Alkali ions delivered by the ion source are implanted into a tungsten foil mounted in one endcap of the Penning trap, then surface ionized by heating the foil, trapped [13, 14], and finally ejected by a sudden change of the trapping potential. In case of the study of cluster ions this trap can be abandoned because it is possible to create cluster ions directly in a bunched mode.

Transfer Tube: After leaving the trap the ion bunch is accelerated to 1 keV by electrodes. A system of electrostatic lenses guides the bunch into the fringe field of the superconducting magnet in such a way that the radial energy is not increased. Extensive ion optical calculations have been performed to find an optimum design [15]. This is most important for mass measurements, which require a very small energy spread of the trapped ions. The transmission of the transfer tube has been measured to be 80% (including the transmission through the holes in the endcaps of trap 2). At the entrance of trap 2 the duration of the bunch is shorter than 30 μs.

Precision Trap: Here the bunched ion beam is captured in flight by retarding it to a few eV just before entering the trap. The potential of the lower endcap of trap 2 is switched to the potential of the ring electrode just at the moment when the ion bunch arrives. When the ions pass the center of trap 2 the potential of the endcap is raised again and the ions can no longer escape. More details of the apparatus built for mass measurements are given in [11].

5. Performance

Trapping Efficiency: It has been determined that up to 70% of an ion bunch ejected out of trap 1 can be retarded and captured in flight in trap 2. More details can be found in [11].

Resolution: A resolving power of 500,000 has been observed in a mass measurement of ^{39}K. Here a line width of 4.4 Hz was obtained at a resonance frequency of 2.3 MHz (Fig. 3). Studies to increase further the resolving power are under way.

Accuracy: The ratio of the cyclotron frequencies of, for example, the Rb/K pair coincide with the tabulated mass ratio within 2×10^{-7}. This deviation corresponds approximately to the statistical uncertainty of the centroid of the resonance (Fig. 3) which is about 10% of the line width. Therefore the ratios of unknown to previously-known masses can be determined with an accuracy of the order of 2×10^{-7}.

Storage Time: The storage time of our Penning trap has not yet been investigated systematically. We can give only a lower limit: Up to a delay of 1.5 s with respect to the capture of the ions in trap 2 no decrease of the number of stored ion was observed. It can be expected that the life time of the charged particles is much longer because the ions already survived a large number of oscillations (see (3)–(5))

during this time of storage. The storage time depends strongly on the background pressure in the apparatus which was $<10^{-9}$ mbar in the experiments reported here.

Sensitivity: Typically 10 to 100 ions were stored in trap 2 at a time. Since the detection scheme applied here is destructive, the trap has to be refilled after each cycle. This involves filling the trap, inducing the RF, ejecting the ions, and measuring the time of flight. The cycle time is about 0.5 s. For the mass measurement of ^{39}K (Fig. 3) 100 cycles per frequency value were performed, corresponding to a total time of 20 min needed for the measurement of the ^{39}K mass.

6. Discussion

It is very attractive to apply the Penning trap technique to the study of clusters. In comparison to investigations on atomic or ionic beams one gains several orders of magnitude in observation time by using stored cluster ions. The ultra-high mass resolution achievable will even permit investigations of clusters of very large size. These advantages, together with the possibility of preparing pure samples of one cluster size only, lead us to expect that new fields in cluster spectroscopy will open up. All those techniques developed for the study of atomic ions can, in principle, be applied also to a study of cluster ions. These methods include cooling of the ions in the trap, detection of their motion, investigations using single ions, ultra-high resolution spectroscopy, and spectroscopy in the microwave as well as in the optical region. A quite comprehensive overview on the status of trapping atomic ions can be found in the contributions to the 1984 International Conference on Atomic Physics [16–18].

We intend to combine a Penning trap with a cold metal cluster beam produced in a supersonic expansion. Bunched ions can be obtained by pulsed photoionization. Appropriate timing of the opening of the trap can be used to pre-purify the cluster bunch with respect to the mass before the clusters are captured in the Penning trap. It should also be possible to accumulate a larger number in the trap than obtained by a single injection. For this purpose the potential of the ring electrode can be continuously decreased while the lower endcap is pulsed with constant amplitude in order to allow the pass of additional bunches. In this way, stacks of cluster bunches can be accumulated in the trap.

As discussed briefly in the Introduction, a large number of experiments can be performed with stored clusters. The most challenging might be to investigate Na clusters by laser spectroscopy and to search for optical resonances. Recently, Knight et al. [19] found peaks in the mass spectra of Na clusters at cluster sizes of $n=8$, 20, 40, 58, and 92. These "magic" numbers can be explained in a one-electron shell model in which independent delocalized atomic $3s$ electrons are bound in a spherically symmetric potential well [19]. The calculation yields discrete electronic energy levels and a shell structure which reproduces the peaks observed in the mass spectra. The authors conclude that the good correspondence between the experimental results and the model calculations suggests that there are no perturbations large enough to distort the main features of the level structure. Hence, discrete resonances should also be observable by laser spectroscopy of stored Na cluster ions. Such an experiment which presents a stringent test of the model seems only feasible using the trapping technique.

This work was supported by the Bundesministerium für Forschung und Technologie. We would like to thank G. Bollen, P. Dabkiewicz, P. Egelhof, F. Kern, H. Kalinowsky, H. Stolzenberg and H. Stürmer for their collaboration in realizing the Penning trap designed to measure the masses of short-lived nuclei.

References

1. Dehmelt, H.G.: Adv. At. Mol. Phys. **3**, 53 (1967)
2. Brown, L.S., Gabrielse, G.: Rev. Mod. Phys. **58**, 233 (1986)
3. Wineland, D.J., Itano, W.M.: Adv. At. Mol. Phys. **19**, 135 (1983)
4. Neuhauser, W., et al.: Phys. Rev. A **22**, 1137 (1980)
5. Nagourney, W., Janik, G., Dehmelt, H.: Proc. Natl. Acad. Sci. USA **80**, 643 (1983)
6. Gräff, G., Kalinowsky, H., Traut, J.: Z. Phys. A – Atoms and Nuclei **297**, 35 (1980)
7. Van Dyck, R.S. Jr., et al.: Int. J. Mass Spectrosc. Ion Proc. **66**, 327 (1985)
8. Blatt, R., Schnatz, H., Werth, G.: Phys. Rev. Lett. **48**, 1601 (1982)
9. Schwinberg, P.B., Van Dyck, R.S. Jr., Dehmelt, H.G.: Phys. Rev. Lett. **47**, 1679 (1981)
10. Recknagel, E.: Proceedings of the 9th International Conference Atomic Physics, Seattle 1984. Dyck, Jr., R.S. van, Fortson, E.N. (eds.). Atomic Physics 9. Singapore: World Scientific 1984
11. Schnatz, H., et al.: Nucl. Instrum. Methods (submitted for publication)
12. Alford, J.M., et al.: Int. J. Mass Spectrosc. Ion Proc. (submitted for publication)
13. Coutandin, J., Werth, G.: Appl. Phys. B **29**, 89 (1983)
14. Bonn, J., et al.: Appl. Phys. B **30**, 83 (1983)
15. Stürmer, H.: Diploma work, Mainz, unpublished (1984)
16. Dyck Jr., R.S. van, Church, D.A.: Proceedings of the 9th International Conference Atomic Physics, Seattle 1984. Dyck Jr., R.S. van, Fortson, E.N. (eds.). Atomic Physics 9. Singapore: World Scientific 1984
17. Werth, G.: Proceedings of the 9th International Conference Atomic Physics, Seattle 1984. Dyck Jr., R.S. van, Fortson,

E.N. (eds.). Atomic Physics 9. Singapore: World Scientific 1984
18. Wineland, D.J., et al.: Proceedings of the 9th International Conference Atomic Physics, Seattle 1984. Dyck Jr., R.S. van, Fortson, E.N. (eds.). Atomic Physics 9. Singapore: World Scientific 1984
19. Knight, W.D., et al.: Phys. Rev. Lett. **52**, 2141 (1984)

H.-J. Kluge
CERN
EP-Division/Isolde
CH-1211 Geneva 23
Switzerland
and
Institut für Physik
Universität Mainz
Jakob-Welder-Weg 11
D-6500 Mainz
Federal Republic of Germany

H. Schnatz
L. Schweikhard
Institut für Physik
Universität Mainz
Jakob-Welder-Weg 11
D-6500 Mainz
Federal Republic of Germany

The Chemistry and Physics of Molecular Surfaces

A. Kaldor, D.M. Cox, D.J. Trevor, and M.R. Zakin
Corporate Research Laboratory Exxon Research and Engineering Company, Annandale, New Jersey, USA

Received June 18, 1986

This article reviews the results of several recent experiments performed in our laboratory designed to elucidate the fundamental chemical and physical properties of clusters of both transition metals and other refractory elements containing from one to several hundred atoms. The gas-phase reactivity of clusters towards a variety of reagents is explored using a fast-flow reactor system. Strong cluster size-dependent variations in reactivity are observed, especially for the case of hydrogen chemisorption. Measurement of cluster photoionization thresholds (IPs) provides a sensitive probe of the evolution of cluster electronic structure as a function of the number of constituent atoms.

Cluster ionization potentials are observed to exhibit fluctuations about the smooth global falloff predicted by the classical drop model, indicating the non-bulk-like behavior of small clusters. Measurement of shifts in IP induced by chemisorption of different reagents provides insight into the nature of adsorbate-cluster bonding. The formation and properties of bare and metal-doped carbon clusters are explored, with particular emphasis on elucidating the photophysics and photochemistry of the postulated ultra-stable larger clusters. The results suggest that further work is required to prove soccer ball-like structures for C_{50}, C_{60}, etc. Finally, infrared multiple-photon dissociation (IR-MPD) is demonstrated to be a viable technique for obtaining infrared spectra of absorbate-cluster complexes. This technique is an important new tool for obtaining information about the molecularity of gas-phase reactions beyond that currently available from mass spectrometric analysis. As an illustration of the method, IR-MPD spectra of methanol chemisorbed on small iron clusters are obtained.

PACS: 36.40.+d; 33.80.Rv; 35.20.V_f

Introduction

The physical and chemical properties of materials in the transition to ever smaller units of constituent atoms are of substantial scientific and technological interest. Change manifests itself in a variety of different forms, from electronic structure [1] to details of the physical arrangement of the atoms [2]. Perhaps one of the most significant changes occurs in the relative contribution of surface atoms to those below the surface as a function of cluster size, including the development of long range order in the "bulk", or interior atoms. To put this in perspective consider that for clusters of fewer than twelve atoms all are exposed surface atoms; a globular Pt_{12} forms a 7 Å diameter cluster. In comparison, for a 20 Å platinum cluster of about 280 atoms, about 120 are "bulk" atoms [3].

There have been few experimental methods available to investigate the properties of materials as a function of the number of constituent atoms. Most techniques, such as supported catalyst preparations, produce a distribution of cluster sizes [4]. Some, such as synthesis of organometallic cluster compounds [5], produces materials of single cluster size, but these are surrounded by ligating agents to satisfy the coordination requirements of the cluster. Much of what we know about the structure and bonding of metal clusters comes from these systems. Synthesis of materials from the thermal decomposition of single cluster size organometallic complexes deposited on supports has only produced a distribution of cluster sizes [6]. Other, presumably more gentle preparation techniques in which only a single size cluster would be produced have not been demonstrated yet. Matrix isolation techniques also produce a distribution of clusters, and due to the nature of the experiments have primarily focused on collecting and studying materials consisting of four or fewer atoms [6]. Pioneering experiments in molecular beams have made it possible to study the electronic properties of Na, K, and a few other elements which are relatively easy to vaporize [7]. Recently, utilizing laser vaporization in a pulsed molecular beam, gas phase cluster studies have been extended to virtually every element, or combination of elements in the periodic table, including such refractories as carbon and tungsten [8]. As a result, there is a virtual explosion in the data base on many aspects of unsupported gas phase clusters of metals and non-metals alike. In this paper we review recent findings in our laboratory on the ionization potential of bare clusters and cluster molecule adducts, with special emphasis on transition metals, on cluster chemical reactivities, and on infrared laser-driven dissociation of cluster adducts to provide insight to their vibrational spectroscopy. The richness of these findings only reveal the tip of the iceberg that awaits us as we explore this scientific frontier.

Experimental

In recent papers we reported the techniques we use to produce, detect, and react clusters [1, 9, 10]. Here we only provide a brief description of the operation and refer the reader interested in the specific conditions used to the original publications which describe the work. A schematic of the nozzle-reactor section of the experimental apparatus is shown in Fig. 1. A one inch rod of 0.125 inch diameter of the particular metal studied is positioned near the exit of a high pressure pulsed valve. Metal is vaporized by focusing a high power pulsed laser to 1/2 mm diameter spot on the rod. The rod is rotated and translated to continually expose fresh material to the laser. The vaporization laser is timed to fire so as to entrain the vaporized metal in the high pressure helium gas released by the pulsed valve into a 0.2 cm diameter, 4.5 cm long tube. In this tube the vapor condenses forming a broad distribution of clusters. At the end of the tube the entrained clusters flow into a 1 cm diameter, 5 or 7.6 cm long reactor tube into which either pure helium, or a reactant/helium mixture is pulsed. This approach enables one to separate chemical and physical effects on the cluster beam. The residence time in the 7.6 cm long reactor is estimated to be about 160 ± 50 μs; the total pressure 5 to 25 torr. The gas temperature is estimated to be between 300 to 500 K[10].

Upon exiting the reactor tube, the cluster/gas mixture expands rapidly into vacuum; it is cooled and subsequently collimated by a series of skimmers into a molecular beam. Figuratively, as well as literally, all reaction is frozen after the expansion. Downstream, 85 cm from the reactor, the beam is intercepted by an ionizing laser pulse fired at a pre-selected delay time. The ions produced are analyzed by time-of-flight mass spectrometry. Scanning of the delay time provides a temporal profile of the cluster beam.

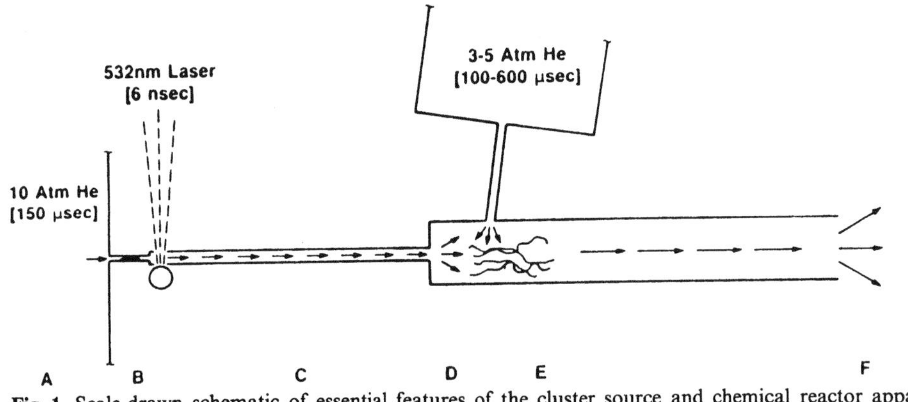

Fig. 1. Scale-drawn schematic of essential features of the cluster source and chemical reactor apparatus. Moving from left to right the method works as follows: **A** The primary nozzle opens and He expands from 3–10 atm stagnation pressure into a 1 mm diameter channel. **B** Near the peak of the primary He pulse, a focused frequency-doubled Nd:YAG laser strikes a rotating-translating rod positioned where the 1 mm channel opens to 2 mm. **C** Sputtered metal atom vapor rapidly condenses and cools while flowing in the 4.5 cm long condensation channel. **D** The primary He pulse and clusters expand into the 1.0 cm diameter reactor. **E** The secondary nozzle opens at a variable time with respect to the primary nozzle and laser, and the resulting expansion across the primary stream creates a turbulent zone of intimate mixing; this is the reaction zone, extending downstream toward the end of the reactor. **F** The mixed flow expands into vacuum ($\sim 10^{-5}$ torr), and a portion is collimated by conical skimmers into a molecular beam. Clusters and products are detected $\sim$85 cm downstream by photoionization mass spectrometry

Ionization is achieved by either a tunable laser system, or by an excimer laser operating on any one of several lasing mixtures, such as F_2 at 157 nm (7.87 eV), ArF at 193 nm (6.42 eV), or KrF at 249 nm (5 eV).

Infrared spectra are obtained by using a pulsed, line tunable CO_2 laser, aligned coaxial with the molecular beam. The infrared laser output energy is attenuated with a dimethylether gas cell. The TEM_{00} laser output is focused to a spot of 0.8 mm diameter, 65 cm upstream from the ionization region. The laser is timed to produce the maximum depletion in the photoionization mass spectrum of specific masses as a result of IR multiple-photon laser-induced dissociation. Consistent with our observations, calculations also show that the photofragments leave the photoionization detection zone. The frequency dependence of the depletion in the mass spectrum is related to the infrared absorption spectrum of the species.

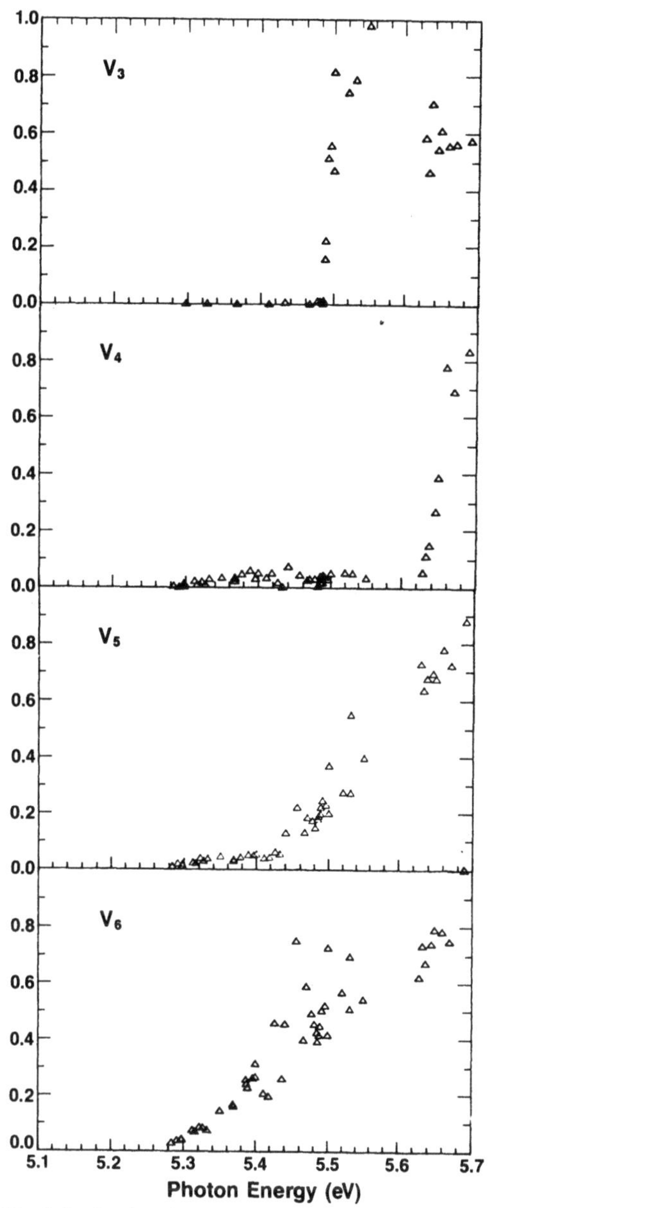

Fig. 2. Ionization thresholds for V_{3-6} clusters. Each curve is normalized first to the laser energy and then to its maximum value. The photon intensity is 25 μJ/cm² in a 4 ns pulse

Discussion

a. Ionization Potential

We have studied the variation of ionization potential (IP) with cluster size for a number of transition metals [1, 10]. Figure 2 shows the onset of photoionization for V_{3-6}. As the cluster size increases from V_3, the threshold shape changes from a rather sharp onset for V_3 to a more gradual one for V_6; these shapes are typical for a number of metal systems. Both V_4 and V_5 display measurable photoion yield below what we are calling the threshold; these ions may be produced by a number of different mechanisms. These include ionization from vibrational hotbands, ionization from thermally occupied low lying electronic states, ionization of structural isomers, or multiple photon ionization processes. Ionization threshold data are summarized in Fig. 3 for Nb and V. The IP values as a function of cluster size for transition metal systems exhibit significant oscillations about the monotonic $1/R$ fall-off predicted by the classical spherical metal drop model [11],

$$W_{\text{cluster}} = W_{\text{bulk}} + (3/8)e^2/R \qquad (1)$$

where R is the radius, W_{cluster} is the IP, and W_{bulk} is the infinite flat plane bulk work function [11]. The correction term $(3/8)e^2/R$ arises from two effects. One effect takes into account the fact that the image potential differs between a metal sphere and a flat metal surface. This effect leads to a *lowering* of the work function for a metal sphere relative to a flat plane by an amount $(5/8)e^2/R$. The second effect accounts for the coulomb attraction between the departing electron and the positive hole (for an infinitely conducting sphere this hole charge may be assumed to reside at the center of the sphere). This effect *increases* the work function of the isolated cluster relative to a grounded flat plane by an amount e^2/R. Thus the work function of the small metallic sphere is predicted to be larger than that of the flat plane by $(3/8)e^2/R$. This relationship holds surprisingly well for clusters of free electron metals such as Na and K[12]. When the model is applied to transition metals e.g. Fe, V, Nb, the overall $1/R$ falloff is obeyed quite nicely. However, for the solid curves in Fig. 3a value of 3.4 is assumed for the work function of the bulk ($R=\infty$), 0.9 eV below the bulk polycrystalline value. It has

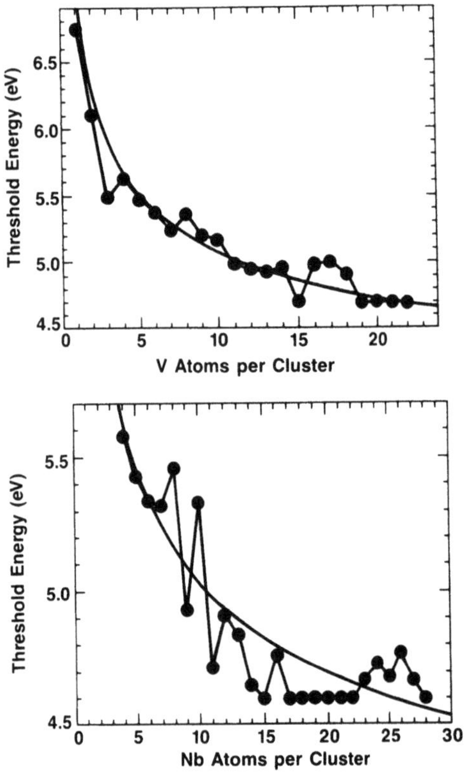

Fig. 3. Ionization thresholds for V and Nb as a function of the number of metal atoms in the cluster. The solid line is the evaluation of equation 1 assuming $W_{bulk} = 3.4$ eV and bulk density. The reported work function for both polycrystalline V and Nb is 4.3 eV

been observed that the work function for rough surfaces with substantial defect and step density can vary by 0.5 eV from that of the most dense packing. This suggests qualitatively that clusters in this small size range may mimic such defected, disordered surfaces.

The oscillations about the $1/R$ fall-off also closely correlate with strong variations in reactivity of these clusters. The exact reason for this non-monotonic behavior is not yet resolved. We suspect that large, discontinuous changes in the measured IP's may reflect major geometric and/or electronic structural changes in the clusters. Quantification of such effects requires much more extensive theoretical calculations than are available today.

One important issue that needs to be resolved is whether we are dealing with globular, structureless materials, or if unique stable structures for specific clusters exist. One could imagine that several structures with nearly the same energetics coexist, but with low barriers among them they will be fluctional. The observation of significant magnetic deflections of iron clusters suggest however that well defined electronic structures exist, at least on the scale of the transit time through the inhomogeneous magnetic field [1 b, c].

Table 1. Upper (U) and lower (L) bounds for increase in ionization threshold for $V_x(H_2)_y$ species

metal cluster	H_2 L−U	$(H_2)_2$ L−U	$(H_2)_3$ L−U	$(H_2)_{\geq 4}$ L−U
V_3	0.9–2.4	0.9–2.4		
V_4		0.8–2.2	0.8–2.2	0.8–2.2 ($y=4$)
V_5			1.0–2.4	1.0–2.4 ($y=4$)
V_{11}				0.6–1.4 ($y=6,7$)

We have also examined the effect of chemisorption on the IP of some clusters, i.e. I.P. of cluster complexes [13]. Specifically, we have observed that the IPs of vanadium, niobium, and iron clusters increase upon chemisorption of hydrogen (D_2), and that a stepwise increase is evident as the hydrogen (D_2) "coverage" increases. Table 1 summarizes these observations for V_x. Though not quantified, O_2 adsorption on clusters of Pd show similar behavior as a function of cluster size. In other cases the IPs drop; for example when benzene is chemisorbed on Pt and Pt_2. The measurement of IPs may be a very sensitive tool to determine the nature of adduct bonding. More detailed studies in conjunction with theoretical analysis of the electronic structure of cluster adducts offer powerful, new insight into the bonding chemistry and perhaps even the structure of these new materials.

Carbon clusters are one example where consideration of IP variation with cluster adducts is useful for the interpretation of photoionization mass spectra and in discussions about cluster structure. Heath et al. [14] recently reported the detection of LaC_{60}^+ complexes and postulated, based both on its intensity in the photoionization mass spectra, and the fact that $La_2C_{60}^+$ was not observed that the La was likely to be bound within a spheroidal C_{60} cage supporting the hypothesized Buckminsterfuller structure [15]. Based on a careful analysis of the ionization behavior of LaC_{60} and C_{60}, we found [16] the cluster complex IP to be significantly lower than for C_{60}. This suggests that the intensity of the LaC_{60}^+ and absence of $La_2C_{60}^+$ may be attributed to limited La loading rather than to a superstable structure. To test the idea we synthesized K doped carbon clusters. The observed photoionization mass spectrum is shown in Fig. 4. Up to 3 potassium atoms on the C_{60} species can be identified, demonstrating that there is not a sharp restriction in the number of metal atoms that will attach to a C_{60} as might be expected if they only bonded inside the cage, but that the metal atoms are most likely on the exterior of the cluster. Both La and K may reside on the edges or faces of the carbon clusters. Further measurements are needed to clarify metal loading, aggregation, and bonding sites of the metal on the carbon clusters.

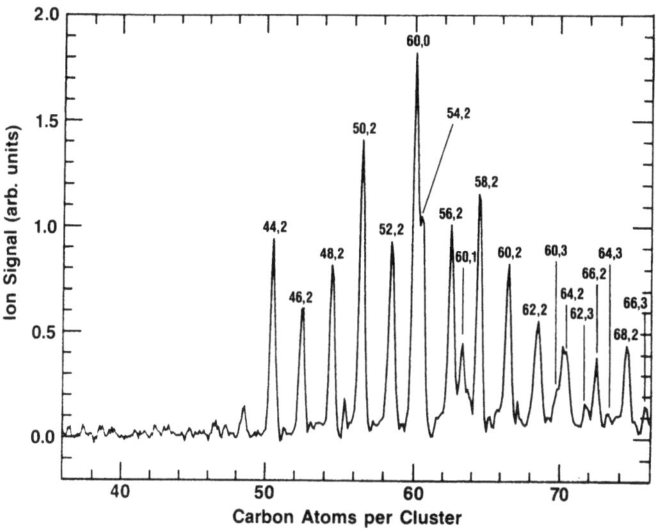

Fig. 4. Time-of-flight mass spectrum for clusters detected when a graphite rod treated with potassium hydroxide is vaporized. The numbers *n*, *y* refer to the number of carbon atoms, *n*, and potassium atoms, *y*, in the cluster. The major peaks are all $C_nK_2^+$, except for the bare C_{60}^+ peak. The ArF laser at ~1 mJ per pulse was used for ionization. Mass peaks corresponding to C_nK^+ and $C_nK_3^+$, although less intense, are also observed

b. Chemical Reactivity

Many fundamental chemical reactions on surfaces and catalysts have been investigated during the past two years on unsupported gas phase clusters. These include H_2 [17], N_2 [17c], and C−H [18] bond activation, as well as chemisorption of CO [19], O_2 [9, 20], H_2O [17a], CH_3OH [17a], and H_2S [9]. The bond activation reactions of H_2, N_2, and C−H exhibit cluster size dependence, while the initial chemisorption proceeds with more limited size dependence for reagents that form strong (>14 kcal/mole) bonds. Before we go too deeply into this subject it may be useful to define the meaning of reaction in the context used here.

We measure reaction by two means. The first is depletion of the metal ion signal as a result of the addition of the reactant into the helium diluent in the pulsed reactor. The other is the detection of stable products. Depletion is a good overall measure that a reaction has taken place, and because there is little uncertainty in the detectivity it is a good means to obtain quantitative reaction rates. However it does not shed any light on the reaction pathway. Detection of products does offer the additional benefit that a reaction pathway is identified, but it does not mean that it is the only, or even a major pathway. In fact it most likely identifies only the stable products with sufficiently low IPs to be detected.

In order to observe a reaction it has to be fast enough to occur during transit in the reactor tube described in the previous section; in order to obtain kinetics it has to be slow enough not to approach equilibrium. Similarly, the activation energy of the reaction has to be relatively low since at this point we are limited to 300–500 K as operating temperature. The reaction products have to be stable enough to survive transit through the reactor, and in order to observe products they have to survive transit to the photoionization detection zone. To detect products they have to survive the ionization process as well.

In order to detect dissociative processes it is useful if some of the reaction products leave the cluster some time during the process. Recent vibrational spectroscopy results, discussed in the next section, show that it is possible to observe reaction pathways on the cluster itself even when all fragments remain bound. In many ways cluster science replicates the developments in surface science; as the tools and techniques develop significant progress in thermodynamics, structure, and kinetics/dynamics will emerge.

Structure sensitivity in reactions cannot be reported at this point because of a real gap in the measurement capabilities of the field; we cannot determine the structures of clusters. Instead we discuss the relationship of reaction selectivity/sensitivity to cluster size. It is important to appreciate that we use kinetic criteria for reaction selectivity rather than equilibrium criteria. It should be noted that we can only estimate the internal temperature of the cluster prior to reaction. This we assume to be 300–500 K, because of the large number of collisions with the helium carrier added in the reactor. The relatively high pressure environment in the reactor tube helps to stabilize collision complexes and thus remove the exothermicity of the cluster reactions. But there may remain temperature variations with cluster size. Clusters with a larger number of internal vibrational modes can accommodate more energy, and also more efficiently transfer energy to the heat bath. Often no reaction is observed for the atoms, dimers, and trimers consistent with smaller heat capacity due to fewer internal vibrational modes. The considerations necessary to sort out these problems are identical to calculations of the fall-off curve in chemical activation of unimolecular decomposition. The selectivity this may lead to in cluster reactions is similar to what is encountered in catalyst studies where heat transfer is not adequate and "hot spots" develop. Of course where a specific cluster size is reactive, but those around it are relatively inert, other, more interesting explanations have to be sought. We also see evidence for subsequent reactions of the initial chemisorption products; these may complicate any detailed understanding of the observed selectivity. In the kinetic studies reported to date only the initial chemisorption step is probed. Reaction pathways, molecular vs. dissociative chemisorption, and isomerization reactions are important processes still waiting to be elucidated.

A model that describes our observed reactivity pattern predicts that H_2, N_2, and CH_4 are activated because they act as acceptors of electrons donated by the transition metal cluster, while CO, O_2, D_2O, CH_3OH, and C_6H_6 are non-activated because they can form a reasonably strong chemisorbed state before bond activation occurs. These qualitative trends are consistent with changes in ionization potentials. Specifics of H_2 chemisorption on metal clusters will be discussed below in more detail.

It has been observed that clusters of Fe [17c–e], Nb [17b, c, 10c], V [10d], Co [17b, c], and Al [21] exhibit strong size dependent reactivity with molecular hydrogen, specifically for the absorption of the first molecule. These trends are shown in Fig. 5 for Nb and V and in Ref. 17c–e for Fe. For example Nb_x show little propensity for reaction for $x=8$, 10, and 16. We find a reasonable correspondence between variations in ionization potential and the reactivity of most clusters containing more than 8 atoms. This anticorrelation suggests an activation barrier to reaction that is controlled by the energy of the HOMO of the metal cluster, which in a one electron picture relates directly to the IP. Although variations in electron binding energy qualitatively explain most of the individual cluster reactivity results, we expect that effects such as geometrical packing and/or electronic shell structures may also make important contributions. Such effects have recently been proposed by Upton [22] to explain the reactivity of D_2 with Al_x clusters shown in Fig. 6. Specifically Upton was able to show that Al_6 was the first cluster large enough to stabilize the D_2 adduct. Further calculations are needed to explain the loss of reactivity shown in the figure for $x>7$.

Dihydrogen reacts slowly with platinum clusters (Pt_x; $x<12$) and only a low coverage is observed even at our highest H_2 partial pressures. This turns out to be important for it permits observation of dehy-

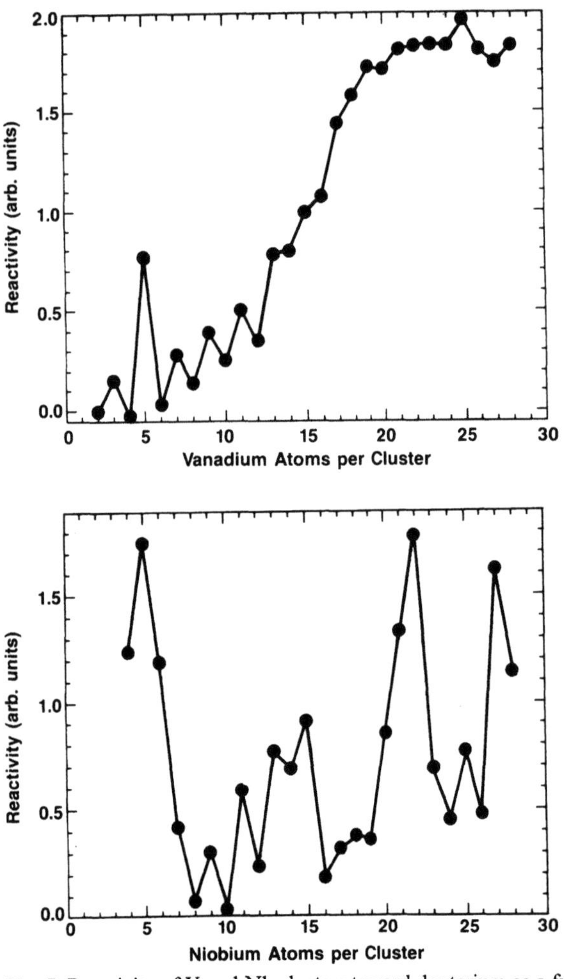

Fig. 5. Reactivity of V and Nb clusters toward deuterium as a function of cluster size

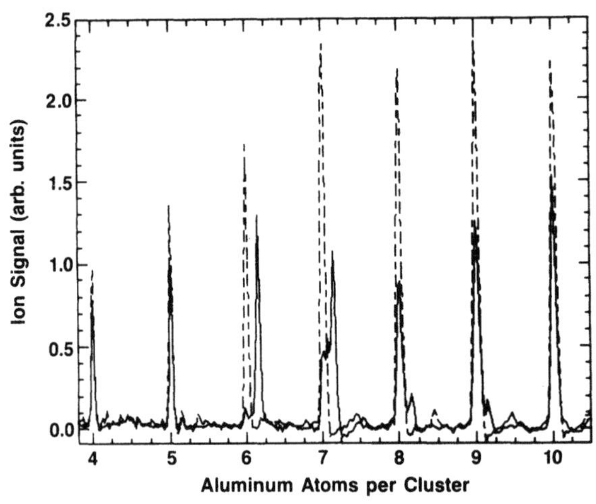

Fig. 6. Time-of-flight mass spectra for aluminum clusters. The dashed line is the reference spectrum in which helium only is pulsed into the reactor (0.45 cm dia. × 5 cm long) and the solid line the result when a deuterium/helium mix is used. Ionization is with the F_2 laser line at 7.87 eV. Note that for Al_6 the only product ion observed is $Al_6D_2^+$, but that on larger reactive clusters both $Al_nD_2^+$ and Al_nD^+ are observed. In a larger diameter (1 cm dia.) reactor only $Al_nD_2^+$ product ions are detected on those clusters which react. This effect is due to differences in the cluster growth mechanisms between the small and large diameter reactor; cluster growth is more rapidly quenched once clusters enter the large diameter reactor. In the small reactor Al_nD_2 species probably react with residual Al atoms and abstraction reactions such as $Al_nD_2+Al \rightarrow Al_{n-1}D+AlD$ may occur. Subsequent ionization would result in $Al_{n-1}D^+$ ions. Such mechanisms are discussed in Ref. 21

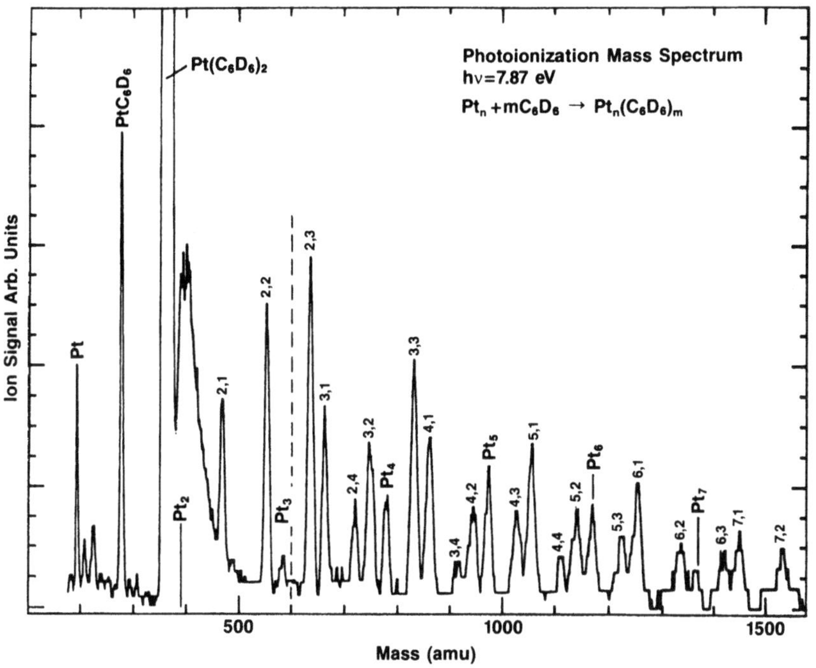

Fig. 7. Time-of-flight mass spectrum for platinum clusters after reaction with deuterobenzene. The numbers (n, y) refer to the number of metal atoms, n, and benzene adducts, y. The vertical dashed line indicates a change in mass spectrometer transmission function and gain

drogenation reactions on the cluster surface by mass spectrometric detection alone. Perhaps one of the most interesting observations is that a single Pt atom can dehydrogenate the first chemisorbed cyclohexane to a C/H ratio of approximately one [18]. The second chemisorbed cyclohexane requires a six atom cluster to reach the same level of dehydrogenation, though there is progressive loss of hydrogen toward $C/H=1$ as the number of metal atoms increases from 1 to 6. Normal hexane also dehydrogenates to $C/H=1$, though it requires a three atom platinum cluster to reach that point [23]. When benzene reacts with platinum clusters, a number of new chemical species are "synthesized" [18]. Figure 7 demonstrates the observed photoionization mass spectrum using 7.87 eV photons. Pt atom reacts to produce a bisaryl complex, while Pt_2 forms complexes with up to four benzenes. Larger platinum clusters form complexes involving multiple benzene species, though beyond $x=4$ dehydrogenation is observed with 6.4 eV ionization but not to the extent observed for the alkanes.

These experiments, when compared to surface science studies show at least two surprising trends. First, metal atom clusters are used with surprising efficiency to achieve the desired chemistry; 1 to 6 atom clusters can handle several organic molecules at once. Similar C_6 dehydrogenation chemistry on a single crystal surface requires the interaction with about seven metal atoms on the surface to transform a single reactant species. In the temperature range of 300–500 K hydrogen desorbs from a platinum surface and C–H activation is readily observed in chemisorbed hydrocarbon species; the clusters appear to do the same. Second, the similarity of small metal cluster reactivity to highly disordered surfaces, such as stepped surfaces with high kink density, is also consistent with the observed efficiency on clusters. The only technologically significant domain where such small clusters are now found as "stable species" is in the pores of zeolitic structures, but additional factors, such as acidity, etc. complicate the picture of simple size selective chemistry. On other supported cluster systems particle size tends to be much larger, (20 Å and up) involving hundreds to thousands of atoms where surface structures similar to single crystal faces can be defined. This probably is another domain of size/structure sensitivity.

c. Vibrational Spectroscopy

Development of new tools to gain insight and understanding of the structure and bonding, and the dynamics of reactions of metal clusters is an important element in the progress of the field. We have had recent success with using infrared multiple photon dissociation spectroscopy to obtain the vibrational spectra as a function of cluster size of molecules adsorbed on metal clusters [24]. These early experiments have been limited by the availability of broadly tunable infrared lasers with adequate photon flux to cover the important spectral regions. But, even with a line tunable CO_2 laser it has been possible to demonstrate the potential of the approach.

In these experiments we have chosen the reaction of methanol on iron clusters as the system to probe. On iron single crystal surfaces thermal desorption coupled to high resolution electron energy loss spectroscopy has been used to show that near room temperature methanol decomposes to produce a methoxy radical and a surface bound hydrogen atom [25]. Mass spectroscopic analysis alone is not sufficient to demonstrate that analogous dissociative chemisorption is taking place on small iron clusters. Figure 8 shows the observed photoionization mass spectrum of Fe_x and $Fe_x(CH_3OH)_y$ adducts. The data shows that in sequential reactions more than one methanol can react with a cluster. In the same figure the impact of intense infrared laser irradiation is also evident. As a result of absorption of the IR laser photons enough energy has been deposited into the cluster complexes to cause dissociation. Enough momentum is imparted to the fragments to remove them from the molecular beam before they reach the photoionization volume. The lowest energy dissociation pathway in the complex is probably activation of CH_3O-Fe_x to desorb $2H_2+CO$. Other reaction possibilities also abound; one example is loss of a metal atom. From the point of view of the vibrational spectroscopy the only important issue is clear definition of the excitation, or doorway states to multiple photon absorption. As indicated in Fig. 9 a dissociation spectrum is obtained showing both the CO vibrational frequency associated with the methoxy group at about 985 cm^{-1}, and a second transition at about 1,075 cm^{-1}, possibly associated with an H atom bound to the iron atoms directly [26]. The 1,075 cm^{-1} transition disappears when CH_3OD is used, consistent with the above assignment. Recent experiments indicate that the 1,075 cm^{-1} transition is not present when CD_3OH is used; further work is required to decide the assignment of the transition.

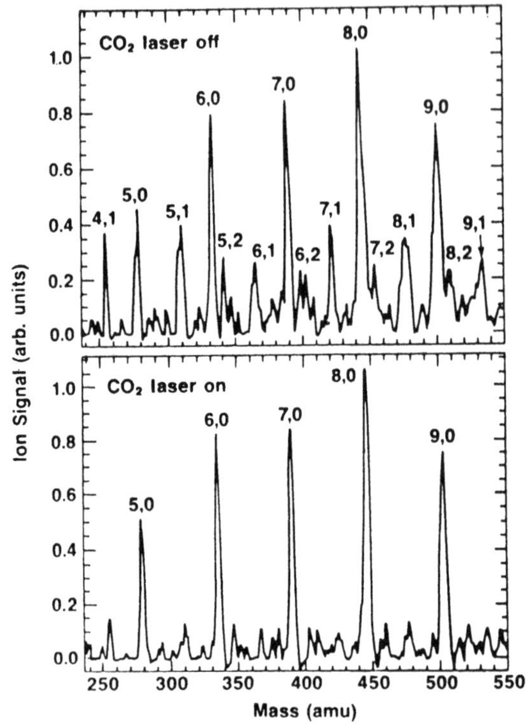

Fig. 8. Time-of-flight mass spectra for iron-methanol adducts. The upper trace shows the reference spectrum with the infrared laser blocked; the lower trace is with the infrared laser unblocked

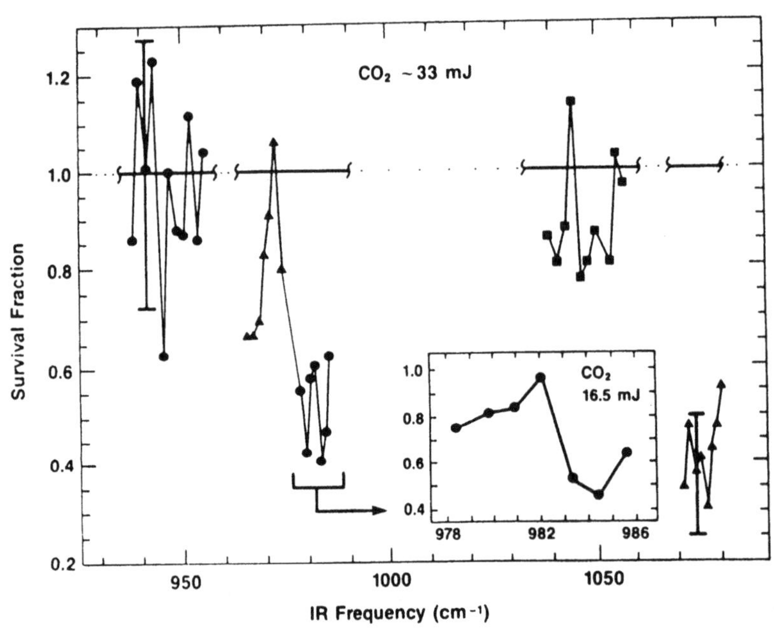

Fig. 9. The dissociation spectrum for $Fe_8(CH_3OH)$. Strong depletion of the $Fe_8(CH_3OH)$ is observed in the 980 and 1,070 cm^{-1} region

The infrared laser induced dissociation of $Fe_x(CH_3OH)_y$ exhibits threshold behavior consistent with a multiple photon absorption process. This is similar to results obtained for large complex molecules with high densities of low lying vibrational levels [27]. Fluence dependent red-shifted line broadening is also observed and provides further evidence for the multiple photon pathway. Additional experiments and modeling studies are underway to better exploit this approach.

Conclusion

This has been a brief summary of some of the recent results obtained in our laboratory. In the process we have not covered adequately many exciting developments in other laboratories. Finding ways to isolate specific cluster sizes and to study their properties in the gas phase and on supports is a major issue. Significant and exciting progress is being made in isolating cluster ions of specific size and performing studies on them [28]. Other thrusts involve the synthesis of isolated cluster species, using ion separation and deposition as one route to such materials. Coincidence photoelectron spectroscopy promises yet another approach to gain electronic structure data. We hope that the flavor that cluster science is robust and exciting has been captured and that important current frontiers have been brought into focus. There are many potential directions to pursue, ranging from physical characterization, to reaction kinetics and mechanisms, to the preparation of new materials. Definition of what these new materials are, clusters of well defined numbers of metal atoms, is difficult. They are surely molecules at some size, but their chemical and physical properties display unique characteristics associated with surfaces. At some intermediate size level defining a cluster as a molecular surface may be appropriate. The search for the most accurate definition of the transition zones should be most exciting.

References

1. a. Rohlfing, E.A., Cox, D.M., Kaldor, A., Johnson, K.H.: J. Chem. Phys. **81**, 3846 (1984)
   b. Cox, D.M., Trevor, D.J., Whetten, R.L., Rohlfing, E.A., Kaldor, A., Phys. Rev. B**32**, 7290 (1985);
   c. Cox, D.M., Trevor, D.J., Whetten, R.L., Rohlfing, E.A., Kaldor, A.: J. Chem. Phys. **84**, 4651, (1986)
2. Martins, J.L., Buttet, J., Car, R.: Phys. Rev. B**31**, 1804 (1985)
3. Van Hardeveld, R., Van Montfort, A.: Surf. Sci. **4**, 396 (1966) and Parks, E.K., Liu, K., Richtsmeier, S.C., Pobo, L.G., Riley, S.J.: J. Chem. Phys. **82**, 5470 (1985) and references therein
4. Colbert, J., Zangwill, A., Strongin, M., Krummacher, S.: Phys. Rev. B**27**, 1378 (1983)
5. a. Muetteries, E.L.: C&EN, 30 Aug. 1982, pp. 28–42;
   b. Cotton, F.A., Chisholm, M.H.: C&EN, 28 June 1982, p. 40
6. a. Ozin, G.A.: Far. Symp. Chem. Soc. **14**, 7 (1980);
   b. Schulze, W., Frank, F., Charle, K.-P., Tesche, B.: Ber. Bunsenges. Phys. Chem. **88**, 263 (1984)
7. Leutwyler, S., Herrmann, A., Wöste, L., Schumacher, E.: Chem. Phys. **48**, 253 (1980); Kappes, M.M., Kunz, R.W., Schumacher, E.: Chem. Phys. Lett. **91**, 413 (1982)
8. See for example:
   a. Dietz, T.G., Duncan, M.A., Powers, D.E., Smalley, R.E.: J. Chem. Phys. **74**, 6511 (1981);
   b. Riley, S.J., Parks, E.K., Mao, C.R., Pobo, L.G., Wexler, S.: J. Phys. Chem. **86**, 3911 (1982);
   c. Rohlfing, E.A., Cox, D.M., Petkovic-Luton, R., Kaldor, A.: J. Phys. Chem. **88**, 6229 (1984);
   d. Rohlfing, E.A., Cox, D.M., Kaldor, A.: J. Chem. Phys. **81**, 3322 (1984)
9. Whetten, R.L., Cox, D.M., Trevor, D.J., Kaldor, A.: J. Phys. Chem. **89**, 566 (1985)
10. a. Rohlfing, E.A., Cox, D.M., Kaldor, A.: Chem. Phys. Lett. **99**, 161 (1983);
    b. Rohlfing, E.A., Cox, D.M., Kaldor, A.: J. Phys. Chem. **88**, 4497 (1984);
    c. Whetten, R.L., Zakin, M.R., Cox, D.M., Trevor, D.J., Kaldor, A.: J. Chem. Phys. (in press);
    d. Cox, D.M., Whetten, R.L., Zakin, M.R., Trevor, D.J., Reichmann, K.C., Kaldor, A.: Proceedings of 1985 International Laser Science Conference (in press)
11. Wood, D.: Phys. Rev. Lett. **46**, 749 (1981)
12. Kappes, M.M., Schar, M., Radi, P., Schumacher, E.: J. Chem. Phys. **84**, 1863 (1986)
13. Zakin, M.R., Cox, D.M., Trevor, D.J., Kaldor, A.: (to be submitted)
14. Heath, J.R., O'Brien, S.C., Zhang, Q., Liu, Y., Curl, R.F., Kroto, H.W., Tittel, F.K., Smalley, R.E.: J. Am. Chem. Soc. **107**, 7779 (1985)
15. Kroto, H.W., Heath, J.R., O'Brien, S.C., Curl, R.F., Smalley, R.E.: Nature **318**, 162 (1985)
16. Cox, D.M., Trevor, D.J., Reichmann, K.C., Kaldor, A.: J. Am. Chem. Soc. **108** 2457 (1986)
17. a. Cox, D.M., Trevor, D.J., Whetten, R.L., Kaldor, A.: (in preparation);
    b. Geusic, M.E., Morse, M.D., Smalley, R.E.: J. Chem. Phys. **82**, 590 (1985);
    c. Morse, M.D., Geusic, M.E., Heath, J.R., Smalley, R.E.: J. Chem. Phys. **83**, 2293 (1985)
    d. Richtsmeier, S.C., Parks, E.K., Liu, K., Pobo, L.G., Riley, S.J.: J. Chem. Phys. **83**, 2882 (1985);
    e. Whetten, R.L., Cox, D.M., Trevor, D.J., Kaldor, A., Phys. Rev. Letter. **54**, 1494 (1985)
18. Trevor, D.J., Whetten, R.L., Cox, D.M., Kaldor, A.: J. Am. Chem. Soc. **107** 518 (1985)
19. Cox, D.M., Reichmann, K.C., Trevor, D.J., Kaldor, A.: (to be submitted)
20. Riley, S.J., Parks, E.K., Nieman, G.C., Pobo, L.G., Wexler, S.: J. Chem. Phys. **80**, 1360 (1984)
21. Cox, D.M., Trevor, D.J., Kaldor, A.: (to be published)
22. Upton, T.H.: Phys. Rev. Lett. **56**, 2168 (1986)
23. Trevor, D.J., Cox, D.M., Kaldor, A.: (in preparation)
24. Zakin, M.R., Brickman, R.O., Cox, D.M., Reichmann, K.C., Trevor, D.J., Kaldor, A.: J. Chem. Phys. **85**, 1198 (1986)

25. McBreen, P.H., Erley, W., Ibach, H.: Surf. Sci. **133**, L469
26. Baro, A.M., Erley, W.: Surf. Sci. **112**, L759 (1981)
27. Dietz, T.G., Duncan, M.A., Smalley, R.E., Cox, D.M., Horsley, J.A., Kaldor, A.: J. Chem. Phys. **77**, 4417 (1982)
28. a. Brucat, P.J., Zhang, L.S., Pettiette, C.L., Yang, S., Smalley, R.E.: J. Chem. Phys. **84**, 3078 (1986);
b. Geusic, M.E., McIlrath, T.J., Jarrold, M.F., Bloomfield, L.A., Freeman, R.R., Brown, W.L.: J. Chem. Phys. **84**, 2421 (1986)
c. Fayet, P., Granzer, F., Hegenhart, G., Moisar, E., Pischel, B., Wöste, L.: Phys. Rev. Lett. **55**, 3002 (1985)

A. Kaldor
D.M. Cox
D.J. Trevor
M.R. Zakin
Corporate Research Laboratory
Exxon Research and Engineering Company
Clinton Township
Route 22 East
Annandale, NJ 98801
USA

Analysis of the Reactivity of Small Cobalt Clusters

A. Rosén and T.T. Rantala *

Department of Physics, Chalmers University of Technology, and University of Göteborg, Sweden

Received April 7, 1986; final version May 30, 1986

The electronic structures of small cobalt clusters have been calculated within the local spin density approximation using the LCAO method. The calculations were done for simple geometries with the optimized number of interatomic bonds, and both for the bond length of the cobalt dimer and the bulk metal. The Fermi energy is found to be smaller for Co_N clusters with $N=3$, 4, 5 and $N>10$ than for the other ones. The variation of the Fermi energy with the cluster size correlates in a striking way with the observed H_2 tendency for chemisorption as found for cobalt clusters in a supersonic beam. Furthermore, the magnetic moments are somewhat smaller for these active clusters. In addition the lowest unoccupied levels of majority spin appear close to the highest occupied levels of minority spin which is not the case for the inert clusters.

PACS: 36.40; 31.20G

I. Introduction

Over the past decades extensive experimental and theoretical research in heterogeneous catalysis have been focused on the understanding of the microscopic behaviour of catalytic reactions [1–7]. In the experiments metal catalysts are dispersed on amorphous carriers or prepared in a form of well defined surfaces or surfaces with different types of defects as kinks, steps, terraces etc. Knowing, the structure of the substrate further studies in surface reaction dynamics have given information about the adsorption and desorption processes of reactants, and the physics behind the formation of intermediate species and final reaction products [8]. The theoretical treatments have been focused on the electronic structures of the substrate and particularly the adsorbate. One of the main objectives in this field has also been the calculation of potential energy surfaces of the combined system and in such a way get the possibility to map out the kinetics and dynamics of catalytic reactions [9–10]. However, without access to theoretical potential energy surfaces knowledge of charge transfer processes close to the Fermi level give useful information about catalytic activity. For example the adsorption of CO on transition metals have to a large extent been analyzed in terms of σ donation and 2π back donation [11–13].

The electronic structures of the solid metal catalysts are strongly related to the catalytic behaviour of small metal clusters [14] and particularly to the activity of a few atom metal clusters as recently observed by a few groups [15–19]. In these experiments metal cluster beams were produced in a supersonic beam and seeded with reactants such as H_2, D_2, O_2, H_2S or CH_4. The Cu_N clusters were found to be quite inert to H_2, while the activity for hydrogen chemisorption changed quite drastically as a function of the number of atoms for cluster beams of Co_N and Nb_N [18]. In the present work calculations of the electronic structure for some of these small cobalt clusters will be used for the analysis of the difference in reactivity for hydrogen chemisorption. The emphasis of the calculations has been to analyze trends of parameters as a function of cluster size such as the Fermi energy, one-electron levels at the Fermi level, magnetic moments, $3d$- and $4s$- contributions to the electronic structure and the energy difference between the highest occupied molecular orbitals (HOMO) and lowest unoccupied molecular orbitals (LUMO).

A short summary of our computational procedure is presented in Sect. II with the results and discussion in Sect. III.

* Present address: Department of Physics, University of Oulu, SF-90570 Oulu 57, Finland

II. Computational Procedure

The electronic structures of the small Co_N-clusters presented in this work were calculated within the local-denisty approximation [20, 21] in spin-polarized form with the exchange correlation potential by von Barth and Hedin [22]. A variational method is used to find the molecular wave functions which are determined within the LCAO method with numerical basis functions. The one-electron equations are solved within the self-consistent multipolar (SCM) method [23] as used in other applications of metal clusters [24–26]. The basis functions have been generated from a free atom Co $3d^8\,4s\,4p^0$ configuration with the addition of $3d$, $4s$ and $4p$ wave functions generated from a free ion Co^{4+} $3d^5\,4s^0\,4p^0$ configuration. The calculations for all the clusters were performed for two different bond-distances: $d(\mathrm{Co-Co}) = 4.2$ a.u. and 4.7 a.u. corresponding to the bond-distance in the Co dimer [27] and bulk [28], respectively. The conformation or the geometry for the clusters bigger than the dimer is not known. Extended X-ray absorption fine structure (EXAFS) measurements for big copper particles have shown a linear decrease of the bond distance from the bulk value for smaller clusters [29]. Calculations for bond distances of the two extremes are therefore expected to show the general trends of the electronic structure for these cobalt clusters. We have done the calculations for clusters with optimum number of bonds of equal length, which can be done within the same symmetry group C_{3v}. Optimization of the geometry by minimization of the total energy as done in other works [25, 30–32] would be highly desirable. However, that would lead to too extensive calculations for the series of clusters treated in this work.

Co_3 is in our calculations treated as a triangle with extension to a pyramid and double pyramid for Co_4 and Co_5, respectively. Co_6 has been calculated as a double pyramid with four atoms in a plane, one below and one above the plane which geometry can also be oriented to C_{3v} symmetry. This geometry has then been extended to the bigger clusters up to Co_7 and Co_8, and finally geometries approaching the *fcc* structure were used for Co_{10} and Co_{13}. The levels close to the Fermi level have been occupied with a Fermi-Dirac distribution corresponding to a temperature of about 160 K or a thermal energy of about 14 meV.

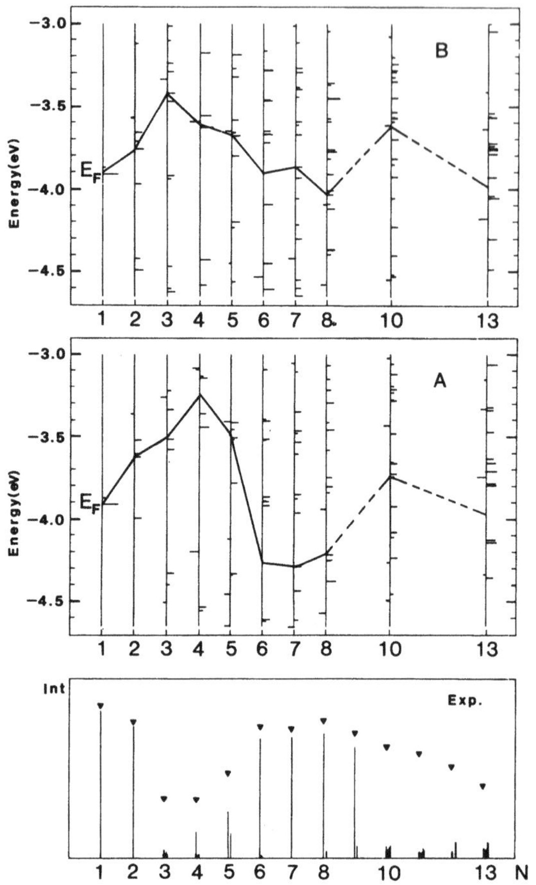

Fig. 1. Energy level diagram of the molecular one-electron levels in the Co_N clusters calculated with the bond distance for the dimer denoted by A and for the bulk denoted by B. The levels to the left for each cluster are those with majority up-spin, while those to the right are those with minority down-spin. As a comparison experimental photoionization spectra for pure cobalt clusters denoted by a black triangle and clusters reacted with D_2 are displayed with vertical lines [18]. The difference between these experiemental spectra shows the reactivity of cobalt clusters

III. Results and Discussion

The trends in hydrogen chemisorption on transition metal surfaces have recently been analyzed by Scillard and Hoffmann [33] and calculated applying the effective medium theory by Nordlander et al. [34]. These results show that the chemisorption of hydrogen is due to sd-hybridization with a reduced occupancy of anti-bonding metal/hydrogen orbitals for the lighter substrate elements. The experimentally found activity of H_2 chemisorption on Co_N metal clusters and inertness for Cu_N clusters follows therefore the general theoretical and experimental trends of H_2 chemisorption on transition metal surfaces. The difference in chemisorption of H_2 on Cu and Ni surfaces have also recently been analyzed by Harris and Andersson [10].

A view of the photoionization mass spectra for Co_N metal clusters without and with seeded D_2 [18] is displayed in the lowest panel of Fig. 1. The most interesting result of this experiment is that clusters with $N = 3$, 4, 5 and $N > 10$ were very active for chemi-

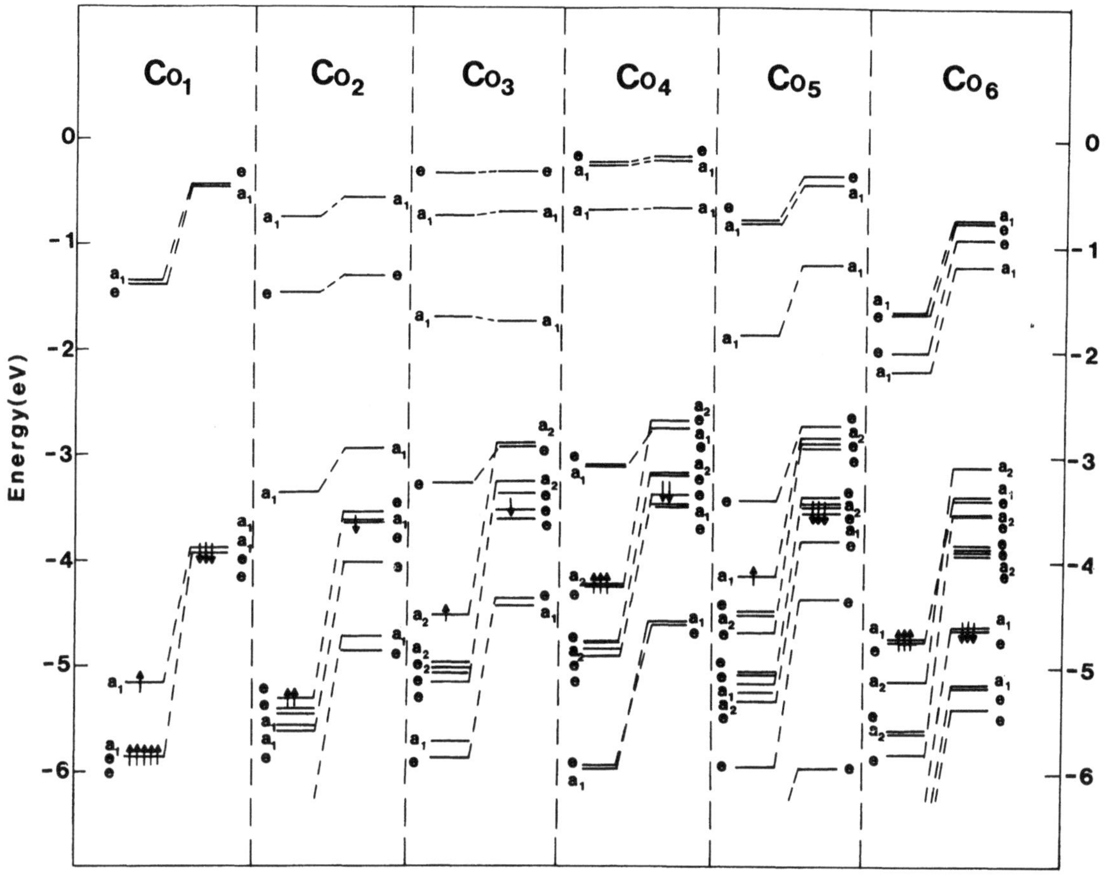

Fig. 2. Energy level diagram of the molecular one-electron levels in the Co_N clusters for $N=1$–6 calculated for the bond distance of the dimer. The labeling of the levels refer to the C_{3v} symmetry representation. The levels to the left are those with majority up-spin while those to the right are those minority down-spin. Occupancies of the last occupied levels are marked with arrows

sorption dissociation of H_2 while the clusters of other sizes were rather inert. The question is if this variation in reactivity with cluster size correlates with the cluster ionization energy as found experimentally for the Fe_N clusters [19]. The results for the Fermi level and the molecular levels close to the Fermi level for two series of our calculations are presented, denoted A and B for the dimer and bulk bond distance, respectively. The calculations show the same qualitative behaviour for the two bond distances as function of cluser size, although the results for the dimer bond distance are more pronounced. We notice a comparatively big change in the position of the Fermi level as a function of cluster size when going from Co_2 to Co_3 and from Co_5 to Co_6 and from Co_8 to Co_{10}, respectively. The value of the Fermi energy is significantly smaller for the reactive clusters and correlate in a striking way with the activity for hydrogen chemisorption of the Co_N clusters.

A more detailed view of the electronic structure for the Co_N clusters with $N=1$–6 calculated with the dimer bond distance is presented in Fig. 2. We notice how the spin-polarized calculations allow the majority up-spin to become stabilized and the minority down-spin to be destabilized relative to their common origin in a spin restricted calculation. The resulting difference in one-electron energies identified as the orbital exchange energy amounts to about 2 and 1.5 eV for the cobalt atom and the dimer, respectively. This exchange splitting is further reduced to less than 1 eV for the Co_3, Co_4 and Co_5 clusters with an increase to 1.5 eV for Co_6. Further examination of the molecular one-electron levels shows how the LUMO levels of majority up-spin are rather close in energy with the HOMO levels of minority down-spin for Co_N with $N=2$–5. This behaviour is changed drastically for the Co_6 cluster for which the energy difference between the same LUMO and HOMO levels is about 2 eV. A similar behaviour exist also for the Co_7 and Co_8 cluster as seen in Fig. 1. Another interesting feature for the Co_6 cluster is that HOMO levels of majority and minority spin from different one electron levels are close in energy. For detailed analysis of the relative importance of the HOMO and LUMO levels spin polarized calculations for clusters with adsorbed H_2 would be desirable. The relative

Table 1. Mulliken orbital and spin populations evaluated for the Co_N clusters. Values are given for calculations with the bond distance of the dimer (first line) and the bulk (second line) denoted by A and B in Fig. 1

	Orbital			Spin diff.			μ_{eff}
	$3d$	$4s$	$4p$	$3d$	$4s$	$4p$	
1	7.91	1.09	0.00	2.09	0.91	0.00	3.0
2	7.94	1.00	0.07	2.02	−0.01	−0.01	2.0
	7.94	1.01	0.06	2.03	−0.02	−0.01	2.0
3	7.99	0.83	0.19	1.87	−0.14	−0.06	1.7
	8.00	0.86	0.15	1.88	−0.17	−0.00	1.7
4	8.01	0.69	0.32	1.75	−0.15	−0.11	1.5
	7.95	0.80	0.28	1.89	0.03	0.01	1.9
5	7.94	0.77	0.34	1.83	−0.01	−0.02	1.9
	7.90	0.83	0.29	1.94	0.04	0.01	2.0
6	7.82	0.82	0.39	2.04	0.25	0.05	2.3
	7.81	0.87	0.34	2.08	0.21	0.04	2.3
7	7.86	0.81	0.36	1.97	0.15	0.02	2.1
	7.55	0.86	0.32	2.01	0.11	0.01	2.2
8	7.89	0.79	0.34	1.88	0.09	0.02	2.0
	7.81	0.89	0.32	2.05	0.15	0.04	2.2
10	7.94	0.80	0.26	1.80	0.03	−0.03	1.9
	7.90	0.85	0.25	1.91	0.06	−0.01	2.0
13*	8.04	0.88	0.29	0.90	−0.04	0.04	
	7.87	0.66	0.55	1.76	−0.07	−0.37	1.6
**	7.95	0.72	0.32	1.76	−0.04	−0.04	2.1
	7.87	0.82	0.31	1.95	0.15	0.04	

* Central atom in the cluster
** Outer atoms in the cluster

importance of metal-hydrogen interaction and the σ- and $\sigma^{\star}$-bonding within the adsorbed H_2 molecule will then determine the dissociation or adsorption of H_2.

A commonly discussed feature for the transition elements is the relative importance of $3d$ and $4s$ electrons [35]. Generally, the atomic ground state configuration for the transition elements is $nd^N(n+1)s^2$ except for a few elements as Cr, Ru, Rh, Mo, Pt for which $nd^{N+1}(n+1)s^1$ appears as the ground state configuration [36]. The configuration for metals is normally closer to $nd^{N+1}(n+1)s^1$ [37], which is also the configuration used in the generation of our basis set. Examination of the molecular orbitals through a Mulliken orbital- and spin-analysis in terms of our basis functions is presented in Table 1. We notice how the $3d^8\,4s$ configuration for the cobalt atom and the dimer is changed for the bigger clusters with redistribution of the $4s$ to the $3d$ and $4p$ occupancy. The most striking effect is the orbital contribution to the spin difference i.e. the difference between the value of up (majority) and down (minority) spins for the reactive clusters with a zero and even negative value for clusters with $N=3$, 4, 5, 10, 13. Summation of the spins gives magnetic moments presented in the last column in Table 1. These values should be compared with the total magnetic moment of 1.72 μ_B for the metal of which 1.59 μ_B represents the spin part [38, 39].

To summarize, there is a striking correlation between the cluster size dependence of the Fermi energy, HOMO/LUMO electronic structure, magnetic moments etc to the experimentally observed reactivity for hydrogen chemisorption. Further work is now in progress for calculations of Nb_N clusters and for the combined system of a Co_N cluster and a H_2 molecule. One of the goals is to analyze the importance of symmetry rules [40] in chemisorption which has been used particularly in the analysis of CO chemisorption [11, 12, 41].

The authors wish to thank S. Larsson, B. Lindgren, U. Landman, P. Siegbahn and various colleagues in the molecular and surface physics groups at Chalmers for enlightful discussions. This work has been supported by the Swedish National Research Council (NFR), Joint Committee of the Nordic National Science Research Council (NOS-N) and Nordic Research Courses (Nordic Council of Ministers).

References

1. Feuerbacher, B., Fitton, B., Willis, R.F.: Photoemission and the electronic properties of surfaces. New York: John Wiley and Sons 1979
2. Rhodin, T.N., Ertl, G.: The nature of the surface chemical bond. Amsterdam: North Holland 1979
3. King, D.A., Woodruff, D.P.: The chemical physics of solid surfaces and hetrogenous catalysis. Vol. 1. Clean Solid Surfaces. Vol 2. Adsorption at Solid Surfaces. Vol. 3. Chemisorption Systems. Vol. 4. Fundamental Studies of Hetrogenous Catalysis. Amsterdam: Elsevier 1983
4. Langerth, D., Suhl, H.: Many-body phenomena at surfaces. Orlando, Florida: Academic Press 1984
5. Somorjai, G.A.: Science **227**, 902 (1985)
6. Shustorovich, E., Baetzhold, R.C.: Science **227**, 876 (1985)
7. Shustorovich, E., Baetzhold, R.C., Muetterties, E.L.: J. Phys. Chem. **87**, 1100 (1983)
8. Kasemo, B., Lundqvist, B.I.: Comm. At. Mol. Phys. **14**, 229 (1984)
9. Nørskov, J.K., Houmøller, A., Johansson, P.K., Lundqvist, B.I.: Phys. Rev. Lett. **46**, 257 (1981)
10. Harris J., Andersson S.: Phys. Rev. Lett. **55**, 1583 (1985)
11. Blyholder, G.: J. Phys. Chem. **68**, 2772 (1964)
12. Plummer, E.W., Eberhardt, W.: Adv. Chem. Phys. **49**, 533 (1982)
13. Sung Shen-Shu, Hoffmann, R.: J. Am. Chem. Soc. **107**, 578 (1985)
14. Hamilton, J.F., Baetzold, R.C.: Science, **205**, 1213 (1979)
15. Riley, S.J., Parks, E.K., Nieman, G.C., Pobo, L.G., Wexler, S.: J. Chem. Phys. **80**, 1360 (1984)
16. Richtsmeier, S.C., Parks, E.K., Lin, K., Pobo, L.G., Riley, S.J.: J. Chem. Phys. **82**, 3659 (1985)
17. Parks, E.K., Lin, K., Richtsmeier, S.C., Pobp, L.G., Riley, J.S.: J. Chem. Phys. **82**, 5470 (1985)
18. Geusic, M.E., Morse, M.D., Smalley, R.E.: J. Chem. Phys. **82**, 590 (1985)
19. Whetten, R.L., Cox, D.M., Trevor, D.J., Kaldor, A.: Phys. Rev. Lett. **54**, 1494 (1985); J. Phys. Chem. 89, 566 (1985)
20. Hohenberg, P., Kohn, W.: Phys. Rev. **136**, B864 (1964)

21. Kohn, W., Sham, L.J.: Phys. Rev. **140**, A 1133 (1965)
22. von Barth, U., Hedin, L.: J. Phys. **C5**, 1629 (1972)
23. Delley, B., Ellis, D.E.: J. Chem. Phys. **76**, 1949 (1982)
24. Delley, B., Freeman, A.J., Ellis, D.E.: Phys. Rev. Lett. **50**, 488 (1983)
25. Ellis, D.E., Delley, B.: In: Local density approximations in quantum chemistry and solid state physics. Dahl, J.P., Avery, J. (eds.). New York: Plenum Publ. Corp. 1984
26. Delley, B., Jarlborg, T., Freeman, A.J., Ellis, D.E.: J. Magn. Magn. Mater **31–34**, 549 (1983)
27. Huber, K.P., Herzberg, G.: Constants of diatomic molecules. New York: Van Nostrand Reinhold Comp. 1979
28. Slater, J.C.: Quantum theory of molecules and solids. Vol. 2, New York: McGraw-Hill Book Comp. 1965
29. Apai, G., Hamilton, J.F., Stohr, T., Thompson, A.: Phys. Rev. Lett. **43**, 165 (1979)
30. Delley, B., Ellis, D.E., Freeman, A.J., Baerends, E.J., Post, D.: Phys. Rev. **B27**, 2132 (1983)
31. Koutecký, J., Paccioni, G., Jeung, G.H., Hass, E.C.: Surf. Sci. **156**, 650 (1984)
32. Martins, J.L., Buttet, J., Car, R.: Phys. Rev. **B31**, 1804 (1985)
33. Scillard, J.E., Hoffmann, R.: .J. Am. Chem. Soc. **106**, 2006 (1984) This work gives references to experimental and theoretical works on metallorganic compounds and surfaces
34. Nordlander, P., Holloway, S., Norskov, J.K.: Surf. Sci **136**, 59 (1984). This work gives references to earlier studies of hydrogen chemisorption on surfaces
35. Siegbahn, P.E.M., Blomberg, M.R.A., Bauschlicher Jr, C.W.: J. Chem. Phys. **81**, 1373 (1984)
36. Moore, C.E.: Atomic Energy Levels. NBS Circ. 467 Washington DC 1971
37. Moruzzi, V.L., Janak, J.F., Williams, A.R.: Calculated electronic properties of metals. New York: Pergamon Press Inc. 1978
38. Myers, H.P., Sucksmith, W.: Proc. R. Soc. London Ser. **A207**, 427 (1951)
39. Rodbell, D.S.: Proceedings of the International Conference Magnetism and Crystallography, Kyoto, 1961; J. Phys. Soc. Jpn, **17**, Suppl. B-I, 313 (1962)
40. Hoffmann, R., Woodward, R.B.: The conservation of orbital symmetry. New York: Academic Press 1970
41. Paul, J., Rosén, A.: Surf. Sci. **127**, 193˙(1983)

A. Rosén
Department of Physics
Chalmers University of Technology
and University of Göteborg
Fack
S-412 96 Göteborg
Sweden

T.T. Rantala
Department of Physics
University of Oulu
SF-90570 Oulu 57
Finland

Compound Clusters

T.P. Martin
Max-Planck-Institut für Festkörperforschung, Stuttgart,
Federal Republic of Germany

Received April 7, 1986; final version May 14, 1986

The cluster beam technique offers promise of becoming a useful tool for chemists in their search for new compounds. Using this technique elemental vapors can be made to react under controlled conditions to form units of condensed matter (clusters) small enough to be analyzed in a mass spectrometer but still large enough to contain several structural building blocks. Periodic patterns in cluster mass spectra often allow these building blocks to be uniquely identified. Of particular interest is the identification of building blocks unknown in the crystalline state.

PACS: 36.40.+d

1. Introduction

Cluster research is still too young a field to speak of tradition. But if this infant field has any unifying activity at all, it is the measurement and interpretation of mass spectra. The mass spectrometer has been used primarily to measure the number of atoms in elemental clusters, and the results have been exciting; Xe_{55} [1–3], Na_{20} [4, 5], Si_{10}, Ge_{10} [6–9], C_{60} [10], to mention a few. In this paper we will emphasize a slightly different aspect of mass spectrometry. The technique can be used to chemically analyze compound clusters.

If the building blocks of a chemical compound can be identified, the character of its bonding and its crystalline structure are quick to follow. Building blocks in clusters are often easily identified by means of mass spectroscopy. Since the same building blocks can be expected to form solid materials, cluster beam techniques should prove to be useful in the preparation and characterization of new chemical compounds. This paper presents some recent results in our search for new compounds.

What happens when two metallic vapors are mixed to form clusters? The results of such an experiment depend on whether the two metals are chemically similar [12, 13] or whether they lie far apart in the periodic table [6, 14]. Chemically similar metals mix in all concentrations to form alloys with properties intermediate to the pure elements. However, when Cs and Au are mixed together something quite surprising happens [15–17]. Although Cs and Au are both model metallic elements, if they are mixed in the ratio of 1:1, the result is a non metallic compound. The reason for this can be found in the large difference in the electronegativity of these two materials. Its almost as large as that between cesium and iodine. In fact Cs and Au combine to form a crystal with the CsCl structure, a structure characteristic of ionic materials. Cesium transfers an electron to gold resulting in a bond with a strong ionic contribution.

In this paper we describe the results of experiments on the systems Cs-Pb and Cs-Sn, for the practical reason that gold has a relatively low vapor pressure. However, the electronegativity difference between the components of these two systems is almost as large as that between Cs and Au, and unusual intermetallic clusters can be observed in the mass spectra. Apparently these clusters do not act as building blocks in the construction of larger clusters. A new building block has been observed in the binary system P-O and this will be described in the second part of the paper.

2. Experiment

Compound clusters, i.e. clusters composed of more than one element, can be produced by allowing the elements to react in the vapor phase [6, 11]. It is then possible to determine the composition of the clusters using a mass spectrometer. By varying the partial pressures of the two elements, the stability of compounds of all conceivable compositions can be tested in just one experiment. Various combinations of elemental vapors (Ge, Sn, Pb, Ga, In, Cs, Rb, Sr, Tm, Eu, Cl, O, S and P) have been allowed to react in the vapor phase and simultaneously thermally quenched in cold He gas. Compound clusters condensed out of the vapor and were identified by using a quadrupole mass spectrometer.

Elemental vapors were created in ovens of various design and allowed to react in a chamber containing from 0.7–5 mbar He gas. The outer mantel of the reaction chamber was cooled with liquid nitrogen. The reaction chamber contained two of these ovens. The vapors from the ovens were mixed and cooled. Compound clusters condensed out of the supersaturated vapor and passed through a 3 mm diameter hole into an intermediate chamber where the excess He gas was pumped off with a 500 l/s turbopump. The cluster beam then passed through a 1 mm diameter hole into a high vacuum chamber containing a quadrupole mass spectrometer pumped with a 360 l/s turbopump. The beam intersected the axis of the spectrometer where the clusters were ionized with 80 eV electrons.

The spectrometer could be operated in the mass range from 1–2,000 amu.

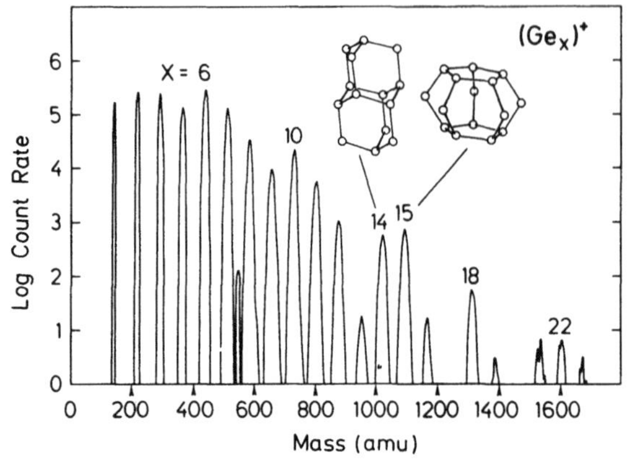

Fig. 1. Mass spectrum of clusters formed by quenching Ge vapor in He gas. Peaks corresponding to clustes containing 6, 10, 14, 15 and 18 Ge atoms are particularly strong. Two structures constructed with bond angles of nearly 109° are indicated

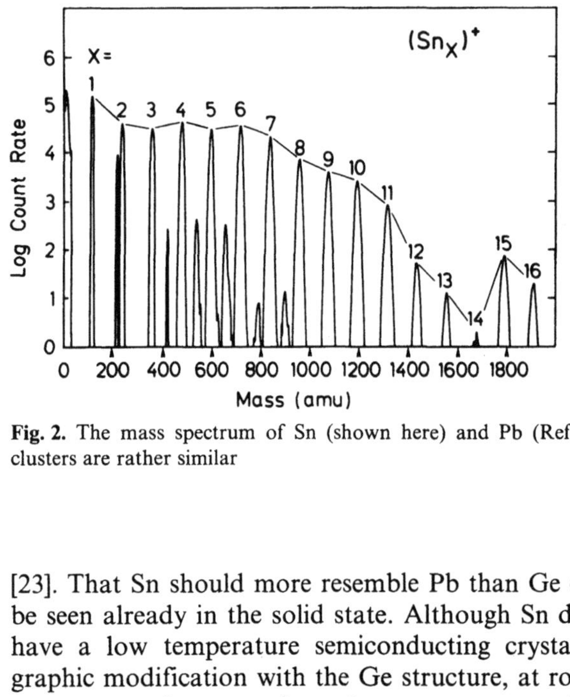

Fig. 2. The mass spectrum of Sn (shown here) and Pb (Ref. 23) clusters are rather similar

3. Group IV Element Clusters

Before combining Cs with elements from group IV, it will be useful to first consider the properties of pure group IV clusters. All of these elements, carbon [18], silicon [19], germanium [20], tin [21], and lead [22], cluster very easily. In fact, polyatomic species can be identified even in equilibrium vapors. As we will see this strong tendency for self-aggregation will also affect in a crucial way the structure of intermetallic compound clusters.

A mass spectrum of clusters formed by quenching Ge vapor in He gas is shown in Fig. 1. Peaks corresponding to clusters containing 6, 10, 14, 15 and 18 Ge atoms are particularly strong. A spectrum of Sn clusters produced under similar conditions is shown in Fig. 2. The unlabelled peaks in Figs. 1 and 2 are oxides. The dominant feature here is the weakness of the peak Sn_{14}^+. In this respect Sn resembles Pb [23]. That Sn should more resemble Pb than Ge can be seen already in the solid state. Although Sn does have a low temperature semiconducting crystallographic modification with the Ge structure, at room temperature the properties and structure are metallic. Because of silicon's low vapor pressure and high reactivity, it is not easily evaporated from an oven. However, clusters can be produced by rf discharge dissociation of SiH_4 gas [8]. The composition of the clusters can be changed from almost pure $(Si)_n$ to $(SiH)_n$ by varying the discharge conditions. A portion of a mass spectrum of such clustrers is shown in Fig. 3.

Each peak is broadened by a distribution of unresolved hydrogen atoms. The hydrogen content clearly changes with cluster size. In particular, there is a well defined transition from silicon-rich clusters to hydrogen-rich clusters near $n = 10$. Notice also that the peak corresponding to pure Si_{10}^+ is particularly strong, just

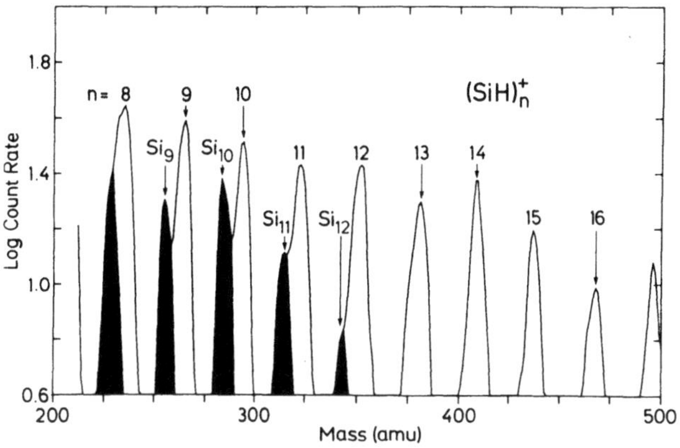

Fig. 3. Mass spectrum of Si-H clusters formed by decomposition and ionization of SiH_4 with an rf discharge. Notice that the peaks separate into almost pure Si (shaded) clusters and cluster with approximate composition $(SiH)_n$ (unshaded). The peak corresponding to Si_{10}^+ is strong

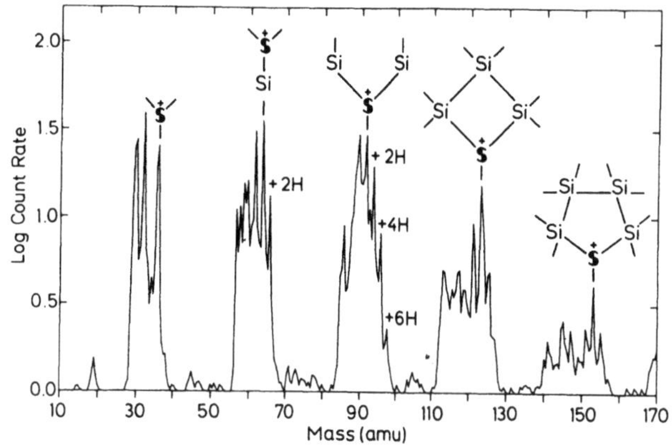

Fig. 5. A sulfur atom can be added to Si-H clusters by applying an rf discharge to a SiH_4-H_2S gas mixture

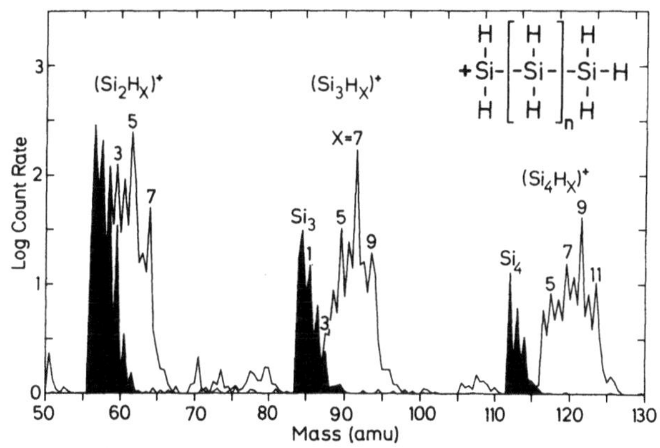

Fig. 4. High resolution mass spectra of Si-H clusters produced with high (shaded) and low (unshaded) rf power and SiH_4 pressure. Positively charged clusters have a strong tendency to bond with an odd number of hydrogen atoms

as it was for Ge. The flux of small Si-H clusters is large enough to allow the mass spectrometer to be operated under conditions for high mass resolution, Fig. 4. Each peak corresponds to the successive addition of a hydrogen atom. Low SiH_4 pressures lead to pure Si clusters, high pressures produce predominantly $[(SiH_2)(SiH_2)_n(SiH_3)]^+$ clusters. Clusters containing an odd number of hydrogen atoms seem to be particularly stable. The reason for this will be discussed later in the paper.

By adding H_2S to the SiH_4 gas it is possible to build one sulfur atom into each of the clusters. As indicated in Fig. 5 the main new peaks correspond to the clusters and molecules $(H_3S)^+$, $(H_3S\ Si)^+$, $(H_3S\ Si_2)^+$, $(H_7S\ Si_3)^+$ and $(H_9S\ Si_4)^+$. The bonding indicated in Fig. 5 follows from the assumption that the sulfur atom has lost an electron making it isoelectronic with phosphorus and thus suggesting a coordination of three.

Excellent theoretical work is beginning to appear on the structure of group IV element clusters [24–33]. It would appear that these clusters, in particular Si, will become test objects for calculational methods.

4. Intermetallic Clusters

Figure 6 shows a mass spectrum of clusters formed by mixing a small amount of Cs vapor with Pb vapor. Mass peaks due to pure Pb_n^+ are plainly evident. Of the mass peaks corresponding to clusters containing both Pb and Cs, two stand out particularly strongly, $(Cs_3Pb_2)^+$ and $(Cs_3Pb_5)^+$. Why are these peaks so strong? What do these clusters look like? In order to answer these questions it is necessary to make a digression and further discuss the properties of pure group IV clusters. Figure 7 shows the results of an extended Hückel calculation assuming specific geometries. The matrix element parameters have been fitted to the atomic ionization potential and the well-known band structure of Si. Notice that for a pentamer with the trigonal bipyramid structure it is possible to add two bonding electrons to form a very stable M_5^{2-} polyanion. In addition, the calculation indicates that the tetrahedral M_4^{4-} and the M_2^{2-} polyanions should also demonstrate unusually high stability.

How can these results concerning pure group IV element clusters be applied to the Cs-Pb system and in particular to the cluster $(Cs_3Pb_5)^+$? Could Pb_5 be a trigonal bipyramidal polyanion? If so, where does it get its electrons? We believe that two Cs atoms transfer their electrons to Pb allowing it to condense

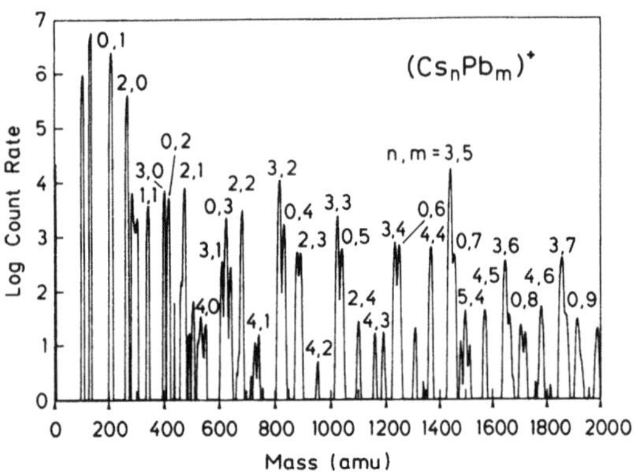

Fig. 6. Mass spectrum of Cs-Pb clusters. Peaks corresponding to $(Cs_3Pb_2)^+$ and $(Cs_3Pb_5)^+$ are particularly strong

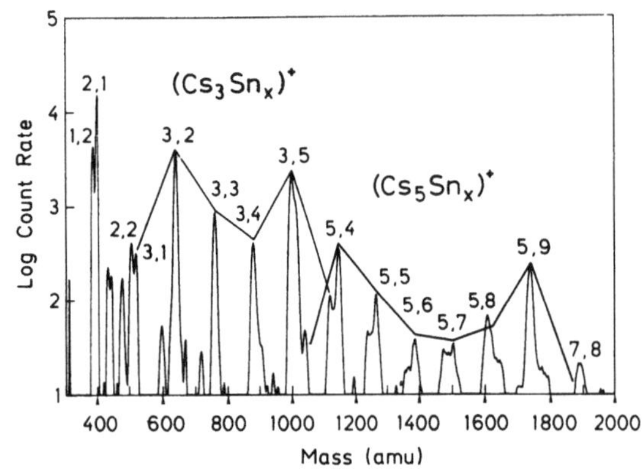

Fig. 8. Mass spectrum of Cs-Sn clusters. The peaks are broad due to the large number of Sn isotopes

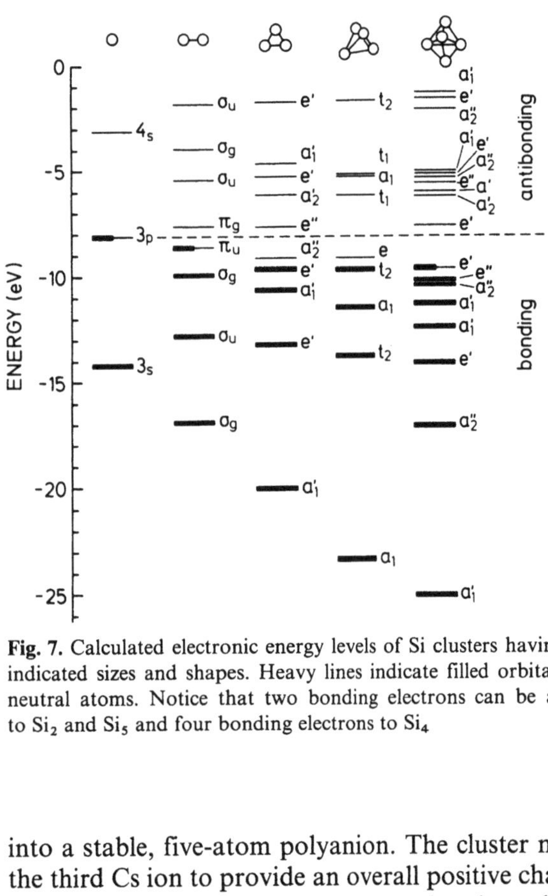

Fig. 7. Calculated electronic energy levels of Si clusters having the indicated sizes and shapes. Heavy lines indicate filled orbitals for neutral atoms. Notice that two bonding electrons can be added to Si_2 and Si_5 and four bonding electrons to Si_4

into a stable, five-atom polyanion. The cluster needs the third Cs ion to provide an overall positive charge.

Additional evidence exists concerning the stability of Pb_5^{2-}. Zintl and co-workers [34, 35] made an early study of a Cs-Pb solution in liquid NH_3. They were able to identify many polyanions in liquid solution, for example Pb_5^{2-} and Pb_9^{4-}. However, these ions did not prove to be sufficiently stable to act as building blocks in the construction of crystals when the solvent was evaporated. This problem was solved by Corbett and co-workers who stabilized the polyions in crystals by complexation with 2,2,2-crypt [36, 37]. In this way they were not only able to grow the salt [crypt $Na^+]_2$ Pb_5^{2-} but also to determine the structure of the P_5^{2-} polyanion: it is a trigonal bipyramid.

Lead is so heavy that the stability of two important clusters cannot be checked with our spectrometer, the cluster $(Cs_5Pb_9)^+$ also expected from Zintl's work and the cluster $(Cs_5Pb_{10})^+$. What is the significance of this second cluster? We stated in the introduction of this paper that two types of information concerning stability can be obtained from cluster mass spectra. A single intense isolated mass peak indicates a highly stable cluster, however, a periodicity in the mass spectrum means that this cluster can be used as a building block to construct larger clusters. We have noticed that $[Cs(Cs_2Pb_5)]^+$ is stable. Is the next cluster in the series stable, $[Cs(Cs_2Pb_5)_2]^+$? In order to test the stability of these two clusters we turn to the system containing the lighter element Sn. This change in material has, however, the disadvantage that Sn has many more isotopes and the mass peaks will be broader.

A mass spectrum of Cs-Sn clusters is shown in Fig. 8. Two series of peaks emerge, one serie for clusters containing three Cs atoms and the other series for clusters containing five Cs atoms. This selectivity reflects the strong tendency to build clusters with paired electrons only. In Fig. 8 peaks corresponding to the clusters $(Cs_3Sn_5)^+$, $(Cs_5Sn_4)^+$ and $(Cs_5Sn_9)^+$ are particularly strong. The first cluster was already observed and discussed for the Cs-Pb system. In order to discuss the structure of $(Cs_5Sn_4)^+$ it is useful to return to the results of the Hückel calculation shown in Fig. 7. If a neutral group IV element tetramer is assumed to be a tetrahedron, it can be seen that the unoccupied orbital offers a place for four additional bonding electrons, resulting in the formation of a highly stable polyanion M_4^{4-}. Four Cs atoms are necessary to provide the electrons. In our case a fifth Cs ion is necessary to provide the net positive charge on the cluster, $(Cs_5M_4)^+$. Actually, it is quite plausi-

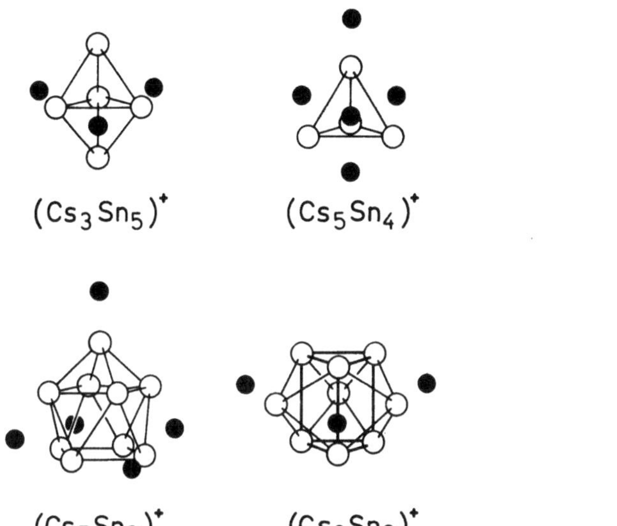

Fig. 9. The known structure of several polyanions (Refs. 36–41) and the probable distribution of alkali metal ions (black) in several Cs-Sn clusters

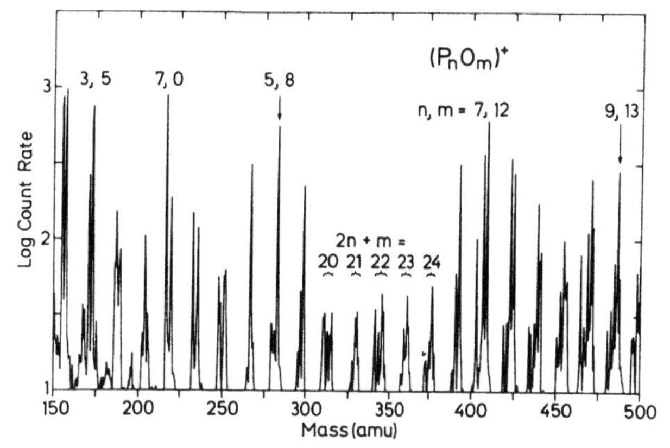

Fig. 10. Mass spectrum of clusters formed by quenching the vapor of red phosphorus in a mixture of 1% oxygen in He gas. Each group of peaks corresponds to clusters with composition $(P_xO_y)^+$ where $2X+Y=const$

ble that such a polyanion should be tetrahedral. Si_4^{4-} and Ge_4^{4-} are isoelectronic to neutral P_4 and As_4, both well-known tetrahedral molecules. In fact, tetrahedral Ge_4^{4-} has been observed in Li- and Na-germanide crystals [38–41].

Clearly, Zintl clusters exist in the vapor phase. These clusters can probably be described as a core-cluster within a cluster, Fig. 9. However, the core-cluster does not appear to act as a building block for the construction of larger units. In order to illustrate the aggregation of building blocks we turn now to the system P-O.

5. P-O System

Phosphorus and oxygen combine to form a rich variety of molecules and crystalline compounds. In fact, the chemistry of phosphorus and oxygen is one of the best known among complex binary systems. For this reason it would appear to be an unlikely place to look for new compounds. However, in this section we hope to show that the cluster beam technique can yield new information on even this well-studied system.

Five distinct molecules containing these elements have been identified. All can be thought to be based on the tetrahedron composed of four phosphorus atoms. In P_4O_6 each of the six edges of the tetrahedron are bridged with oxygen atoms. In the molecules P_4O_7, P_4O_8, and P_4O_{10} additional, terminal, double-bonded oxygen atoms are successively added to each phosphorus atom.

The most common form of the crystalline material is phosphorus pentoxide, normally obtained by burning phosphorus in excess oxygen. It should be remembered, however, that this compound can occur either as a molecular crystal composed of weakly bound P_4O_{10} units or with a two dimensional sheet structure or with a three-dimensional structure or with a glassy structure. What can we expect of P-O clusters?

White phosphorus (thermodynamically unstable below 560 °C) has an inconveniently high vapor pressure. It tends to evaporate at room temperature before the system can be evacuated. For this reason we used red phosphorus which required an oven temperature of at least 300 °C. Figure 10 shows a mass spectrum of clusters obtained by evaporating phosphorus into a 1 mbar He gas containing 1% O_2. Although the spectrum is complicated, indicating the formation of a large variety of oxides, it has one simplifying feature: The mass peaks occur in bunches. The reason for this bunching is that one phosphorus atom (31 amu) has almost the same mass as two oxygen atoms (32 amu). Therefore, each bunch is separated by about 16 mass units. More explicitly, the oxide clusters $(P_nO_m)^+$ in each bunch satisfy the relation, $2n+m$ is equal to a given integer.

Useful information can be obtained from this complex spectrum if we look at it under two extreme conditions of viewing, *1)* a large mass interval under low mass resolution and *2)* a small mass interval under high resolution. These two cases are shown in Figs. 11 and 12. In Fig. 11 the structure within a bunch is not resolved. However, mass peaks corresponding to pure phosphorus clusters can always be identified and are connected with a solid line. Notice the even-odd alternation. With the possible exception of P_{11}^+, pure phosphorus mass peaks are strong for

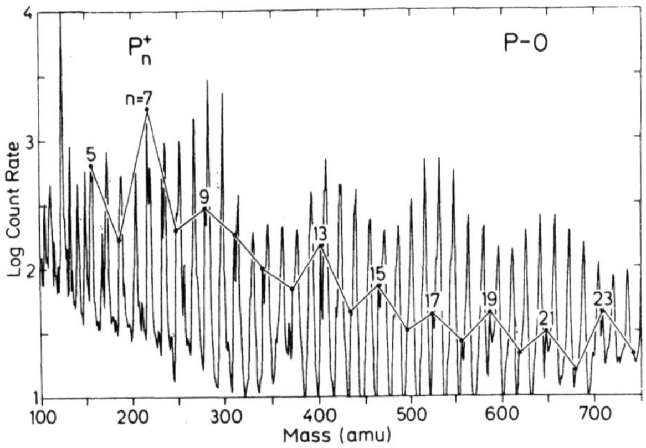

Fig. 11. Low resolution mass spectrum of phosphorus-rich P-O clusters. Peaks corresponding to pure phosphorus clusters are connected with a solid line

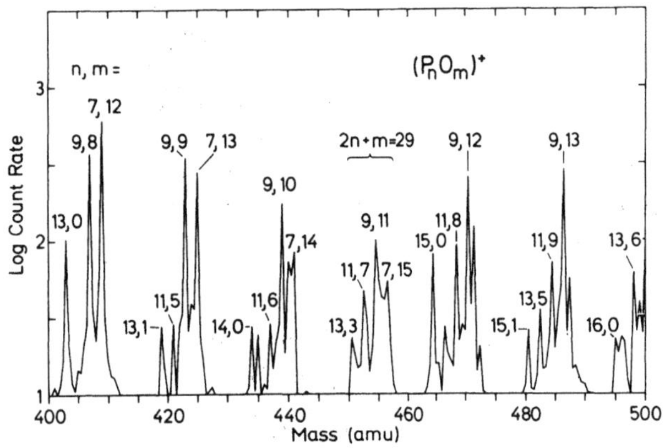

Fig. 12. High resolution mass spectrum of phosphorus-rich P-O clusters. Strong peaks correspond to clusters containing an odd number of phosphorus atoms

clusters containing an odd number of atoms. Now look at a small portion of the spectrum under high resolution. In Fig. 12 only six bunches are shown. A rather surprising result emerges out of the complexity of the earlier spectra. With only a few exceptions *all oxide clusters contain an odd number of phosphorus atoms*.

These observations give further support to a general rule concerning the stability of compound clusters. This rule is applicable to all clusters having elemental components with an odd number of electrons, e.g. containing elements from the I, III, V or VII groups (odd-electron-elements). The rule reads as follows: *Positively charged compound clusters are particularly stable when they contain an odd number of odd-electron-elements*. Stated in less practical but more easily understood terms, this rule states that highly stable clusters contain an even number of valence electrons.

Complex as these low oxide spectra are, they contain no periodicity that might indicate an oxide building block. This situation changes dramatically if the

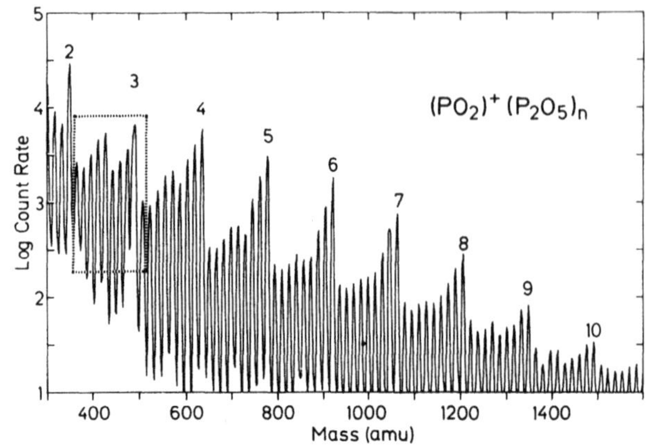

Fig. 13. Low-resolution mass spectrum of oxygen-rich P-O clusters. The periodic pattern indicates the presence of a building block

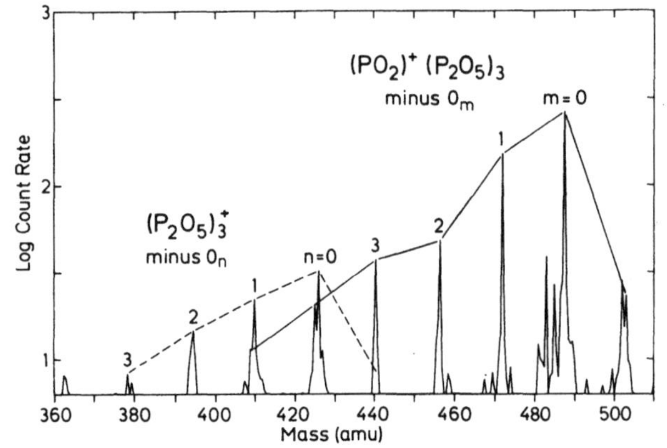

Fig. 14. High resolution mass spectrum of oxygen rich P-O clusters. The building block is here identified as P_2O_5..

oxygen content in He is increased to 10%, Fig. 13. The most characteristic feature of this spectrum is that every ninth bunch is strong. Formally such a series can be represented as $A^+(P_xO_y)_n$ where $2x+y=9$. That is, the building block is P_xO_y. In addition, there is a charge carrying ion A^+. Although this spectrum does allow us to assign a constraint on the values of x and y, the resolution is too low to uniquely identify either the building block or the charge carrying ion. One period of this spectrum is shown with high mass resolution in Fig. 14. Now all the necessary information is available. The building block is P_2O_5 and the charge carrying ion $(PO_2)^+$. Also present is a secondary periodicity corresponding to the same building block but the charge carrying ion $(P_2O_5)^+$. The lines filling in the periods are ionization fragments corresponding to the loss of 1, 2, ... oxygen atoms.

The unusual aspect of our findings is that no one has ever observed a P_2O_5 molecule either in the gas phase or in a solid. In this situation it is helpful to examine oxides of the related element nitrogen. N_2O_5

is well-known not only as a molecule, but also as a low temperature crystal. The crystal, however, is not a molecular solid. It is an ionic solid composed of nitronium $(NO_2)^+$ cations and nitrate $(NO_3)^-$ anions. Therefore, it is difficult to decide whether our clusters are composed of P_2O_5 molecules weakly bonded with each other or if they are ionic in nature composed of $(PO_2)^+$ and $(PO_3)^-$ pairs. In either case it would appear that a new P-O building block has been observed.

6. Concluding Remarks

A question that always arises in mass spectrometry investigations is: Are we studying the properties of neutral clusters or of charged clusters? This question cannot be ignored nor can it be conclusively answered. Many of the observations we have discussed are clearly attributable to the properties of charged clusters. For example, Cs-Pb clusters always contain an extra Cs ion to carry the cluster charge. Similarly, the P-O clusters contain an extra $(PO_2)^+$ unit. The even-odd alternations in Si-H, P-O, Cs-Sn can be attributed to an even number of electrons in *charged* clusters. This alternation seems to be a universal characteristic of charged clusters containing atoms from any of the groups I, III, V or VII [6, 11, 42–47].

Other observations we have made probably allow conclusions concerning the properties of neutral clusters. If P_2O_5 is a building block in the series $(PO_2)^+(P_2O_5)_n$ then it would be surprising if it were *not* a building block for neutral clusters. If a two-body potential [48] works for $Na^+(NaCl)_n$, then it would be very surprising if it were inappropriate for $(NaCl)_n$ (notice we do not generalize this conclusion to non-ionically bonded clusters). The polyanion Sn_5^{2-} does not seem to act as a building block for larger clusters and is therefore less likely to be found as a unit in neutral clusters and in pure crystals.

In those cases where mass spectroscopy indicates the existence of a new building block for neutral clusters, the next step is to collect and characterize the material *in situ*. That is one of the directions we are now following.

References

1. Echt, O., Sattler, K., Recknagel, E.: Phys. Rev., Lett. **47**, 1121 (1981)
2. Ding, A., Hesslich, J.: Chem. Phys. Lett. **94**, 54 (1983)
3. Birkhofer, H.P., Haberland, H., Winterer, M., Worksnop, D.R.: Ber. Bunsenges. Phys. Chem. **88**, 207 (1984)
4. Knight, W.D., Clemenger, K., Heer, W.A. de, Saunders, W.A., Chou, M.Y., Cohen, M.L.: Phys. Rev. Lett. **52**, 2141 (1984)
5. Kappes, M.M., Kunz, R.W., Schumacher, E.: Chem. Phys. Lett. **91**, 413 (1982)
6. Martin, T.P.: J. Chem. Phys. **83**, 78 (1985)
7. Bloomfield, L.A., Freeman, R.R., Brown, W.L.: Phys. Rev. Lett. **20**, 2246 (1985)
8. Martin, T.P., Schaber, H.: J. Chem. Phys. **83**, 855 (1985); Z. Phys. B – Condensed Matter **35**, 61 (1979)
9. Heath, J.R., Yuan Liu, O'Brien, S.C., Qing-Ling Zhang, Curl, R.F., Smalley, R.E., Tittle, F.K.: J. Chem. Phys. (to be published)
10. Kroto, H.W., Heath, J.R., O'Brien, S.C., Curl, R.F., Smalley, R.E.: Nature **318**, 162 (1985)
11. Martin, T.P.: J. Chem. Phys. **81**, 4426 (1984)
12. Rohlfing, E.A., Cox, D.M., Petkovic-Luton, R., Kaldor, A.: J. Phys. Chem. **88**, 6227 (1984)
13. Kappes, M.M., Radi, P., Schär, M., Schumacher, E.: Chem. Phys. Lett. **119**, 11 (1985)
14. Scheuring, T., Weil, K.G.: Surf. Sci. **156**, 457 (1985)
15. van der Lugt, W., Geertsma, W.: J. Non-Cryst. Solids **61**, 187 (1984)
16. Hoshino, H., Schumtzler, R.W., Hensel, F.: Phys. Lett. A, **51**, 7 (1975)
17. Christensen, N.E., Kollar, S.: Solid State Commun. **46**, 727 (1983)
18. Honig, R.E.: J. Chem. Phys. **22**, 126 (1954)
19. Honig, R.E.: J. Chem. Phys. **22**, 1610 (1954)
20. Kant, A., Strauss, B.H.: J. Chem. Phys. **45**, 822 (1966)
21. Ginerich, K.A., Desideri, A., Cocke, D.L.: J. Chem. Phys. **62**, 731 (1976)
22. Ginerich, K.A., Cocke, D.L., Miller, F.: J. Chem. Phys. **64**, 4027 (1976)
23. Sattler, K., Mühlbach, J., Recknagel, E.: Phys. Rev. Lett. **45**, 821 (1980)
24. Pacchioni, G., Koutecky, J.: Ber. Bunsenges. Phys. Chem. **88**, 242 (1984)
25. Weinert, C.M.: Surf. Sci. **156**, 641 (1985)
26. Raghavachari, K.: J. Chem. Phys. **83**, 3520 (1985); **84** 5672 (1986)
27. Grev, R.S., Schaefer, H.E.: Chem. Phys. Lett. **119**, 111 (1985)
28. Diercksen, G.H.F., Grüner, N.E., Oddershede, J., Sabin, J.R.: Chem. Phys. Lett. **117**, 29 (1985)
29. Phillips, J.C.: J. Chem. Phys. **83**, 3330 (1985)
30. Jones, R.O.: Phys. Rev. A, **32**, 2589 (1985)
31. Blaisten-Barojas, E., Levesque, D.: (to be published)
32. Tomanek, D., Schluter, M.A.: (to be published)
33. Saito, S., Ohnishi, S., Satoka, C., Sugano, S.: (to be published)
34. Zintl, E., Brauer, G.: Z. Elektrochem. **41**, 297 (1935)
35. Zintl, E., Kaiser, H.: Z. Anorg. Allg. Chem. **211**, 113 (1933)
36. Edwards, P.A., Corbett, J.D.: Inorg. Chem. **16**, 903 (1977)
37. Belin, C.H.E., Corbett, J.D., Cisar, A.: J. Am. Chem. Soc. **99**, 7163 (1977)
38. Schnering, H.-G. von: Nova Acta Leopoldina **59**, 165 (1986)
39. Witte, J., von Schnering, H.G.: Z. Anorg. Chem. **327**, 260 (1964)
40. Llanos, J., Nesper, R., Schnering, H.G. von: Angew. Chem. **95**, 1016 (1983)
41. Niecke, E., Rüger, R., Krebs, B.: Angew. Chem. **94**, 553 (1982)
42. Leleyter, M., Joyes, P.: J. Phys. (Paris) **36**, 343 (1975)
43. Hoareau, A., Cabaud, B., Melinon, P.: Surf. Sci. **106**, 195 (1981)
44. Hermann, A., Leutwyler, S., Schumacher, E., Wöste, L.: Helv. Chim. Acta **61**, 4542 (1978)
45. Martins, J.L., Buttet, J., Carr, R.: Phys. Rev. B, **31**, 1804 (1985)
46. O'Brien, S.C., Liu, Y., Zhang, Q., Heath, J.R., Tittel, F.K., Curl, R.F., Smalley, R.E.: J. Chem. Phys. (to be published)
47. Ghatek, S.K., Bennemann, K.H.: (to be published)
48. Martin, T.P.: Phys. Rep. **95**, 167 (1983)

T.P. Martin
Max-Planck-Institut
für Festkörperforschung
Heisenbergstrasse 1
D-7000 Stuttgart 80
Federal Republic of Germany

Structural and Electronic Properties of Compound Metal Clusters

B.K. Rao, S.N. Khanna, and P. Jena
Physics Department, Virginia Commonwealth University, Richmond, Virginia, USA

Received April 10, 1986; final version May 20, 1986

The equilibrium geometries, relative stabilities, and vertical ionization potentials of compound clusters involving Li_n, Na, Mg, and Al atoms have been calculated using ab initio self-consistent field linear combination of atomic orbitals – molecular orbital (SCF-LCAO-MO) method. The exchange energies are calculated exactly using the unrestricted Hartree-Fock (UHF) method whereas the correlation correction is included within the framework of configuration interaction involving pair excitations of valence electrons. While the later correction has no significant effect on the equilibrium geometries of clusters, it is essential for the understanding of relative stabilities. Clusters with even numbers of electrons are found to be more stable than those with odd numbers of electrons regardless of their charge state and atomic composition. The equilibrium geometries of homo-nuclear clusters can be significantly altered by replacing one of its constituent atoms with a hetero-nuclear atom. The role of electronic structure on the geometries and stabilities of compound clusters is discussed.

PACS: 31.20Tz; 36.40+d; 61.55.−x

The success of the electronic shell model [1] in explaining the "magic numbers" in the mass spectra of homo-nuclear alkali metal clusters has attracted a great deal of attention. That a fundamental understanding of observed abundances requires a detailed analysis of nucleation and fragmentation kinetics [2] has been overshadowed by the fact that the energetics of the ground state alone can account for much of the experimental data. To understand the extent to which electronic structure governs the occurrence of magic numbers, experiments [3] have been carried out in compound clusters consisting of free-electron-like metal atoms. Two interesting observations have emerged. First, clusters consisting of hetero-nuclear alkali atoms ($Li_{n-1}Na$, $K_{n-1}Li$,...) exhibit magic numbers for $n=2, 8, 20,\ldots$ as predicted by the shell model [1], since these correspond to complete shell filling by the valence electrons. However, when the heteronuclear atom is a divalent atom (such as Mg, Zn), the clusters $K_{n-1}Mg$, $K_{n-1}Zn$ do not exhibit "magic numbers" for values of n for which the total numbers of valence electrons in the clusters are 2, 8, 20,.... For example, K_6Zn and K_6Mg which have 8 valence electrons each should be "magic". Experimentally, K_8Mg and K_8Zn correspond to magic numbers in K_nMg and K_nZn spectra. This clear disagreement between experimental observation and the predictions of the "shell model" combined with the fact that the "shell model" also fails to account for the odd-even alternation in stability [1] of alkali metal clusters implies that more realistic calculations are necessary to assess the role of ground state energetics on cluster stability.

In this paper, we provide such an analysis. We begin by emphasizing that the equilibrium geometries of clusters must be obtained before attempting an interpretation of the electronic properties of clusters. This is necessary since the relative changes in the total energies between two neighboring clusters may be of the same order as that associated with the different structures of the same cluster. We have determined the equilibrium geometry of a cluster of n-atoms by minimizing the total energy of the clusters with respect to all possible configurations and by ensuring that the force at every constituent atom of the cluster is zero. The total energy of the cluster is calculated in the SCF-LCAO-MO method by including exchange contribution exactly within the unrestricted Hartree-Fock framework and correlation contribution within the configuration-interaction involving pair excitations of valence electrons. The atomic orbitals were represented by Gaussian functions. We have used the well-known STO-6G basis set [4] for $1s$, $2s$, $2p$ orbitals of Li and $1s$, $2s$, $2p$, $3s$, and $3p$ orbitals of Na, Mg, and Al-atoms. We begin with the atoms of a cluster located at random positions. The SCF-energy for this particular cluster is used to provide the gradient forces at individual atomic sites. The atoms are then moved along the direction of the forces and the SCF-procedure is repeated for the new configuration. The above two steps are repeated till the forces vanish. To insure that the atoms are not stuck at a local minimum, this process is repeated with a new random starting point. After a few iterations (typically 5), a global minimum is assumed to have been reached. In searching for the equilibrium geometry, we also consider the various possible spin states (singlet and triplet for odd-electron clusters and doublets and quartets for even-electron clusters). In this paper we present equilibrium structures where both spin and topological parameters are optimized. We refer the reader to a recent paper of ours [2] for more details in the numerical procedure.

In Fig. 1 we present the equilibrium geometries of $Li_{n-1}Mg$ and Li_n clusters. Replacing a Li atom by a Mg atom has a very significant effect on the topology. This clearly indicates that the Mg atom interacts rather strongly with the Li-atoms. This can be further substantiated by noting that the Li atoms in all the clusters (with the possible exception of Li_3Mg where Mg is not at the centroid of the triangular Li_3 cage) studied are symmetrically situated around Mg-atom. This observation is interesting since it goes against what can be expected from the Jahn-Teller theorem which states that a structure can lower its energy by lifting the degeneracy through the lowering of symmetry. The disappearance of the structural distortion around Mg can be understood by comparing the relative strengths of the Jahn-Teller effect (which is typically few tenths of volts) and the electronic bonding (which is typically an eV and is enhanced by a symmetric placement of Li atoms around the impurity). In Fig. 2 we plot the equilibrium geometries of Li_n Al clusters. The structures show remarkable similarity with the Li_n Mg clusters. Note that, unlike Li_3 Mg clusters, the structure of Li_3Al is an equilateral triangle arrangement of Li atoms with Al at its centroid. However, as discussed in the following, the relative stabilities of Li_nAl clusters are

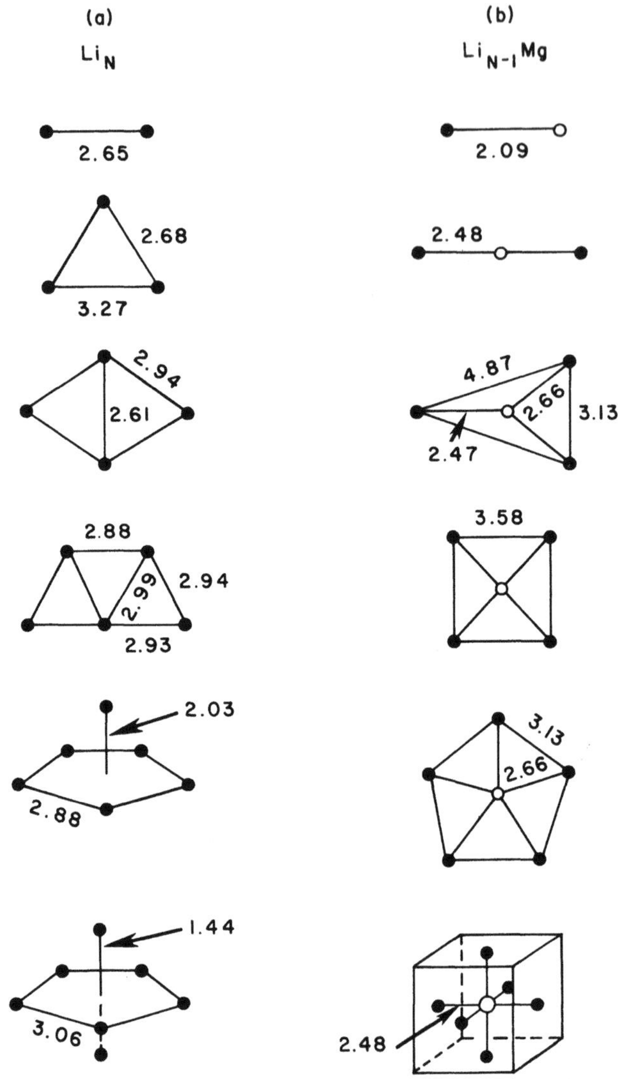

Fig. 1 a and b. Equilibrium geometries and bond lengths of (a) Li_n clusters and (b) Li_{n-1}Mg clusters. The solid (open) circle represents Li(Mg) atoms

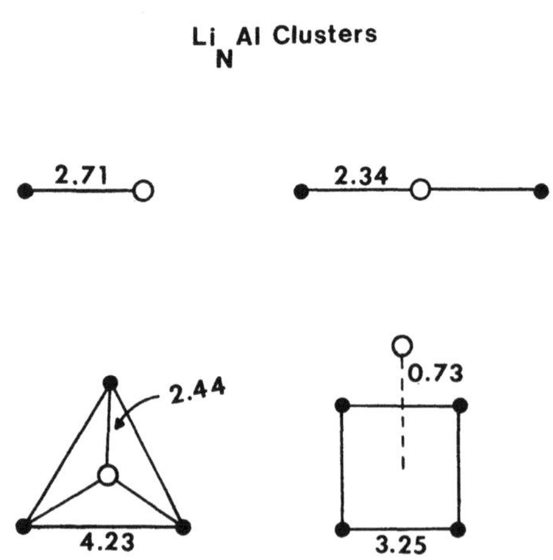

Fig. 2. Equilibrium geometries and bond-lengths of Li_nAl clusters. Al and Li atoms are represented by open and filled circles respectively

very different from those of Li_nMg clusters. We have also studied the structure of Li_nNa clusters. Na is an alkali atom like Li and possesses a single valence electron in the $3s$ shell whose radial distribution is not much different from that of the Li $2s$ electron. Thus, replacing a Li atom by a Na atom in the Li_n cluster is not expected to alter the topology of the Li_n cluster. This, indeed, is the case; the equilibrium topology of Li_2Na is virtually identical to that of the Li_3 cluster, including the bond lengths and bond angles. We have not explored larger clusters of Li_{n-1} Na, but remain confident that these structures will be the same as the Li_n clusters.

The strong influence of Al and Mg atoms on Li_n topology may be taken to indicate that the relative stabilities of Li_nAl and Li_nMg clusters may be quite different from Li_n clusters. To analyze this, we have calculated the energy gained in adding a Li atom to an existing cluster of Li_n and Li_nMg, namely,

$$\Delta E_n{}^0 = E(Li_n) - E(Li_{n-1}) - E(Li) \quad (1)$$

$$\Delta E_n = E(Li_nMg) - E(Li_{n-1}Mg) - E(Li). \quad (2)$$

Here $E(Li_n)$ and $E(Li_nMg)$ are total energies corresponding to the equilibrium geometries in Fig. 1. The results for ΔE_n^0 and ΔE_n are plotted in Fig. 3. ΔE_1 in Fig. 3 is a measure of the atomization energy of the LiMg dimer. The dashed curves correspond to Hartree-Fock values. The clear disagreement between the dashed and solid curves (that include correlation) should serve as a reminder that for a reliable interpretation based upon energetics, inclusion of correlation contribution in the total energy is essential. It is, however, appropriate to mention that correlation has little effect on the equilibrium geometries.

The minima in Fig. 3 represent the increased stability of the corresponding clusters over their neighbors. Two interesting observations can be made from Fig. 3. First, the energetics of Li_n and Li_nMg clusters are very similar indicating that the relative stabilities of Li_n clusters are not affected by introducing Mg. This is, of course, in agreement with the experimental data in K_nMg clusters. Second, there is an odd-even alternation in the cluster stability; even-n clusters being more stable than odd-n ones. The origin of this behavior can be traced to the filling of molecular levels. Clusters with even number of electrons can fill each available molecular level with electrons of opposite spins. For an odd-n cluster, the last electron has to proceed to the next higher molecular level, requiring an additional energy cost. That the odd-even alternation can be evident in other clusters is confirmed by plotting,

$$\Delta E^+{}_n = E(Li_n^+) - E(Li_{n-1}) - E(Li) \quad (3)$$

$$\Delta E_n = E(Li_nAl) - E(Li_{n-1}Al) - E(Li) \quad (4)$$

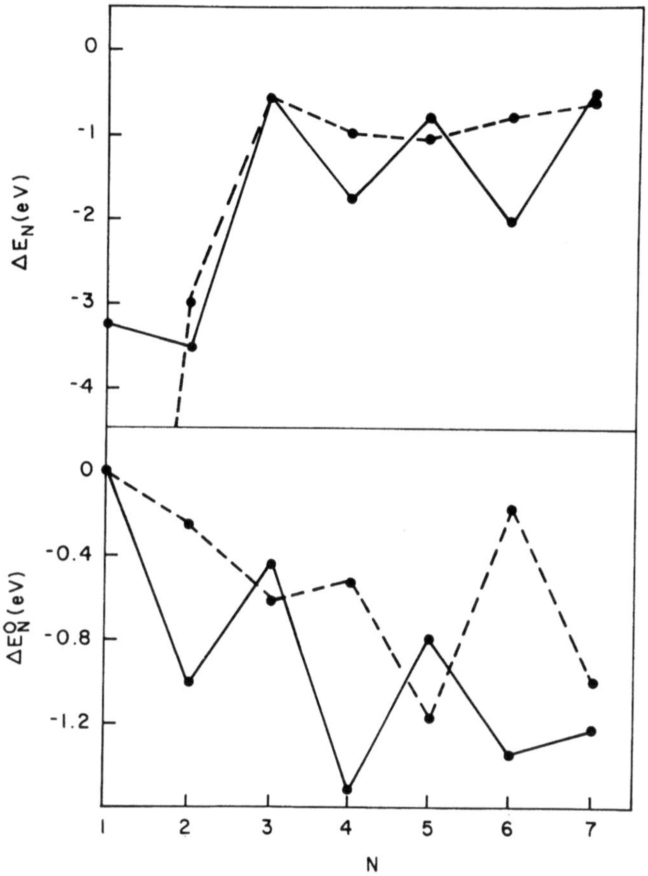

Fig. 3. ΔE_n^0 (for Li_n clusters) and ΔE_n (for Li_nMg clusters) as a function of n. The dashed curves correspond to energies obtained in the UHF procedure whereas the solid curve includes effects of correlation. See Eqs. (1) and (2)

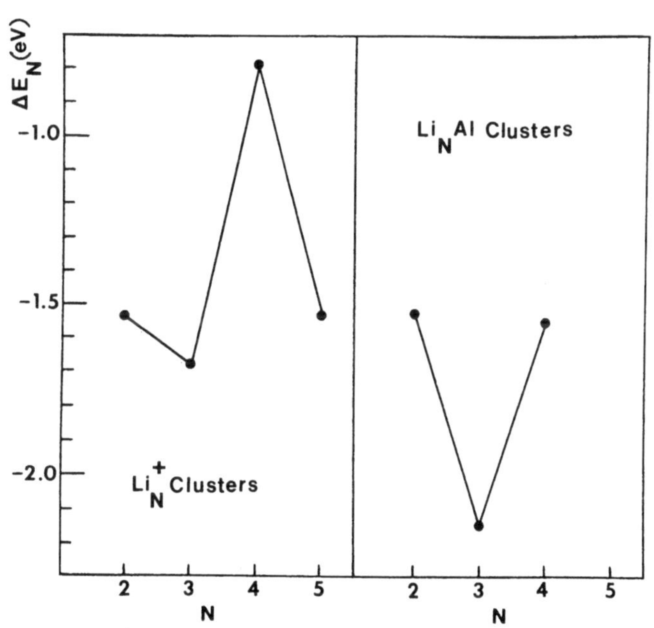

Fig. 4. ΔE_n^+ (for Li_n clusters) and ΔE_n (for Li_nAl clusters) as a function of n. See Eqs. (3) and (4). Only results including correlation effects are shown

in Fig. 4. For clusters in (3) and (4) n-odd clusters have even-number of electrons. Thus, singly charged clusters of Li_n and neutral clusters of Li_nAl would exhibit larger stability for n-odd clusters than n-even clusters. The results in Fig. 4 confirm this assertion.

The vertical ionization potential, defined as the energy necessary to singly ionize a cluster without allowing the cluster to relax to a new geometry, is given in Table 1. The table also lists the total energy of Li_nMg and Li_nMg^+ cluster for reference. The dependence of the ionization potential on cluster size is nonmonotonic in agreement with experimental trend.

In summary, we have performed ab initio calculations of the energetics of compound clusters. While the effect of a defect atom may have substantial effect on the topology of homo-nuclear clusters, its effect on relative stabilities depends on the valence structure of the defect atom. For a defect with an even number of electrons, the relative stabilities of host alkali clusters will remain unaffected. However, defects with an odd-number of valence electrons can change the overall stability pattern. It is hoped that experiments on alkali clusters with Al atom will be soon carried out to verify this prediction.

Table 1. Total energies of Li_nMg and Li_nMg^+ clusters and vertical ionization potentials including correlation at the pair excitations of the valence electron level in the configuration interaction scheme

n	$E(Li_nMg)$ (Hartree)	$E(Li_nMg^+)$ (Hartree)	I.P. (eV)
1	−206.21591	−206.15326	1.70
2	−213.74480	−213.60976	3.67
3	−221.16506	−221.04907	3.16
4	−228.62832	−228.49587	3.60
5	−236.05588	−235.98262	1.99
6	−243.52969	−243.39299	3.72
7	−250.94909	−250.87389	2.05

This work is supported in part by grants from NSF (INT-8306590) and ARO (DAAG 29-85-K-0244). One of the authors (BKR) also acknowledges support from the Grants-in-Aid program for faculty of Virginia Commonwealth University.

References

1. Knight, W.D., Clemenger, K., Heer, W.A. de, Saunders, W.A., Chou, M.Y., Cohen, M.L.: Phys. Rev. Lett. **52**, 2141 (1984)
2. Rao, B.K., Jena, P.: Phys. Rev. B **32**, 2058 (1985)
3. Kappes, M.M., Radi, P., Schar, M., Schumacher, E.: Chem. Phys. Lett. **119**, 11 (1985)
4. Hehre, W.J., Stewart, R.F., Pople, J.A.: J. Chem. Phys. **51**,·2657 (1969)

B.K. Rao
S.N. Khanna
P. Jena
Physics Department
Virginia Commonwealth University
Richmond, Virginia 23284-0001
USA

Binary Metal Alloy Clusters

Klaus Sattler*
Department of Physics, University of California, Berkeley, California, USA

Received April 7, 1986; final version May 21, 1986

Binary alloy clusters $A_n B_m$ (A, B: Bi, Sb, Pb, Ag, in different compositions) have been generated by simultaneous inert gas condensation of the two vapours. Under suitable condensation conditions the adsorption of single atoms A to clusters B_m can be studied. The adsorption probabilities are found to vary with cluster size. The cluster reactivities depend on the number of absorbed foreign atoms. Lead and bismuth clusters are found to be highly reactive in cracking the strongly bound Sb_4-tetrahedron. One single interaction channel is found for Pb_n/Sb_4 (symmetrical dissociation) and three interaction channels are found for Bi_n/Sb_4 (symmetrical and nonsymmetrical dissociation). Pb- and Bi-clusters have different onset thresholds for dissociative reactivity.

PACS: 36.40; 82.65; 81.20.6

1. Introduction

Since the development of new cluster sources during the last few years several studies on reactions of molecules with clusters have been reported. Most of these measurements focus on the reactivity of transition metal clusters for example: $Fe_n + H_2$ [1–3]; $Fe_n + O_2$, H_2S, methane [4]; Pt_n + benzene and several hexanes [5]; Fe_n, Co_n, Ni_n, Nb_n with H_2, D_2, N_2, or CO [6]. Binary clusters have also been produced by combining the atoms or molecules of two elements in the vapour phase: Cs vapour has been allowed to react with O_2, Cl_2, S_8, H_2O, and P_4 [7].

Metal alloy clusters Ni_n/Cr_m and Ni_n/Al_m have been studied after vapourization of the corresponding bulk alloys followed by a free jet expansion of a He/cluster mixture [8]. For clusters larger than about six atoms a statistical behaviour for the interaction between the two components is observed. For smaller clusters departure from a statistical behaviour is found, i.e. their size distributions are not correlated to the initial atomic abundances of the two components.

Different generation methods for binary metal clusters can be applied:

* On leave from Universität Konstanz, Fakultät für Physik, D-7750 Konstanz, FRG

(i) Langmuir or Knudsen evaporation of binary materials

(ii) clustering of binary molecules, *(a)* from a gas, *(b)* evaporated from a solid

(iii) simultaneous growth of the two separate components, *(a)* atom A – atom B interactions, *(b)* cluster A – cluster B interactions

(iv) cluster growth followed by single atom (or molecule) addition.

In this paper we mostly deal with point *(iv)*. Binary clusters from Bi, Sb, Pb, and Ag atoms are considered [9, 10]. The clusters are generated in an inert gas condensation source with two separate ovens being applied.

The clusters are ionized (by electron bombardment) for time of flight mass analysis. Despite the charged clusters are detected we interpret the data in terms of the neutral clusters. This seems to be valid for low enough ionizing energies (soft ionization).

2. Experimental

The clusters are generated by simultaneous condensation of the two vapours in cold He gas. The source is shown in Fig. 1. The inert gas condensation method has been described earlier [11]. Two ovens with resis-

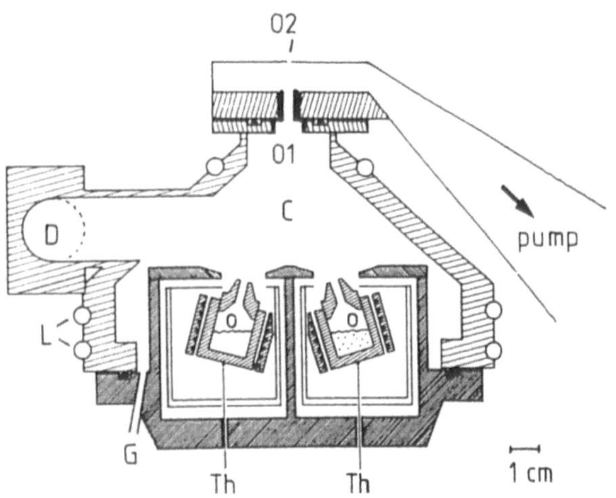

Fig. 1. Cluster source: Two metals are separately evaporated from two ovens, respectively. The ovens are resistantly heated, and the temperatures are measured with thermoelements at the bottom. Thermal shielding is provided by Ta-cylinders and Cu-walls. Condensation occurs in cold He-gas in the region *C*. The He-gas is cooled by the walls of the cell (*G*, gas inlet; *D*, differential pressure monitor; *L*, tubings for the cooling fluid). A differential pumping section is applied above the aperture 01. The clusters enter the mass spectrometer by aperture 02 [10]

tance heating are shielded by Ta-cylinders and surrounded by LN_2-cooled Cu-walls. The vapours of the two components *A* and *B*, respectively, are directed towards the condensation region of the cell there being supersaturated in cold He-gas (temperature 80 K, pressure 1 mbar). By variation of the vapour pressures relative to each other different conditions for cluster growth are obtained:

(a) $p_A \sim p_B$; clusters with a broad variety of combinations n/m are produced

(b) $p_A \gg p_B$; clusters A_n are produced, single atoms *B* are adsorbed at A_n.

With the conditions under *(b)* size dependent adsorption probabilities can be studied. If the component *B* is a molecular unit (for example Sb_4) dissociative reaction probabilities are gained.

After formation, the clusters leave the condensation cell. Then they are ionized by a pulsed electron beam and their masses are analysed by electronic time of flight spectrometry.

3. "Soft" Ionization

The aim of these studies is to get information about the formation of the neutral compound clusters. For mass spectrometry, however, the clusters have to be ionized. At electron ionizing energies close enough to the ionization threshold we suppose that the ionized-cluster spectrum reflects the neutral-cluster spectrum.

This becomes obvious if cluster spectra are compared being taken with different ionizing energies. Antimony is an interesting model substance because the vapour itself already contains cluster units. These are highly stable Sb_4-clusters with all valence bonds saturated in a tetrahedron structure (bond energy ~ 3 eV). A modulo 4 sequence is expected in the mass spectrum if no intramolecular fragmentation occurs.

In common we have to consider two size distributions: *(i)* the distribution of the neutrals, and *(ii)* the distribution of the cluster ions. In the mass spectrum we analyse the ions alone and one would think that the spectrum reflects their stability distribution. In the case of Sb-clusters, however, this expectation does not hold at all. The high tendency for covalent bond formation with this material is expected to lead to geometrical structures for the cluster ions being completely different from that of the neutrals. Both, the neutral and the charged clusters should have their characteristic stability distributions (high tendency for chemical inertness for all cluster sizes) and the two distributions are expected to be fundamentally different. Despite the charged clusters are detected we will show in the following section that in the case of Sb-clusters the stability distribution of the charged clusters does not show up.

Figure 2a shows a mass spectrum of antimony clusters with a sequence of Sb_{4n}-peaks alone. For a discussion we first compare the neutral and the charged Sb_4-cluster: The Sb-atom in the Sb_4-complex provides 3 $5p$-electrons for chemical bonds. The six bonds in the tetrahedron structure are fully saturated by paired electrons the Sb_4 therefore being a highly stable inert unit. The total number of valence electrons is 12 and thus the high stability is explained by the "even electron rule". The ionized cluster Sb_4^+ contains 11 valence electrons and does not fulfill the even electron rule while Sb_3^+ (8 valence electrons) does. From Fig. 2a however we see that Sb_3^+ is not detected in the mass spectrum. This result is found for low electron energies $E_i < 10$ eV. For E_i just above the ionization threshold the energy transferred to the cluster is not high enough for fragmentation processes to occur.

We conclude that ionization alone does not necessarily lead to fragmentation even if the charged clusters have a size distribution different from that of the neutral clusters. This is illustrated in the case of Sb_{4n}-cluster-spectra. The ionized clusters Sb_{4n}^+ do not fulfill the "even electron rule". A stability distribution of the singly charged Sb_{4n}-clusters would show preference for clusters with odd numbers of atoms. At low electron ionizing energy the stability distribution of

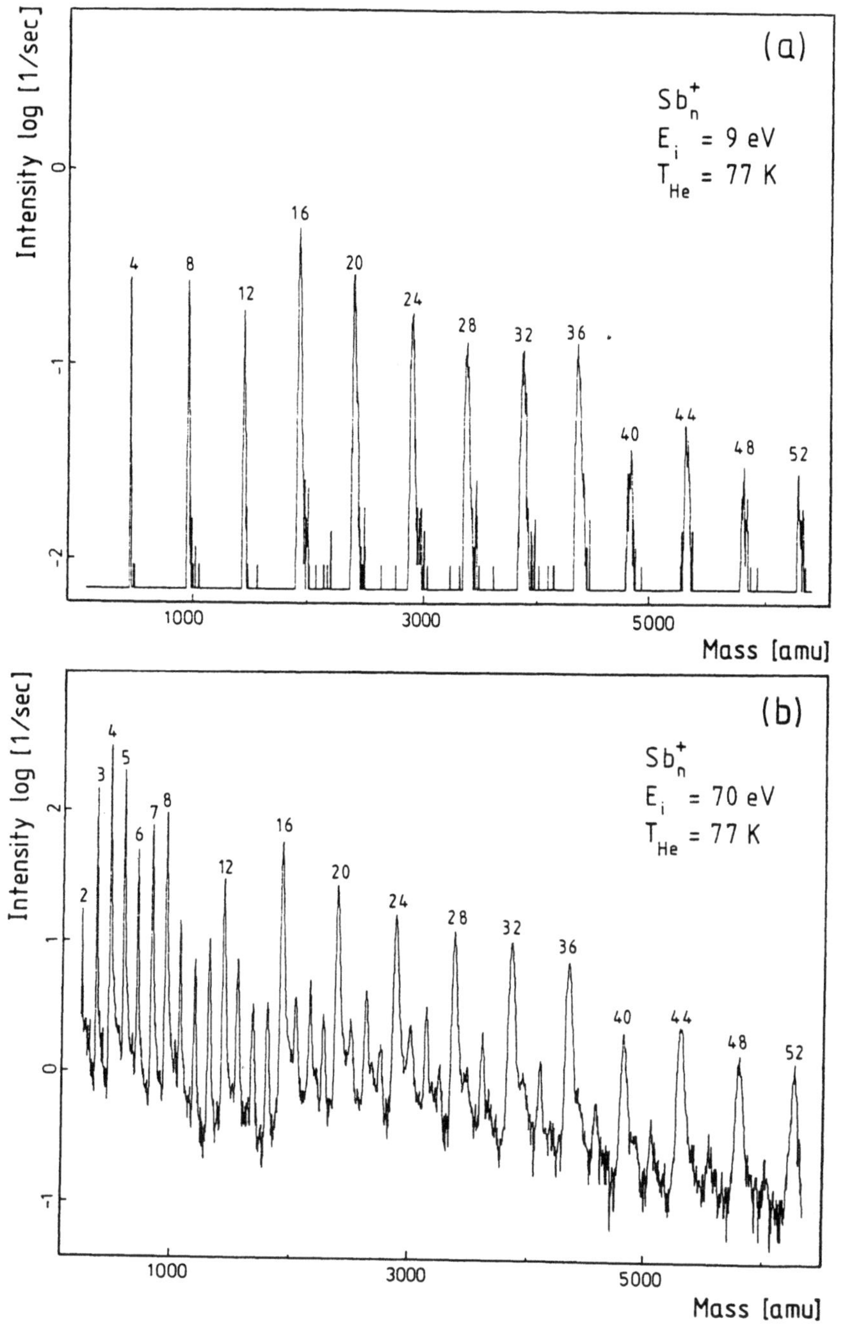

Fig. 2a and b. Mass spectra of Sb-clusters for two different electron ionizing energies E_i: **a** $E_i = 9$ eV, **b** $E_i = 70$ eV. The temperature of the He-gas in the source is $T_{He} = 77$ K [10]

the neutral clusters however cannot be transferred into the stability distribution of the charged clusters because there is a redistribution barrier for this process to happen. Only if E_i is increased above the various appearance potentials for fragmentation, highly stable ionized clusters are observed (for example, Sb_3^+, Sb_5^+, see Fig. 2b).

Fragmentation after electron impact has been studied in detail for the antimony tetramer [12]. The ionization potential (IP) of Sb_4 is 7.7 eV and the lowest appearance potential AP = 10.4 eV is found for the reaction $Sb_4 + e^- \rightarrow Sb_3^+ + Sb + 2e^-$. This shows that at least (AP − IP) = 2.7 eV has to be transferred to a Sb_4-molecule to fragment. This is about the same as the bond dissociation energy of the neutral Sb_4-molecule $D(Sb_3 - Sb) = 2.9$ eV.

The bond dissociation energies of small charged Sb cluster ions are high as well ($D(Sb_3^+ - Sb) = 2.7$ eV, $D(Sb_2^+ - Sb_2) = 3.8$ eV, $D(Sb^+ - Sb) = 3.2$ eV [12]), being derived from ionization potential measurements of Sb_3 (IP = 7.5 eV) and Sb_2 (IP = 8.6 eV). These species are present in the vapour at high enough temperature of the Knudsen cell. For the studies reported here Sb-vapour has been produced at temperatures low enough to consist predominantly of Sb_4-molecules.

A recent study on Bi-clusters showed similar results: onset for fragmentation in the vicinity of 10.5 eV [13]. This is 3.2 eV above the ionization potential of Bi_1 (7.7 eV). The bond dissociation energies for dimers of Bi, Sb, Pb, and Ag atoms lie between 0.8 eV and 3.1 eV: Bi–Bi, 2.1 eV; Sb–Sb, 3.1 eV; Pb–Pb, 0.8 eV; Ag–Ag, 1.7 eV; Bi–Ag, 2.0 eV; Bi–Sb, 2.6 eV; Bi–Pb, 1.5 eV (taken from phys. Tables). Therefore *AB*-bonds could be expected to have similar strength in the clusters as *AA* or *BB* bonds. This however does not hold anymore if "chemically inert" subunits can build up in the cluster. Then a situation is created where the same kind of atoms is bound under different conditions and with strongly differing bond strengths.

So far we did not discuss intermolecular fragmentation $Sb_{4n}^{+*} \rightarrow Sb_{4(n-1)}^{+} + Sb_4$. This would be important if the inert Sb_4-units would be packed in Sb_{4n} with loose intermolecular Van der Waals bond. From the data now in hand concerning their reactivity, however, we doubt if the Sb_4 units retain their identity after being clustered.

For interpretation of the data a further point has to be considered. For ionizing energies E_i close enough to the ionization potentials IP of the clusters one expects that the cross section for ion formation depends on the difference (IP-E_i) which is a function of n for IP(n). This would lead to the detection of different cluster intensities for different n being generated during ionization. In this case the measured mass distribution would neither reflect the equilibrium intensities of the neutral nor that of the ionized clusters. In order to check this possibility we varied E_i but the main features in the spectra being concerned with these studies did not change.

4. Adsorption of Single Atoms

We now consider the adsorption of single Ag-atoms on Sb_{4n}-clusters. Figure 3 shows a mass spectrum of Sb/Ag-clusters and the corresponding histogram for the integrated intensities. Two features can be seen: *(i)* up to three Ag-atoms are adsorbed and *(ii)* for most cluster sizes the intensity $I(Sb_{4n}Ag_m)$, $m = 1$–3, decreases fast with increasing m. Physisorption seams to be relevant for these cases. However, Fig. 3 shows two exceptions, Sb_8 and Sb_{12}: For these clusters the probability for one Ag-atom to be adsorbed is especially high.

The adsorption probability can be defined as follows:

$$P_a(n) = I(Sb_nAg_m^+)/\{I(Sb_n^+) + \sum_m I(Sb_nAg_m^+)\};$$
$m = 1$–3

In Fig. 4, P_a is plotted as a function of n for $m = 1$, 2, 3. The curve for one Ag atom adsorption shows pronounced structure, the ($m = 2$)- and ($m = 3$)-curves are relatively flat. For $m = 1$, maxima are found for $n = 8$, 12, 24, and 32. Sb-clusters with these sizes are more reactive than others in that size region.

It is difficult to understand why the adsorption of the second and third adatom at the cluster surface should be so much different from the adsorption of the first adatom. It could be explained in two ways: *(i)* saturation of the dangling cluster bonds by the first adatom. The clusters Sb_nAg_1 then would be inert for other adatoms, i.e. chemisorption for $m = 1$, and physisorption for $m = 2$ and 3; *(ii)* The clusters have nutshell structures with empty inner cavities and one Ag atom is situated inside the cavity. The structure in the reactivity curve then would be caused by the ability of some clusters to let an adatom move to the center or, to rearrange and incorporate the foreign atom. In this case the outstanding high intensities of Sb_8Ag_1 and $Sb_{12}Ag_1$ could be understood in terms of bcc (8 nearest neighbors) and fcc (12 nearest neighbors) structures, respectively.

For both explanations discussed above, however, geometrical structures for Sb-clusters have to be assumed which are fundamentally different from the structures assumed so far. The expectation so far was that clustering of tightly bound units like Sb_4 leads to packing of these units. They contain their individuality in the cluster. A packing model for tetrahedra explained the observed magic numbers in the Sb_{4n} mass spectra [14]. If the chemically inert Sb_4-units are packed to clusters, the ensemble should be inert as well. No variation in the adsorption probability for foreign atoms is expected in this case. The experimental data, which show strong cluster size dependence for P_a, however, seem to indicate that collective structures build up. The Sb_4-units loose their idendity after being packed and new energetically favoured atomic arrangements are generated. There is no direct proof for this statement but it could explain the pronounced structure in the $P_a(n, m)$-plots.

We cannot completely rule out the possibility that the difference between $m = 1$, 2, and 3 is generated after ionization but this is not likely to occur. For a discussion of this point we first ask which one of the two components (Ag or Sb_{4n}) is ionized: In the case of Van der Waals coupling between Sb_4-units the IP's of both components (IP(Ag) = 7.57 eV, IP (Sb_4) = 7.64 eV, IP(Sb_{4n}) ~ IP(Sb_4) for small n) are nearly the same. In the case of collective Sb_n-structures IP(n) would decrease. For both

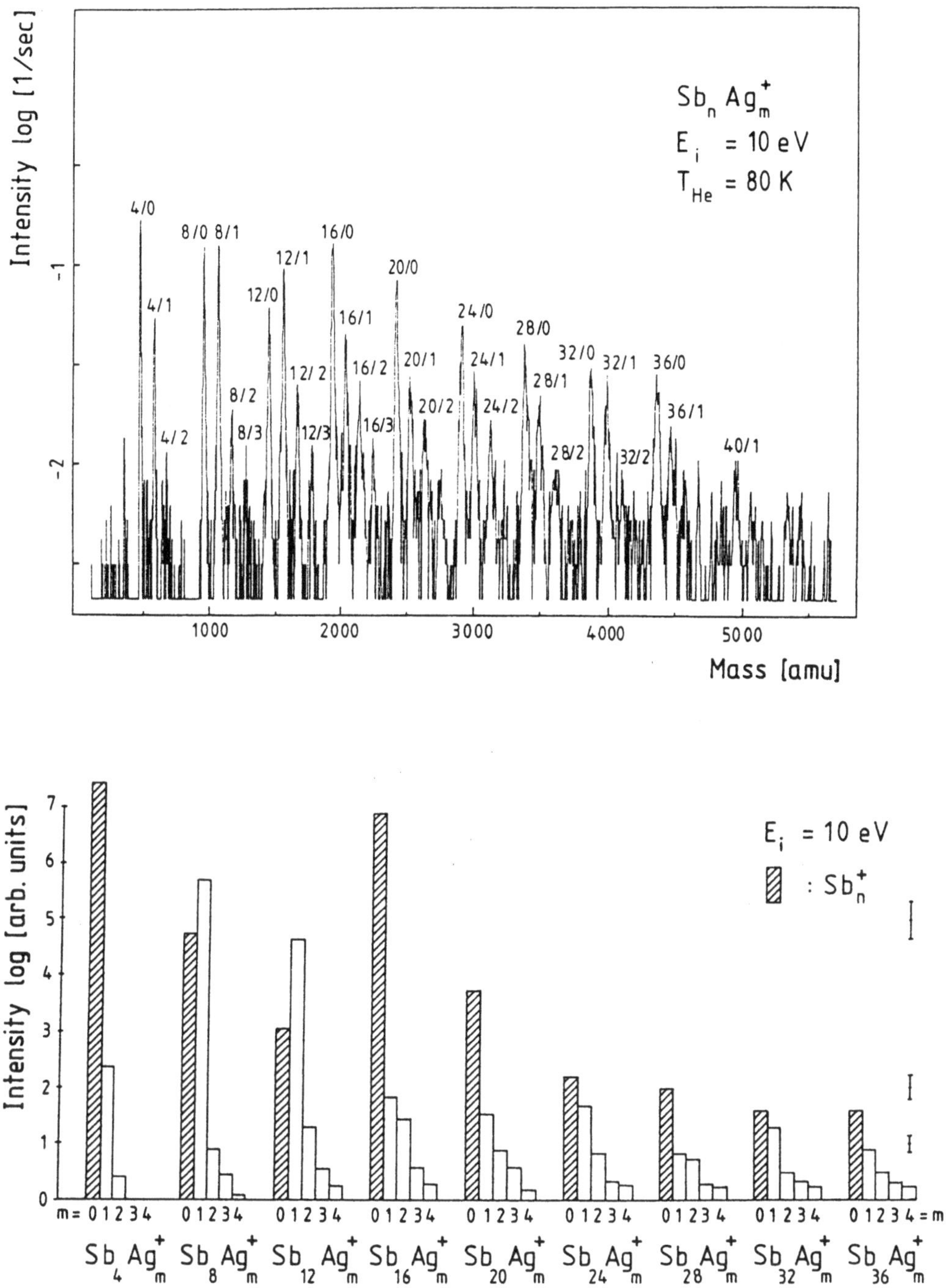

Fig. 3. Mass spectrum of Sb_nAg_m-clusters (top) and the corresponding histogram with integrated intensities (bottom); $E_i = 10$ eV [10]

Sb_n-structures the electron is not removed from the silver atom and therefore solvation of Ag^+ can be ruled out. The other possibility namely redistribution of the whole alloy cluster ion is unlikely because of the relatively high redistribution barriers of a few electron volts. As mentioned before stable cluster ions are only found predominantly for high electron ionizing energies ($E_i \gg 10$ eV) which in the case of alloy clusters will be discussed in another publication [10].

Figure 5 shows the P_a-plot for adsorption of Pb-atoms on Sb-clusters. A strong size dependence is found for $m = 1$ and 2. There is a maximum at $n = 12$, but the absolute P_a-values for $n = 8$ and 12 are much lower than in the case of Ag-adsorption. A silver atom fits into the center of an antimony cluster because the metallic radii r_0 of $Sb(r_0 = 1.39\,Å)$ and $Ag(r_0 = 1.34\,Å)$ are nearly the same. Lead atoms are much larger in size $Pb(r_0 = 1.5\,Å)$. Therefore, a center

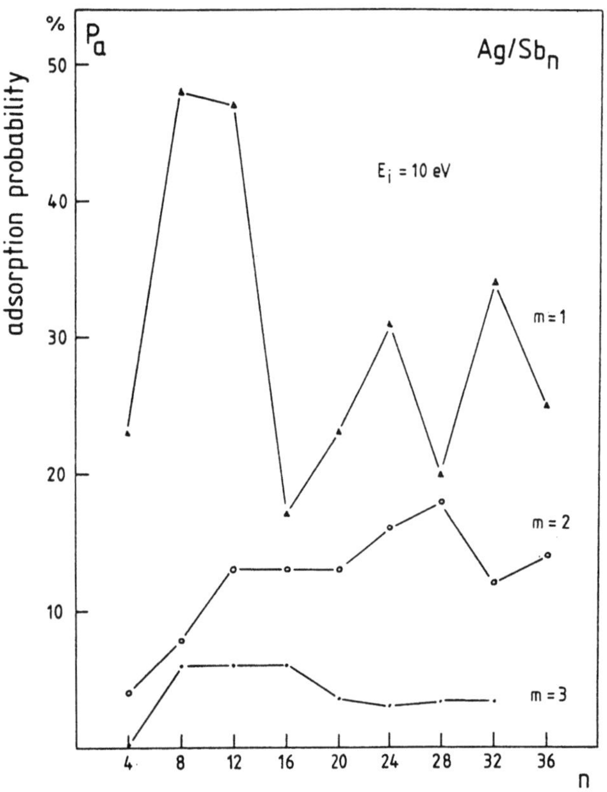

Fig. 4. Adsorption probabilities for one, two, or three Ag-adatoms on Sb-clusters (definition in the text) [10]

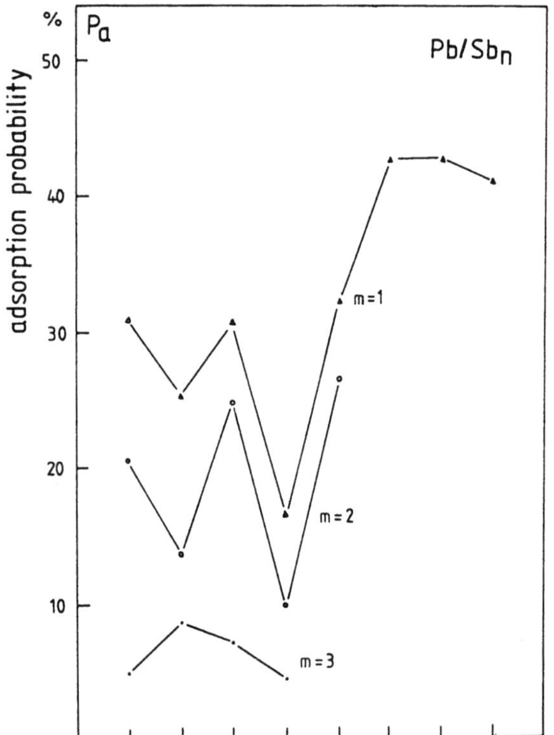

Fig. 5. Adsorption probabilities for one, two, or three Pb-adatoms on Sb-clusters [10]

position seems to be convenient for a Ag-atom but unconvenient for a Pb-atom.

5. Dissociative Chemisorption

With source conditions where Sb_4-units are adsorbed by Pb-clusters the spectrum in Fig. 6 is measured. The adsorption probabilities are extremely high. The histogram shows that one or two lead atoms do not influence the Sb_4-units. However, from Pb_3 on, very effective dissociation of Sb_4 occurs; the molecule is split into two halfs: $Pb_n + Sb_4 \rightarrow Pb_nSb_2 + Sb_2$. Figure 7 shows the dissociation probabilities

$$P_d(n) = I(Pb_nSb_2^+)/\{I(Pb_nSb_2^+) + I(Pb_nSb_4^+)\}.$$

The corresponding mass spectra have been taken at three different electron ionizing energies. The plot at 8 eV shows the dissociation onset at $n=3$. Then, P_d increases fast with n and gets $\sim 90\%$ for $n=10$. Three local maxima ($n=4$, 8, 10) are found. The outstanding high reactivity can also be seen from spectra taken at 10 eV and 80 eV.

It is interesting to note that the adsorption probabilities for Sb_4 on Pb_n-clusters are extremely high (see Fig. 6). The corresponding Pb–Sb bulk phase diagram is simply eutectic. We conclude that two materials which are insoluble in the bulk phase can be soluble in clusters. It is open to question if this is due to the rapid quenching of the two vapours possibly yielding metastable complexes (with life times of at least a few microseconds, the time for mass analysis), to the nonperiodic character of the cluster geometries or to the quantized electronic structure of the two components.

If one or two Sb_4-units are adsorbed by Bi-clusters the histogram in Fig. 8 is obtained. Like Pb_n, Bi-clusters are also reactive in respect to dissociation of Sb_4. In this case however, symmetric and antisymmetric dissociation channels are populated. Bi-clusters have no size threshold for the reaction onset. The single Bi-atom already affects dissociation of Sb_4.

The ability of Pb- and Bi-clusters to crack the covalent Sb-bonds is surprising. The dissociation energies are high for both dissociation paths: $Sb_3 - Sb$, 2.9 eV; $Sb_2 - Sb_2$, 2.9 eV [12]. One possible explanation is that electronic charge is transferred from the clusters into the antibonding states of Sb_4. For the single atoms we would not expect that this occurs: the electronegativities of the atoms Sb(1.9), Pb(1.8) and Bi(1.9) are nearly the same, and the ionization potentials as well (Sb_4, 7.7 eV; Pb_1, 7.42 eV; Bi_1, 7.29 eV). IP's for clusters however decrease fast with increasing size ($\sim n^{-1/3}$ for metals and $\sim n^{-1}$ for insulators) towards the bulk work functions (4.04 eV for

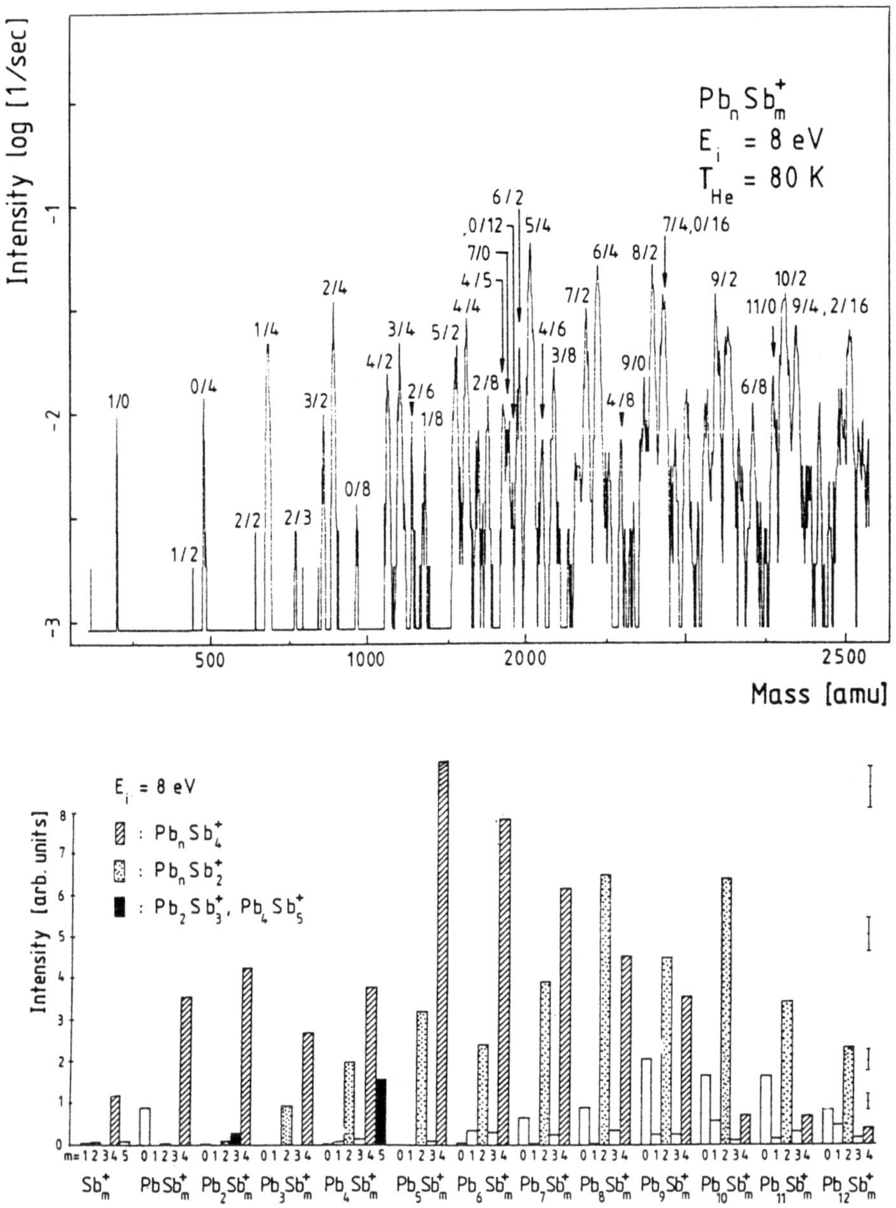

Fig. 6. Mass spectrum of Pb_nSb_m clusters and the corresponding histogram. After adsorption of one Sb_4 molecule at a Pb-clusters the reaction $Pb_n + Sb_4 \rightarrow Pb_nSb_2 + Sb_2$ occurs; dissociation onset at $n=3$ [9]. The two black lines in the histogram increase fast with increasing electron energy and exceed at high E_i by far the other lines in the distribution. This fact is discussed in Ref. 10

Pb(solid) and 4.34 eV for Bi(solid). Therefore, the contact potential difference (considered in analogy to bulk metal-metal contacts) between the atomic adsorbates and the cluster-substrates increases with cluster size. The amount of charge transfer depends on this difference.

Besides charge transfer other effects can lead to dissociation. This is shown in the case of Bi_1, which is able to crack with high effectivity the Sb_4-molecule, despite the ionization potentials of Bi_1 and Sb_4 are almost the same.

The author wishes to express his gratitude to D. Schild to present results of his diploma thesis prior forthcoming publication. He also acknowledges continuous coorporation with R. Pflaum and E. Recknagel. This work is supported by the Deutsche Forschungsgemeinschaft.

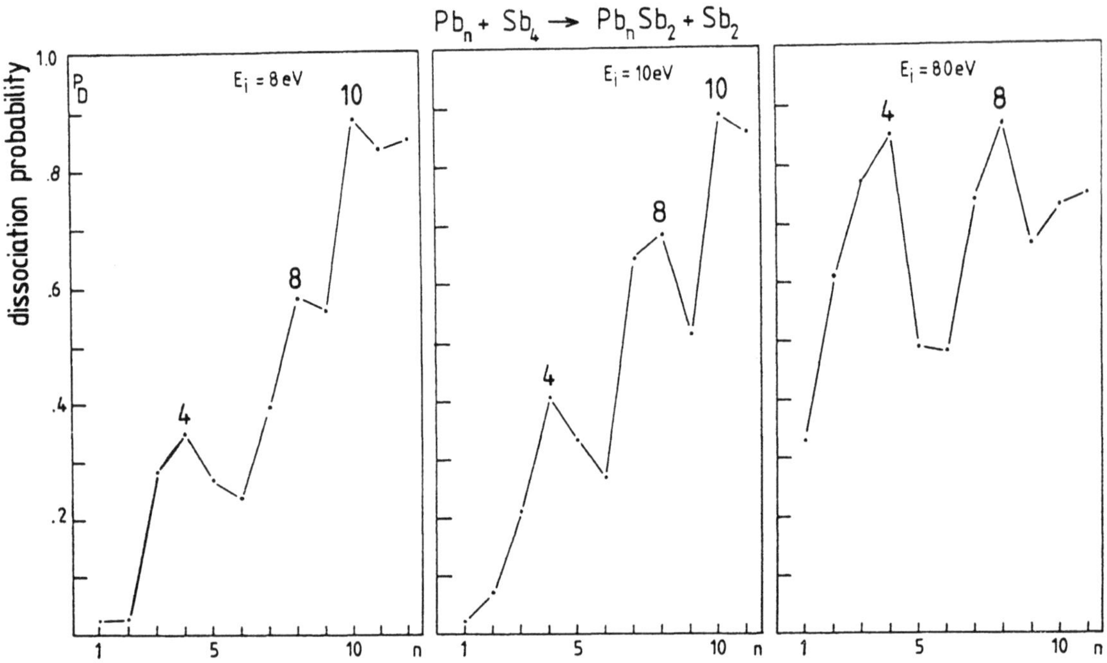

Fig. 7. Dissociation probabilities for the reaction $Pb_n + Sb_4 \rightarrow Pb_nSb_2 + Sb_2$. The corresponding mass spectra have been taken at three different ionizing energies 8 eV, 10 eV, and 80 eV [10]

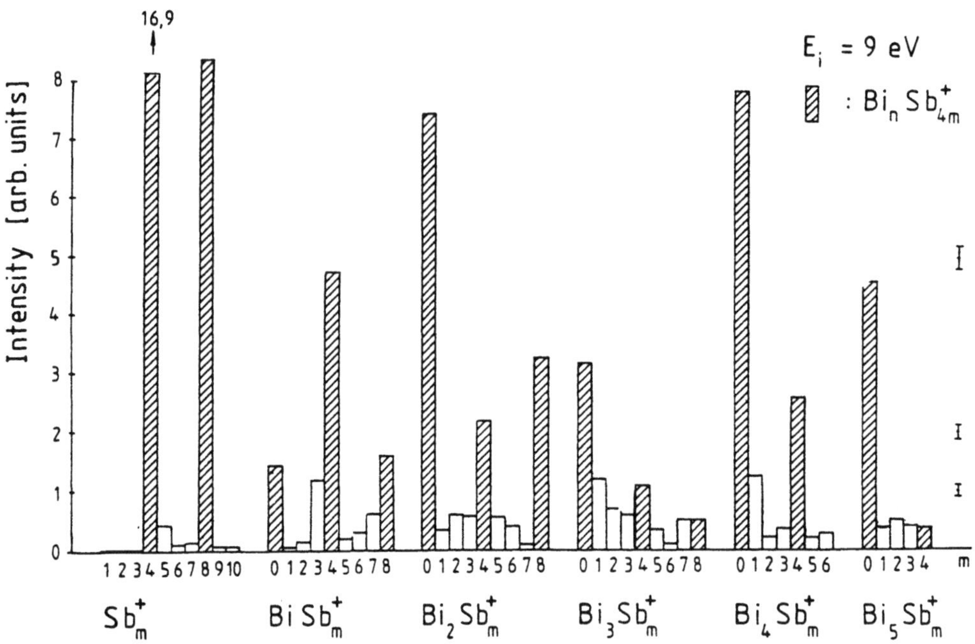

Fig. 8. Histogram for Bi_nSb_m-clusters. One or two Sb_4-units are adsorbed by the Bi-clusters. Symmetrical and nonsymmetrical dissociation channels are observed; dissociation onset at $n=1$ [9]

References

1. Richtsmeier, S.C., Parks, E.K., Liu, K., Pobo, L.G., Riley, S.J.: J. Chem. Phys. **82**, 3659 (1985)
2. Parks, E.K., Liu, K., Richtsmeier, S.C., Pobo, L.G., Riley, S.J.: J. Chem. Phys. **82**, 5470 (1985)
3. Liu, K., Parks, E.K., Richtsmeier, S.C., Pobo, L.G., Riley, S.J.: J. Chem. Phys. **83**, 2282 (1985)
4. Whetten, R.L., Cox, D.M., Trevor, D.J., Kaldor, A.: J. Phys. Chem. **89**, 566 (1985)
5. Trevor, D.J., Whetten, R.L., Cox, D.M., Kaldor, A.: J. Am. Chem. Soc. **107**, 518 (1985)

6. Morse, M.D., Geusic, M.E., Heath, J.R., Smalley, R.E.: J. Chem. Phys. **83**, 2293 (1985)
7. Martin, T.P. Surf. Sci. **156**, 584 (1985)
8. Rohlfing, E.A., Cox, D.M., Petkovic-Luton, R., Kaldor, A.: J. Phys. Chem. **88**, 6227 (1984)
9. Schild, D., Pflaum, R., Sattler, K., Recknagel, E.: Proceedings of the Symposium on Atomic and Surface Physics (SAPS), Obertraun, Austria 1986, p. 300
10. Schild, D., Pflaum, R., Sattler, K., Recknagel, E.: J. Phys. Chem. (in press)
11. Sattler, K., Muehlbach, J., Recknagel, E.: Phys. Rev. Lett. **45**, 821 (1980)
12. Cabaud, B., Hoareau, A., Nounou, P., Uzan, R.: Int. J. Mass Spectrosc. Ion Phys. **11**, 157 (1973)
13. Walstedt, R.E., Bell, R.F.: Phys. Rev. **A33**, 2830 (1986)
14. Sattler, K., Muehlbach, J., Pfau, P., Recknagel, E.: Phys. Lett. **87A**, 418 (1982)

K. Sattler
Department of Physics
University of California
Berkeley, CA 94720
USA

Cluster Compounds Help to Bridge the Gap between Atom and Solid

H. Müller
Sektion Chemie der Friedrich-Schiller-Universität Jena,
German Democratic Republic

Received April 7, 1986; final version May 23, 1986

We discuss in this paper the possibilities to calculate electronic structure of the metal atom cluster compounds (CC) and present some interesting results from binuclear and hexanuclear cluster prototypes.

The aim of this work is to show that quantum chemistry of CC has an extraordinary potential to provide: *(1)* interesting examples of the highly dispersed metallic state (only quantum chemistry has developed a model to bridge the whole gap between atom and solid, the analytic cluster model), *(2)* simple models of local bulk situations (e.g. behaviour of impurity atoms), developing cluster-bulk analogy, *(3)* simple models of local surface phenomena (e.g. chemisorption processes), developing cluster-surface analogy.

The most important point is to initiate knowledge transfer between complex chemistry, surface chemistry, field of highly dispersed metals and metal physics.

PACS: 3640; 3120G; 3120P

1. Introduction

We discuss in this paper the possibilities of calculating electronic structure of one particular class of compounds containing metal-metal bonds, so-called metal atom cluster compounds (Chap. 2.). These cluster compounds (CC) contain polyhedral groups of metal atoms Me_N (with or without interstitial atom in the polyhedron centre), which are surrounded by some ligand sphere [1, 2]. Up to now chemists have synthesized CC containing metal islands Me_N up to $N=55$ [3].

We present some interesting results of quantum chemical calculations from cluster prototypes ($N=2$ and 6) in Chap. 3. The aim of this work is to show that quantum chemistry of CC has an extraordinary potential to provide: *(1)* interesting examples of the highly dispersed metallic state (Chap. 4.1.), *(2)* simple models of local bulk situations (e.g. behaviour of impurity atoms), developing an analogy between bulk solid and clusters (Chap. 4.2.), *(3)* simple models of local surface phenomena (e.g. chemisorption and catalytic processes), developing an analogy between solid surfaces and clusters (Chap. 4.3.). From this point of view CC help to bridge the gap between metal atoms and metals.

2. Cluster Compounds – Scientific Challenge to Quantum Chemistry

Often CC exhibits remarkable structures and properties. Permanently new structure patterns are discovered. The development of bonding models, which can explain the structures, detailed electronic properties and reactivities of CC, has represented a considerable intellectual challenge during the last two decades, particularly since these molecules are some of the most complex chemical species currently known. In spite of considerable difficulties up to now quantum chemists have achieved interesting results.

The difficulties mentioned are mainly caused by *(1)* the large number of metal atoms (moreover, heavy atoms require relativistic corrections), *(2)* number and variety of ligands (often connected with strong lowering of symmetry) and *(3)* additional problems like fluxionality, interstitial atoms etc. As a result [2] more sophisticated quantum chemical methods (like ab initio HF including CI) are only applicable to the smallest CC (binuclear CC) and only $SW-X_\alpha$ approach and above all semi-empirical methods (e.g. EHMO) are suitable in describing electronic structure of larger CC (about 5–10 metal atoms).

3. Results of Quantum Chemical Calculations on CC-Examples

3.1. Electronic Structure of Binuclear CC

This prototype (the simplest type in principle) permits the full quantum chemical improvement, which should be illustrated with the cluster ion $Re_2Cl_8^{2-}$ (Fig. 1). The most important features of its structure are: *(1)* the very short Re–Re distance, *(2)* the eclipsed configuration and *(3)* diamagnetism. The results of the first quantum chemical calculations (Cotton [4], simple *d*-overlap model) agree completely with our own computation (Müller [5], free electron model). The main results are: *(1)* quadruple Re to Re bond (ground state configuration $\sigma^2\pi^4\delta^2$ in correspondence with the very short Re-Re distance; *(2)* δ-bond provides the explanation for the occurrence of the eclipsed structure, although it is the weakest of the Re to Re bonds (in the staggered configuration the corresponding overlap would be zero); *(3)* closed shell electronic structure (since there are 8 electrons to occupy the 4 bonding orbitals) in agreement with the observed diamagnetism of the ion.

The present situation is given by a paper of Cotton [6] who recently has redone $SW-X_\alpha$ calculations, explicitly employing relativistic mass-velocity, Darwin and spin-orbit corrections; it will be sufficient here to state that relativistic corrections dramatically improve the spectral agreement compared to the previous non-relativistic calculations [7]. It is important to state, too, that we can find immediately the quadruple bond scheme from the converged relativistic $SW-X_\alpha$ energy levels for $Re_2Cl_8^{2-}$, although in this calculation the ligands are explicitly involved; therefore we can conclude: *(1)* the qualitative essentials of the bonding are apparent even in the simplest treatments, but *(2)* the more sophisticated methods are required to interpret spectra and other experimental data quantitatively.

3.2. Electronic Structure of Hexanuclear CC

Nowadays the hexanuclear CC are – besides certain exceptions – the actual field of investigation. The main structure of this prototype consists of an octahedral arrangement of the metal atoms with 12 edge-bridging (Me_6X_{12}) or 8 face-bridging (Me_6X_8) ligands (Fig. 1). Well-known examples for these so-called "Münster-Cluster" are given by the halides (X = F, Cl, Br, I):

$Me_6(Vb)X_{12}^{2+}$ with 16 cluster electrons and
$Me_6(VIb)_8^{4+}$ with 24 cluster electrons.

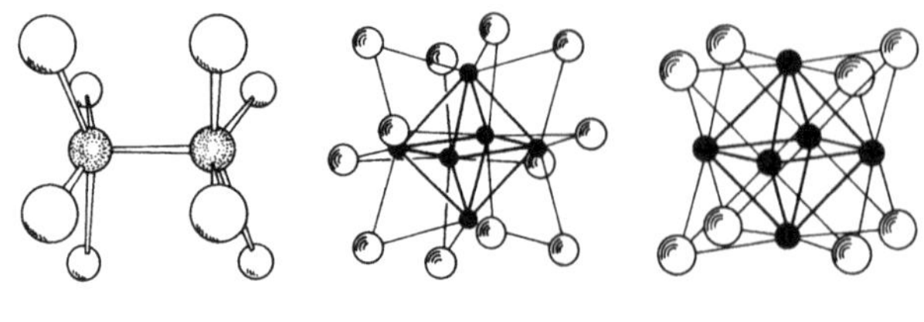

Fig. 1a–c. Structures of selected cluster compounds (CC): **a**: Re_2Cl_8, **b**: Me_6X_{12}, **c**: Me_6X_8 (**b**, **c**: so-called Münster cluster)

Again there is a good agreement between the early quantum chemical calculations (Cotton [8], simple *d*-overlap model) and our computations (Müller [9], free electron network model). In both models the calculated number of bonding orbitals are 8 or 12, respectively, for the two CC, just sufficient to fill all these bonding orbitals with the cluster electrons mentioned above (in agreement with the observed diamagnetism); here the bond orders are less than one.

Nowadays the results are characterized by $SW-X_\alpha$ calculations including explicitly the influence of the ligands [6, 10, 11]. Cotton [6] discussed on this level the ligand destabilization of special cluster MO's in $Mo_6Cl_{14}^{2-}$. If we compare the results of various quantum chemical methods, it will be quite gratifying that the simple method has correctly predicted the properly filled metal-metal MO's, but it is important to state those simple methods certainly cannot predict ligand destabilization. We have discussed the stability within the systems $Mo_6X_8^{4+}$ (X = F, Cl, Br, I) [10] and compared measured and calculated X-ray photoelectron spectra of $[Mo_6Cl_8]Cl_4$ [11], which illustrates the importance of covalent Mo–Cl bonding for the stability of the cluster.

4. The Extraordinary Potential of the CC to Bridge the Gap between Atom and Solid

If one analyses such calculations as discussed in Chap. 3. two steps will be found: *(1)* calculation of the pure metallic island Me_N (e.g. Mo_6 from $[Mo_6Cl_8]Cl_4$) in order to get information concerning the metal to metal bonding scheme, *(2)* calculation including the ligands in an explicit manner to get information concerning its influence.

On this occasion it is important to state that the first step possesses an independent scientific meaning: investigation of properties of a very little piece of metal. This correspondence becomes more evident when one considers the high nuclearity clusters (like $Rh_{12}Rh(CO)_{24}H_3^{2-}$ [12], $Pt_{19}(CO)_{28}^{4-}$ [13],

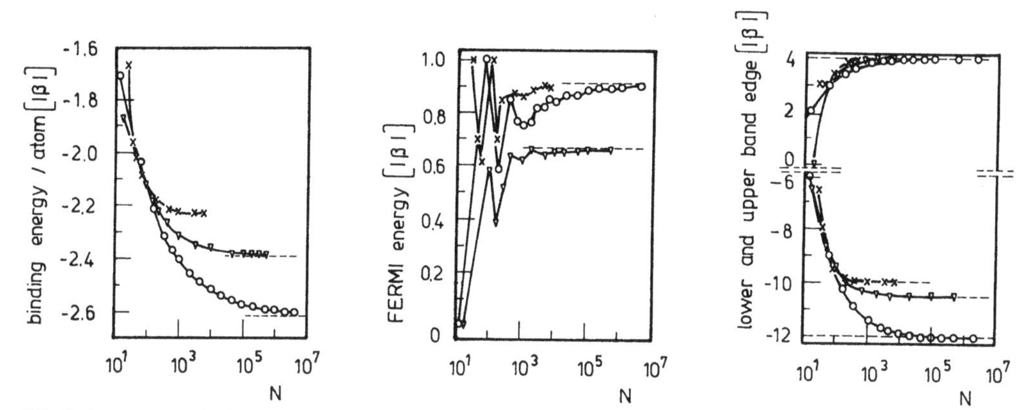

Fig. 2. Some numerical results concerning the electronic structure of three sequences of fcc clusters of rectangular prismatic shape with (100) surface (o cubes, v plates, × rods): *(1)* binding energy per atom, *(2)* Fermi energy and *(3)* lower and upper band edge, plotted against number N of atoms [21]

$Ni_{38}Pt_6(CO)_{48}H_2^{4-}$ [14] and Au_{55} $[(C_6H_5)_3P]_{12}Cl_6$ [3]), clusters which greatly resemble small fragments of bulk metals. On the basis of this simple statement the possibility is open to go from "quantum chemistry of metal atom cluster compounds" into four different areas of natural sciences: *(1)* field of highly dispersed metallic state, *(2)* metal physics, *(3)* surface chemistry and *(4)* complex chemistry (cf. Chap. 3).

4.1. Metallic Islands Me_N and the Highly Dispersed State of Metals

If we only discuss the metallic island Me_N characterized by a given number N of atoms and a definite structure, we are calculating properties of the highly dispersed metallic state. In this field ("fifth state of matter") we find the following situation [15, 16]:

Between the well established microscopic domain of atoms and molecules, mainly ruled by quantum chemistry, and the macroscopic domain of condensed matter, governed by solid state physics, there is an intermediate region dealing with properties of small aggregates or clusters which are neither quite microscopic nor quite macroscopic. This region of highly dispersed matter is very interesting not only from a scientific point of view, but also from a technological one. The main problem in this field is the fact that "giant molecules" are much smaller than "small solids", therefore we have from a theoretical point of view a no man's land in this important region, that means no adequate tools have been developed up to now.

If we try to enter no man's land in order to calculate a certain property $G(N)$ as a function of particle size, after a short distance ($N = N_{\mathrm{critical}}$) all conventional quantum chemical approximation methods will fail (more sophisticated methods earlier, semiempirical methods later); moreover, we have a very similar situation if we try to use the physical approach from the right hand side [17].

In this situation the question arises how to investigate the further development of size effect of the physical or chemical property G if one goes from a single atom via critical cluster to a bulk solid. The answer is unambiguous: there is only one important exception, namely the analytical cluster model (ACM). Only in this model any restriction concerning the number N of atoms is removed completely. The breakthrough concerning N must be paid partly by the shortcomings of this method. If we combine the most simple quantum chemical methods (HMO, FEM) with N atom clusters characterized by specific regular boundaries, we will have the rare possibility to solve Schrödinger equation for this cluster in an analytical manner, i.e. we will get closed formulas for eigen values and eigen functions in dependence on N. Therefore we are able to calculate properties of clusters with arbitrary size.

The literature shows ACM's for simple cubic lattices only [18, 19]. Some years ago we were successful – in cooperation with our colleagues from Charles University Prague – in extending the ACM to fcc- and bcc-clusters [20–22]. The main problem in this context is to overcome the very large secular problems. An effective way is the application of the difference equation formalism.

Figure 2 shows as an example concerning numerical calculations three different properties $G(N)$, plotted versus the number of atoms N: Binding energy per atom, Fermi level, upper and lower band edge. here we use a logarithmic scale because the numerical calculations have been extended up to about ten million atoms (the broken lines at the right of the figures indicate the TB bulk values using periodical boundary conditions). The condensation energy at the linear chain in the framework of free electron model (FEM)

$$\Delta E(N)=0.361+\frac{N^3+\frac{47}{12}N^2+\frac{29}{6}N+\frac{93}{48}}{8\left(N^2+2N+\frac{3}{4}\right)^2}$$

and its asymptotical expansion

$$\Delta E(N)=0.361\left(1+\frac{0{,}346}{N}+\ldots\right)$$

is an example for an analytically calculate property [23]. Three-dimensional clusters lead to the more general formula concerning certain one-electron properties G

$$G(N)=c_1\left(1+\frac{c_2}{N^{1/d}}+\ldots\right) \tag{1a}$$

d = number of dimensions of the cluster

All investigated properties converge to those of the solid at least as quickly as $1/N^{1/3}$ (e.g. bandwidth and energy of LOMO, however, converges more quickly as $1/N^{2/d}$ [24]). It is convenient for larger clusters to convert N into the radius R of the particle

$$G(R)=k_1\left(1+\frac{k_2}{R}+\ldots\right) \tag{1b}$$

We have shown that Eq. (1a, b) are able to describe qualitatively size effects of a great variety of properties. If we fit physically the parameters $c_i(k_i)$ $i=1, 2$

$N\to\infty: c_1=G(\infty)\sim$ bulk value

$N\to n: c_2=[G(n)-G(\infty)]\,G(\infty)^{-1}\,n^{1/d}; n=1, 2,\ldots$

the Eq. (1a, b) can be used as a rule of thumb to predict size effects of various properties of clusters.

We have checked successfully the validity of Eq. (1) for a broad variety of various properties, because the bulk value is known in many cases, and besides this one needs only a single value $G(n)$ (measured or calculated at the cluster Me_n) [25]. Figure 3 gives some examples: ionization potential of Na-clusters [26], binding energy per atom of Li-clusters [27] and melting temperature of gold particles [28].

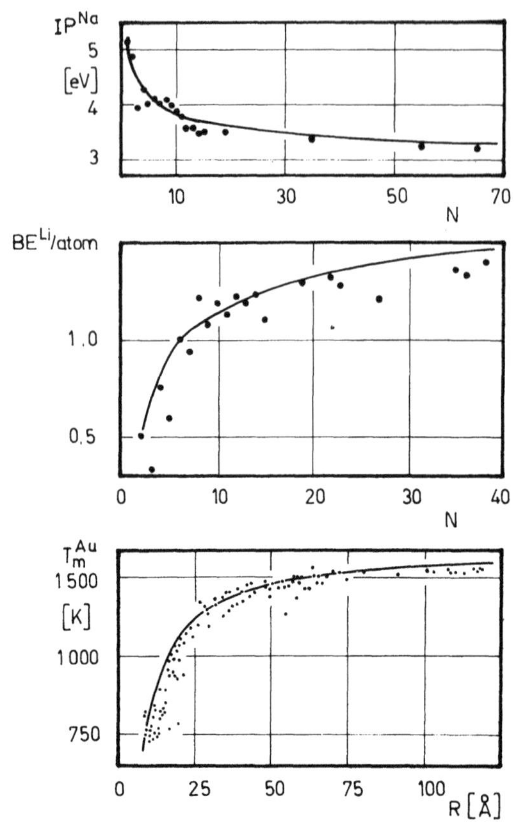

Fig. 3a–c. Description of the size dependency of a few cluster properties (● measured or calculated data) by means of Eq. (1) (—). **a**: photoionization potential measured on sodium clusters Na_N [26]; **b**: binding energy per atom, calculated (CNDO method) on lithium clusters Li_N [27]; **c**: melting temperature measured on gold particles [28]

4.2. *Via Cluster-Bulk Analogy to Metal Physics*

If we discuss the metallic island including an interstitial atom (main group elements like H, C,...) we are calculating – via cluster-bulk analogy – local properties of impurity atoms in bulk metal (corresponding interstitial hydrides, carbides,...).

As an example we consider the hydrido interstitial CC $HNb_6I_{11}(=[HNb_6I^i_8]I^a_{6/2})$ [29, 30, 2], its structure [31] is given in Fig. 1 (Me_6X_8): H Atom is situated in the centre of the Nb_6 octahedron. From calculations of electronic structure of the systems $HNb_6I_8^{3+}/Nb_6I_8^{3+}/HNb_6/Nb_6$ we get information concerning (Fig. 4): *(1)* bonding mechanism (significant lowering of $1a_{1g}$-cluster ($Nb_6I_8^{3+}$)-MO causes a strong covalent H–Nb bond of about 2.5 eV); *(2)* change in magnetism ($Nb_6I_8^{3+}$ [paramagn.] $\to HNb_6I_8^{3+}$ [diamagn.]); *(3)* change in geometry (lattice expansion (Nb–Nb distance) of about 5%, caused by the H atom); *(4)* charge distribution (analysis of the important cluster-MO's shows a negatively charged interstitial H atom); *(5)* penetration mechanism (activation energies (barriers) during H insertion and going from one relatively stable site to another one; estimation of H-vibrations: $\omega_H(0_h)\approx 950\,cm^{-1}$).

A lot of these and similar results, derived from hydrido interstitial CC, leads to useful conclusions for the partner system "H in metals" regarding

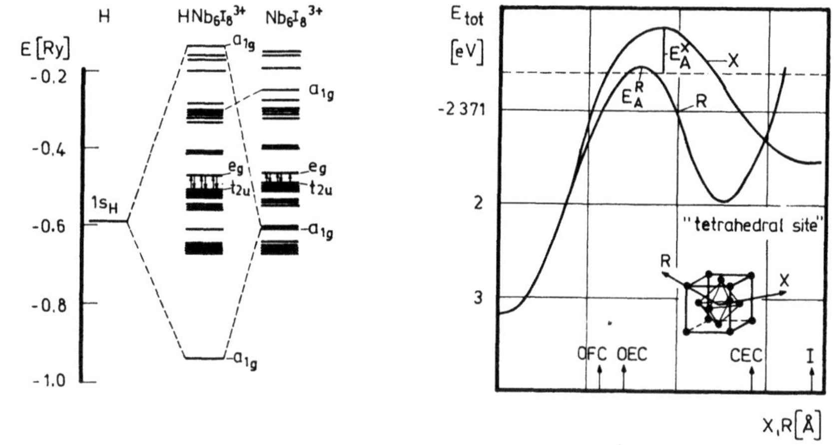

Fig. 4. Electronic structure of the hydrido interstitial cluster compound HNb_6I_{11}. Left side: MO level schemes for $Nb_6I_8{}^{3+}$ and $HNb_6I_8^{3+}$ according to $SW-X_\alpha$ calculations (correlation diagramm); right side: penetration mechanism for hydrogen (total energies of the system $(H+Nb_6I_8^{3+})$ along two "reaction paths", X, R for hydrogen) [29, 30, 2]

bonding models, diffusion, structural changes etc. E.g., in this context: *(1)* we have prepared together with Prof. Wicke/Münster a quantum chemical support for his well-known "two-stage screening model" for H atoms (more general impurity atoms) within transition metals [32]; *(2)* we have studied together with Dr. Seifert/Dresden the interaction of H with Pd and Pt (HPd_6/HPt_6) using a semi-relativistic version of $SW-X_\alpha$ method; the less stability of the H/Pt system – compared with H/Pd – is explained as a relativistic effect [33].

4.3. Via Cluster-Surface Analogy to Surface Chemistry

If we study the "reaction"

$$Me_N + m\,L \rightarrow Me_N L_m$$

we were calculating – via cluster-surface analogy – local properties of chemisorption phenomena. For the cluster carbonyls (ligand L = carbon monoxide CO) this situation is profoundly described in the literature [34–36,1]. Authors have shown that there are a lot of similarities between the CC and chemisorbed systems concerning: *(1)* the photoelectron spectra, *(2)* binding energy, *(3)* infrared spectra and *(4)* mobility of CO. These results demonstrate among other things the localized character of the chemisorption bond and justify cluster approximations for its theoretical description. Some years ago we extended the cluster-surface analogy to metal halogen CC. $Mo_6X_8^{4+}$ (X = F, Cl, Br, I) should act as an example for illustration (as to the structure cf. Fig. 1). The electronic structure of the cluster cations have been calculated using semi-empirical MO method as well as $SW-X_\alpha$-SCF approach [10]. Both quantum chemical methods lead to extensive similarities concerning the ligand bond energies and chemisorption energies, repectively [10].

Further investigations to try to extend cluster-surface analogy to sulfur systems (Mo_6S_8, $Re_6S_8^{2+}$, $Fe_6S_8^{2+}$) [37] have not been finished up to now.

5. Concluding Remarks

The quantum chemistry of CC have an extraordinary potential not only to bridge the gap between atom and solid but also between chemistry and physics (complex chemistry, surface chemistry, field of highly dispersed metals, metal physics). But the most important point is to bridge the gap between chemists and physicists and to initiate knowledge transfer between different areas of natural science.

References

1. Muetterties, E.L.: Science **196**, 839 (1977)
2. Uhlig, E., Müller, H.: Sitzungsber. Akad. Wiss. DDR, Math.-Nat.-Techn. Nr. 12/N (1982)
3. Schmid, G., Pfeil, R., Boese, R., Bandermann, F., Meyer, S., Calis, G.H.M., Velden, J.W.A. van der: Chem. Ber. **114**, 3634 (1981)
4. Cotton, F.A.: Inorg. Chem. **4**, 334 (1965)
5. Müller, H.: Habilitationsschrift, Univ. Jena 1968; Z. Phys. Chem. (Leipzig) **248** 152 (1971)
6. Bursten, B.E., Cotton, F.A., Stanley, G.G.: Israel J. Chem. **19** 132 (1980)
7. Martola, A.P., Moskovitz, J.W., Rösch, N.: Int. J. Quantum Chem. S No **8**, 161 (1974)
8. Cotton, F.A., Haas, T.E.: Inorg. Chem **3**, 10 (1964)
9. Müller, H.: Z. Phys. Chem. (Leipzig) **249**, 1 (1972)
10. Seifert, G., Großmann, G., Müller, H.: J. Mol. Struct. **64**, 93 (1980)
11. Seifert, G., Finster, J., Müller, H.: Chem. Phys. Lett. **75**, 373 (1980)
12. Albano, V.G., Ceriotti, A., Chini, P., Ciani, G., Martinengo, S., Anker, W.N.: J. Chem. Soc. Chem. Commun. 859 (1975)
13. Chini, P.: Gaz. Chim. Italiana **109**, 225 (1979)
14. Ceriotti, A., Demartin, F., Longoni, G., Manassero, M., Marchiana, M., Piva, G., Sansoni, M.: Angew. Chem. **97**, 708 (1985)
15. Müller, H.: Wiss. Z. Humboldt-Univ. Berlin, Math.-Nat. R XXXIV, 78 (1985)
16. Fritsche, H.-G., Kadura, P., Künne, L., Müller, H., Bauwe, E.,

Engels, S., Mörke, W., Rasch, G., Birke, P., Spindler, H., Wilde, M., Lieske, H., Völter, J.: Z. Chem. **24**, 169 (1984)
17. Perenboom, J.A.A.J., Wyder, P., Meier, F.: Phys. Rep. (Rev. Sect. Phys. Lett.) **78**, 173 (1981)
18. Hoffmann, T.A.: Acta Phys. Hungar **1**, 1 (1951); **2**, 97 (1952)
19. Messmer, R.P.: Phys. Rev B**15**, 1811 (1977)
20. Bilek, O., Kadura, P.: Phys. Status Solidi (b) **85**, 225 (1978)
21. Kadura, P., Künne, L.: Phys. Status Solidi (b) **88**, 537 (1978)
22. Künne, L., Bilek, O., Skala, L.: Czech. J. Phys. B**29**, 1030 (1979)
23. Müller, H.: Z. Chem. **12**, 475 (1972)
24. Skala, L.: Phys. Status Solidi (b) **127**, 567 (1985)
25. Müller, H., Strickert, K.: Wiss. Z. Friedrich-Schiller-Univ. Jena/Thür., Math.-Nat. R. (accepted)
26. Schumacher, E., Kappes, M., Marti, K., Radi, P., Schär, M., Schmidhalter, B.: Ber. Bunsenges, Phys. Chem. **88**, 220 (1984)
27. Skala, L.: Phys. Status Solidi (b) **109**, 733 (1982)
28. Buffat, Ph., Borel, J.-P.: Phys. Rev. A**13**, 2287 (1976)
29. Dübler, F., Müller, H., Opitz, Ch.: Chem. Phys. Lett. **88**, 467 (1982); Z. Phys. Chem. (Leipzig) **264**, 936 (1983)
30. Opitz, Ch., Dübler, F.: Phys. Status Solidi (b) **76**, 505 (1983)
31. Simon, A.: Z. Anorg. Allg. Chem. **355**, 311 (1967)
32. Wicke, J., Fritsche, H.-G., Müller, H.: Z. Phys. Chem. Neue Folge (in preparation)
33. Seifert, G., Fritsche, H.-G., Ziesche, P., Heera, V.: Phys. Status Solidi (b) **121** 705 (1984)
34. Conrad, H., Ertl, G., Knözinger, H., Küppers, J., Latta, E.E.: Chem. Phys. Lett. **42**, 115 (1976)
35. Ertl, G.: J. Vac. Sci. Technol. **14**, 435 (1977)
36. Plummer, E.W., Salaneck, W.R., Miller, J.S.: Phys. Rev. B**18**, 1673 (1978)
37. Dübler, F., Müller, H.: Wiss. Z. Friedrich-Schiller-Univ. Jena/Thür., Math.-Nat. R. **31**, 907 (1982)

H. Müller
Sektion Chemie
Friedrich-Schiller-Universität Jena
Am Steiger 3, Haus 3
DDR-6900 Jena
German Democratic Republic

Systems of Small Metal Particles: Optical Properties and their Structure Dependences

U. Kreibig
FB 11 - Physik, Universität des Saarlandes, Saarbrücken,
Federal Republic of Germany

Received April 18, 1986; final version June 2, 1986

Experiments on small particles usually require samples containing large numbers of particles. The properties of such samples are determined both by the properties of the individual particle and by collective effects, if particles are packed closely together. Collective optical effects strongly depend on the topography of the samples. It is shown that they can be classified according to the effective local electromagnetic field. Recent experiments and calculations are presented for optical extinction spectra in the spectral region of plasmon polariton excitations, which clearly show the different behaviour of effective medium-like samples and of samples containing particle aggregates.

PACS: 71.45; 78.20; 81.20; 82.70

0. Introduction

The production and separation of free small metal particles from molecular dimers up to solid state clusters containing several hundreds of atoms has recently been developed to an almost perfect level. The next step, which would be to investigate their physical properties experimentally, is however, hindered by the fact that most experiments require samples containing enormous numbers of such particles, for instance, 10^5 to 10^{10} for conventional optical experiments. Up to now, samples for this purpose have been prepared mainly by storage methods, i.e. collection or production of the particles on a support or in a solid, liquid or gaseous embedding medium. Examples are gas-evaporated particles, island films, matrix-isolated or matrix-embedded particles, colloidal systems, etc. It is often difficult, if not even impossible, to derive physical properties of *individual* particles from experimental results, since in such many-particle systems *collective* properties arising from interactions between the particles may contribute comparable amounts to or even veil these individual properties in the resulting experimental data [1].

In the following, an investigation of these collective effects will be presented, with the aim of classifying them and also explaining them using recent experimental and theoretical results. Both experiments and calculations have been performed for particles in the size regions of microcrystals and bulky particles where solid state concepts can be used (see Table 1), though we would have preferred to perform the investigations on smaller clusters. The reasons are that relevant experiments in these size regions are possible, and that general trends of collective behaviour become conspicuous without being complicated by additional finite-size effects and extra surface effects of the individual particles.

It has to be emphasized, however, that these general results concerning the *collective* properties of densely packed many-particle systems also hold for smaller particles, i.e. clusters and even atoms, provided the interaction is due to dipolar interactions and chemical binding does not occur.

In the first section, the preparation and topographical analysis of samples of noble metal particles used in the experiments are briefly described and some selected examples of the results are discussed. In the second section, the local effective electromagnetic field acting on the particles is used to construct a classification scheme for various collective optical effects of samples with differing topographies. The following four sections treat the Mie case, the effec-

Table 1. Size classification of small metal particles [2]. (It is still an open question how the region $10^1 < N < 10^2$ can belong to the size regions I and II.)

Classification of "Small particles"	Size region I "Molecular clusters"	Size region II "Solid state clusters"	Size region III "Micro-crystals"	Size region IV "Bulky particles"
Number of atoms per aggregate	$N \lesssim 10^1$	$10^2 \lesssim N \lesssim 10^3$	$10^3 \lesssim N \lesssim 10^5$	$N > 10^5$
Number of surface atoms N_s compared to number of atoms of the inner volume (below surface layer) N_v	"surface" and "inner volume" not separable	surface-to-volume ratio: $\frac{N_s}{N_v} \sim 1$	surface-to-volume ratio: $\frac{N_s}{N_v} < 1$	surface-to-volume ratio: $\frac{N_s}{N_v} \ll 1$
Atomic arrangements and electron energies	atomic arrangements and energy spectra vary strongly with N; discrete electron energy spectra	structure of the "volume" may deviate from bulk crystal structure; size dependence of electronic surface and volume properties	bulk-like volume structure; size effects of volume material properties; surface properties similar to those of plane surfaces	bulk-like material properties of the volume; surface only causing polarization effects
Description	molecular theories	solid state theories with size and surface effects	solid state theories with size effects	solid state theories

tive medium case, the separated aggregate case and the general case of sample topography. Results are compiled in the summary.

I. Many-Particle Systems

Metal particles in liquid embedding media, i.e. aqueous colloidal systems, give unique possibilities to separate properties of individual particles from collective sample effects. This is because samples with single particles can be prepared whose topography may be changed subsequently in wide ranges from perfect separation with low filling factors ($f \lesssim 10^{-6}$) to densely packed clusters or aggregates of almost touching particles ($f \sim 0.5$). Hence, the *collective* properties due to interactions between particles may be changed without changing the *individual* properties. However, local arrangements of particles may vary widely in a sample since clustering or aggregation effects usually occur which are governed by complicated statistical laws; hence sizes, shapes and abundances of particular aggregates vary statistically in each sample, as do the distances between and coordination numbers of neighbouring particles within such aggregates. Consequently analyses of optical spectra of such samples are limited, and only a few quantitative ones have been performed [3]. A second consequence is that the preparation of special samples is not reproducible.

The wide field of topographical realizations of many-particle systems may be divided into three gross categories:

Case A	Case C	Case B
Separated single particles in full statistical disorder (parameter: filling factor f)	general case: more or less dense aggregates plus single particles, both in various packing densities	more or less closely connected particles in aggregates; aggregates well separated

To include case C, a weak definition of an aggregate will be used: particle *aggregate* is a section of the sample where all particles have distances to their nearest neighbours that are markedly smaller than the average distance deduced from the filling factor of the whole sample ($\propto R \cdot f^{-1/3}$).

As primary metal particle systems we used Zsigmondy colloids [4] of Ag and Au and subsequently varied their topographies [5, 6]. By changing the pH value of the system, aggregation could be initiated with variable velocity. Special stabilizing agents (e.g. gelatin) were added after some time t to stop the aggregation process, t thus determining the amount and mean size of the aggregates. Aggregation is due to the tendency of the particles to stick together; however thin interface layers between them prevent atomic contact, chemical binding and coalescence. The process is governed by complex cooperation of (electrodynamic) v.d. Waals forces, (electrostatic) Debye-Hückel forces and diffusion processes.

The filling factor of the samples could be varied by extracting water. After stabilization of the aggregates, the extinction of the liquid systems was

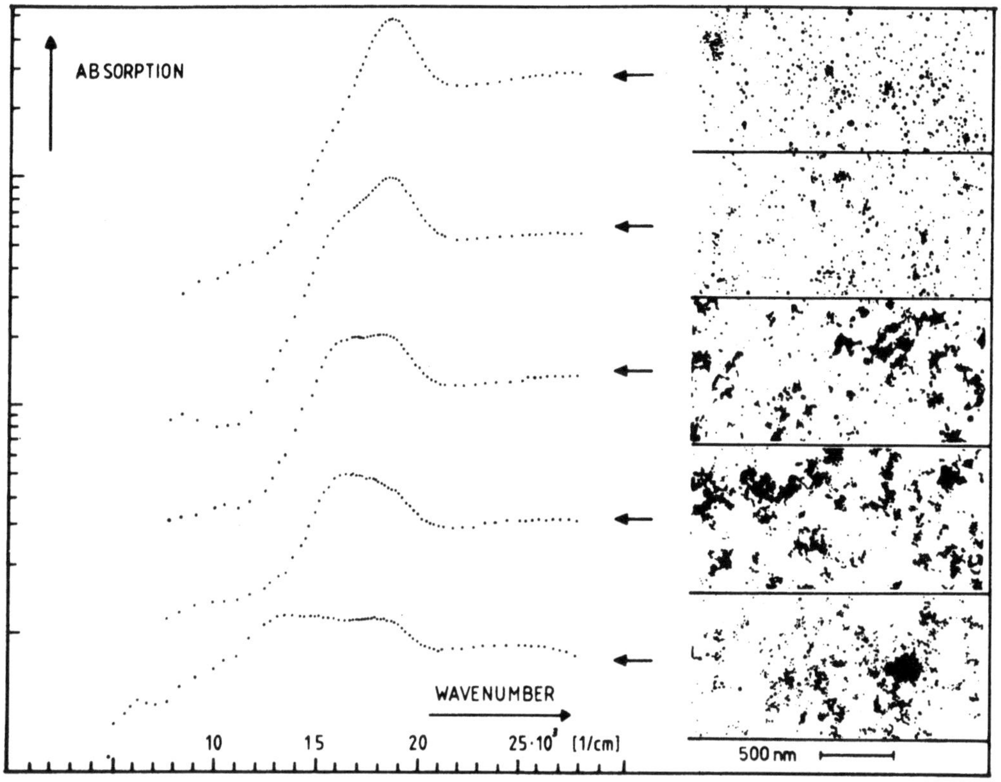

Fig. 1. Extinction spectra and TEM micrographs of one series of samples of Au particles ($2R = 17$ nm) with the amount of aggregation as the *only* varied parameter [7]

measured in a spectrophotometer. In addition, solid samples were prepared by drying on substrates and then analysed by TEM. In fact, only two-dimensional projections were observed by TEM, yet optical spectra of the systems in liquid and solid state show that the stabilization was sufficient to keep the changes due to drying only small (as shown in Fig. 9a).

Figure 1 illustrates one series of Au particle samples with the amount of aggregation as the *only* varied parameter. Drastic changes of the corresponding optical extinction spectra are obvious: we attribute these changes solely to the collective effects discussed in the following sections.

II. The Effective Field

The extinction of the light beam passing through the sample, occurs in part by absorption (producing thermal energy) in the particles and in part by scattering. These extinction losses are determined by the effective electromagnetic field acting on a given particle, and by the individual particle properties, i.e. polarizability α.

In the following, it will be shown that within the framework of Maxwell's theory, this *effective* field includes collective sample effects and their dependence on the sample topography.

For real ε_m (ε_m: dielectric function of the embedding material), the effective electric field can be written as the Lorentz local field

$$E_{\text{local}} = E_{\text{Maxwell}} + \Delta E + \Delta' E, \qquad (1)$$

where E_{Maxwell} is the incident field, ΔE is the correction due to the polarization charges of the Lorentz sphere and $\Delta' E$ is the contribution of polarizable particles within the Lorentz sphere.

The local field of (1), originally introduced as the field at an atom, is now assumed to act on the particle as a whole, thus resembling the field used in effective medium theories.

As discussed in detail in the following, the structure of this local field may be used to classify the three different categories of sample topographies introduced above, namely:

Case A: $\Delta E \neq 0$; $\Delta' E = 0$

Case B: $\Delta E = 0$; $\Delta' E \neq 0$

Case C: $\Delta E \neq 0$; $\Delta' E \neq 0$.

III. Case A Samples

Case A samples contain single particles in full *statistical disorder* with the filling factor f being the only parameter characterizing the topography. Then $\Delta' E \approx 0$.

Here $\Delta' E$ equals zero for particles distributed on a cubic lattice [8]. The situation for random distribution is discussed in detail in [9] on the basis of a lattice gas model, where it is shown that deviations from zero may be important. The following will be restricted to those cases where these deviations are negligible.

If, furthermore, $f \ll 1$, all particles are well separated and there is no interaction. In this limit also $\Delta E \approx 0$. No collective effects are left and the N-particle sample properties equal the N-fold of the individual particle properties. This is the case treated by Mie [10] and Debye, which may be applied to all particle sizes from the solid state cluster (Table 1) upwards, if proper finite size corrections of the material properties are included. For noble metals, this means particles with diameters larger than, say, 2 nm, consisting of more than some hundred atoms. These finite size effects are beyond the context of this paper; some of them are discussed in [2].

The individual particle properties are illustrated in Fig. 2 by comparison of an absorption spectrum with the absorption spectrum of a plane thin film (normal incidence) of the same material, using Ag. At wavelengths below 0.32 μm, the absorption due to interband transitions of the film equals that of a sample containing an adapted number of particles.

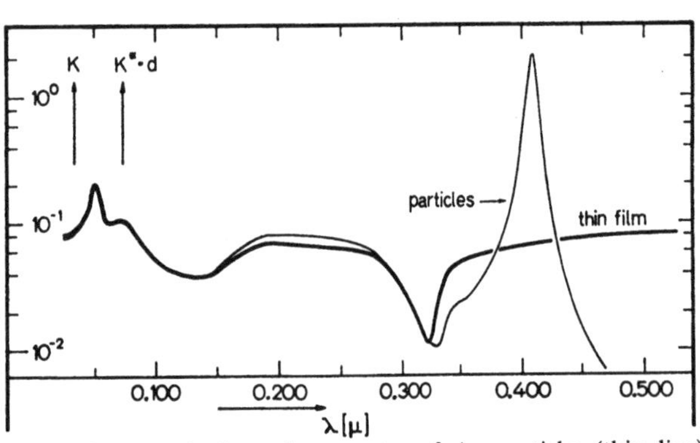

Fig. 2. Computed absorption spectra of Ag particles (thin line) and of a Ag thin film (thick line) of comparable thickness [2]

Above 0.32 μm, however, where the optical properties are mainly due to conduction electrons, both spectra differ markedly. As in the bulk material, the electrons in the film show relaxation behaviour described by the Drude-Sommerfeld theory, since their eigenfrequency (not the plasma frequency!) is zero.

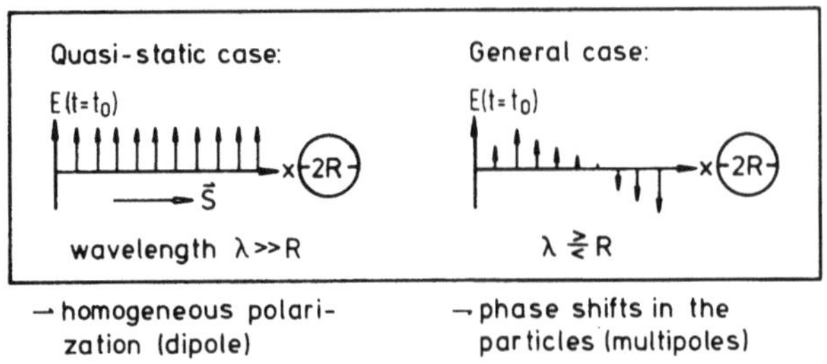

Fig. 3. Definition of quasistatic conditions (E: electric field; $\vec{S}$: Poynting vector; R: particle radius)

As a consequence, the absorption is extended to infinite wavelengths, i.e. zero frequency ("metal absorption"). Small particles, in contrast, are transparent in the low frequency region, since the eigenfrequency differs from zero due to surface charges. The conduction electrons behave like resonators and collective excitations can be excited optically ("plasmon polariton modes").

As long as $\lambda \gg 2R$ (quasistatic condition, Fig. 3), the conduction electron absorption is compressed into a narrow absorption band in the VIS region, which is due to a dipolar excitation of the whole particle. This is shown in Fig. 4a. If, however, the particles are no longer very small compared to wavelength λ, complex resonances due to electric and magnetic multipoles are excited, yielding broader, but still frequency-limited extinction bands. This is shown in Fig. 4b for the extreme example of 200 nm Ag particles. The particle sizes of the samples used for this investigation justify the restriction to dipolar excitations throughout the following. This dipolar excitation mechanism is independent of particle size as long as interior retardation effects do not occur. In this case, the plasmon polariton absorption remains dependent on particle size ex-

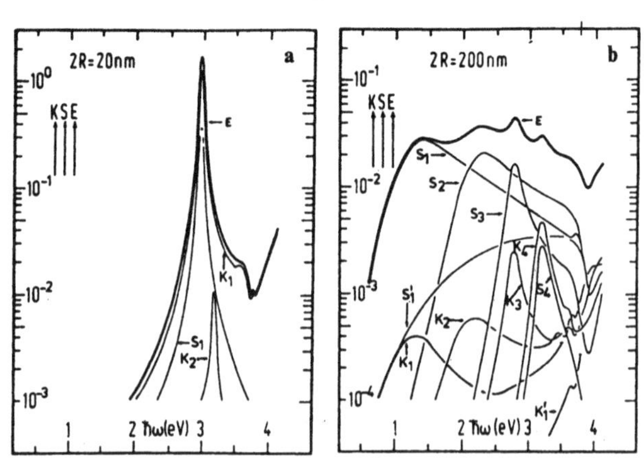

Fig. 4a and b. Extinction spectra (E) of spherical Ag particles with $2R = 20$ nm **a)** and $2R = 20$ nm **b)**, as computed from Mie theory [10]. The spectra are deconvoluted into absorption (K_l) and scattering (S_l) contributions of electric multipoles l and magnetic multipoles (K'_l, S'_l) [2]

clusively via the particle material properties themselves, which for most metals become size dependent due to finite size effects [2], extra surface effects [7] and structural transitions (introducing, for instance, the transition to the molecular cluster region as in Table 1) [2].

These remarks conclude this brief compilation of *individual* particle properties (i.e. the Mie case). The *collective* properties become important for case *A* systems when the filling factor f is increased, in which case the correction term ΔE has to be considered in Eq. (1). If $2R \ll \lambda$ it takes the form

$$\Delta E = P/(3 \cdot \varepsilon_0 \cdot \varepsilon_m), \tag{2}$$

with

$$P = E_{\text{local}} \cdot \Sigma n_j \cdot \alpha_j = (\bar{\varepsilon} - \varepsilon_m) \cdot E_{\text{Maxwell}},$$

where n_j, α_j are number density and polarizability of the jth particle species.

In this case the sample is called an *effective medium* and is described by the effective dielectric function of the whole composite sample $\bar{\varepsilon}(\omega)$, which follows from (2) as

$$\bar{\varepsilon}(\omega) = \varepsilon_m \left(1 + 2\left(\frac{1}{3\varepsilon_0 \varepsilon_m}\right) \cdot \Sigma n_j \alpha_j\right) \Bigg/ \left(1 - \left(\frac{1}{3\varepsilon_0 \varepsilon_m}\right) \cdot \Sigma n_j \alpha_j\right)^{-1}. \tag{3}$$

The explicit absorption spectrum is then given by the absorption constant $K(\omega)$

$$K(\omega) = \omega \cdot (c \cdot \sqrt{\varepsilon_m})^{-1} \cdot \text{Im}\{\bar{\varepsilon}(\omega)\}. \tag{4}$$

Equation (3) is not definitive for effective media; several other effective dielectric functions have been derived [11–13] which deviate from Eq. (3) if percolation effects occur. These are, however, excluded in our experiments because the particles are separated from each other even in densely packed aggregates.

As mentioned in the Introduction, collective effects like those included globally in (3) are only slightly dependent on particle size or can even be independent. The latter holds for (3), since quasistatic conditions are assumed. If the particles are atoms, Eq. (3) is named after Clausius and Mossotti or (for the optical region) after Lorentz and Lorenz. If the particles are molecular clusters, solid state clusters or microcrystals, as in Table 1, then (3) is called the Maxwell Garnett formula. For somewhat larger particles, Gans and Happel extended the formula and added an analogue for effective permeability which in the optical region is determined by eddy currents.

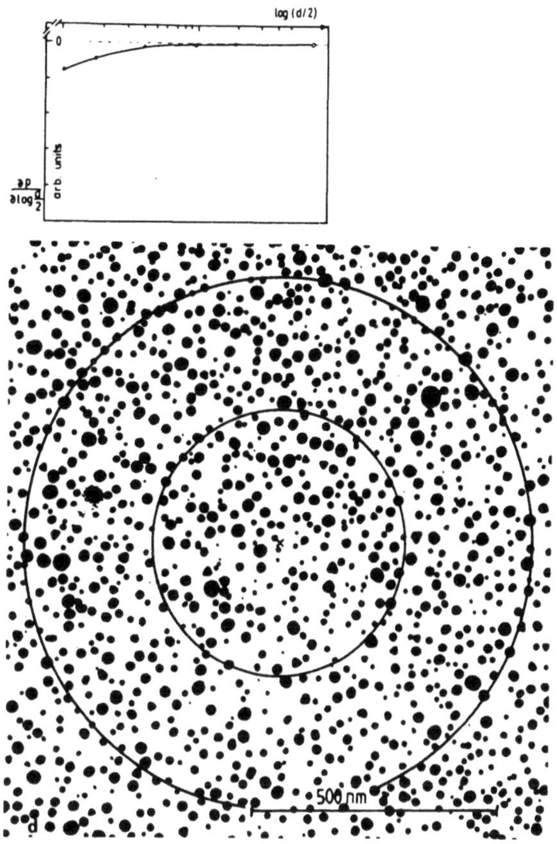

Fig. 5. TEM micrograph of a case A-like sample of Au particles ($2R = 17$ nm). Top inset: deviation from statistical disorder (p: normalized particle-particle distance probability; d: particle-particle distance) [5]

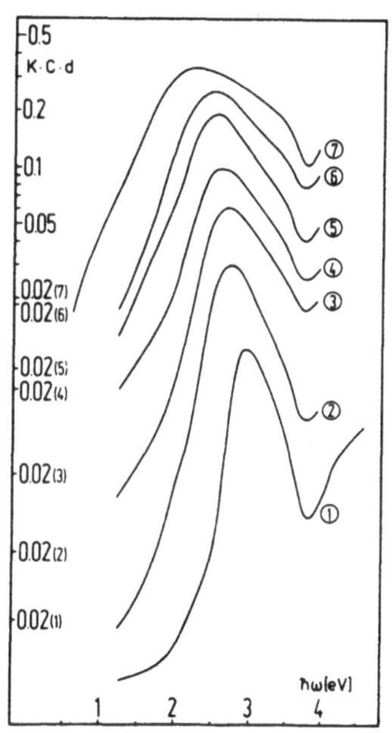

Fig. 6. Extinction spectra of a series of case A samples (Ag particles, $2R = 10$ nm) with the filling factor f as the *only* varied parameter. Curves 1 to 7: $f = 10^{-5}$, 0.003, 0.013, 0.025, 0.12, 0.21, 0.40 [1]

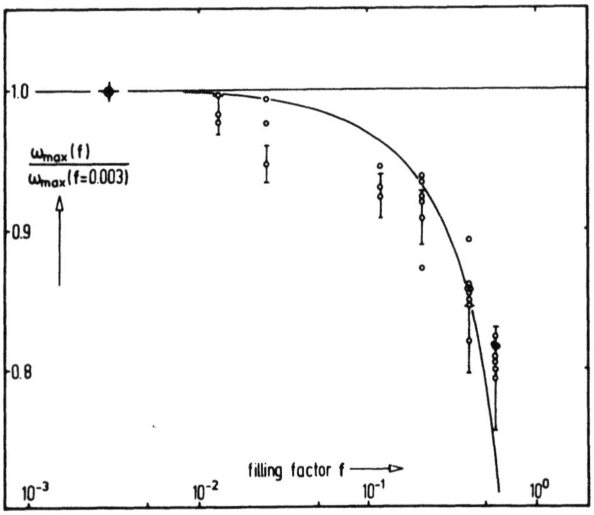

Fig. 7. Relative shifts of the plasmon polariton peak of case A samples due to changes of the filling factor. Solid line: Garnett formula Eq. (3). The bars indicate experimental accuracies

Figure 5 shows a two-dimensional projection of an effective medium-like sample. The normalized particle-particle distance probability $p(d)$ indicates that, in fact, only slight aggregation is present (p should be constant, i.e. $dp/d(\log d)=0$ for full statistical disorder).

Figure 6 examplifies the optical extinction spectra of one series of such Ag particle samples, where, as the only parameter, the filling factor f was varied between 10^{-5} and 0.4 [1]. As a result, the plasmon polariton is shifted towards lower frequencies in quantitative correspondence with (3) (Fig. 7).

The collective effect of case A samples is thus a *shift* of the plasmon polariton peak which remains a *single* peak in contrast to case B samples, as will be shown below. The increasing width in Fig. 6 is, however, in contradiction to Eq. (3) and is partly due to direct particle interaction since the samples are not fully of type A [9], and hence go beyond the effective medium model.

IV. Case B Samples

If the many-particle samples consist of well separated particle *aggregates*, then a local field that differs from that in case A acts upon the particles. Due to the separation of the aggregates $\Delta E \simeq 0$, but since the arrangement of the particles within an aggregate is usually not of full statistical disorder, $\Delta' E$ of (1) is not zero. This correction field may be interpreted as due to the interaction fields among the particles of a given aggregate, which are generated when the particles are irradiated by light. The structure of this correction field is investigated in the following for different aggregate types. But firstly, case B will be further illustrated in discussing some experimental results.

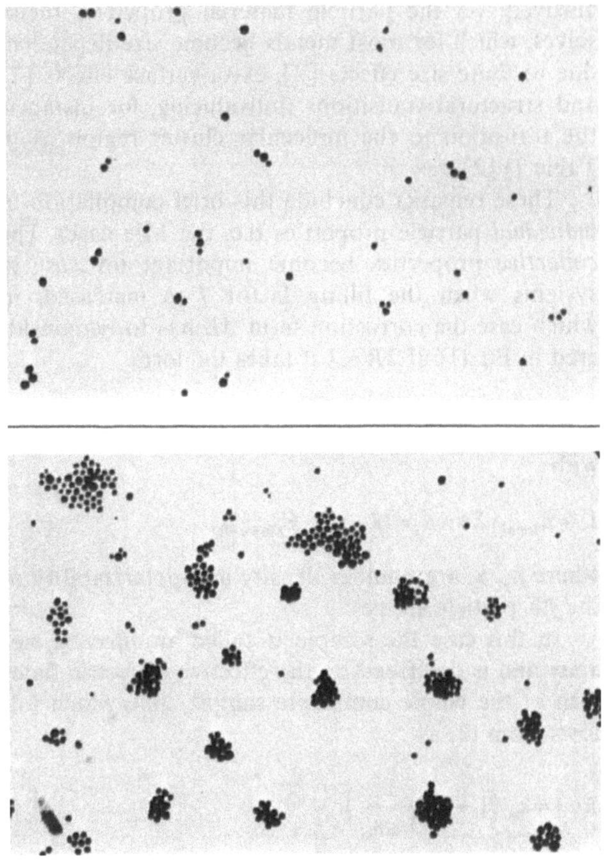

Fig. 8. TEM micrographs of case B samples of Au particles ($2R=56$ nm) with small aggregates [6]

Figure 8 gives two examples of Au particle systems with small, densely packed aggregates. (Corresponding optical extinction spectra are compiled in [6].) Larger aggregates are shown in Fig. 9, together with their extinction spectra. Both samples differ strongly in their topographies - one has quasi-fractal structures, the other contains compact aggregates - yet their spectra show similar broad multipeak structures which are in full contrast to the single plasmon polariton peak of case A samples.

The peak positions prove to depend strongly on the sample topography, as was already demonstrated in Fig. 1. The peak positions from a large number of samples with different degrees of aggregation (but with identical individual particle properties) are compiled in Fig. 10. Obviously, the high frequency peak is rather weakly influenced by the amount of aggregation, while the low frequency peak strongly shifts towards the IR with increasing aggregation. In

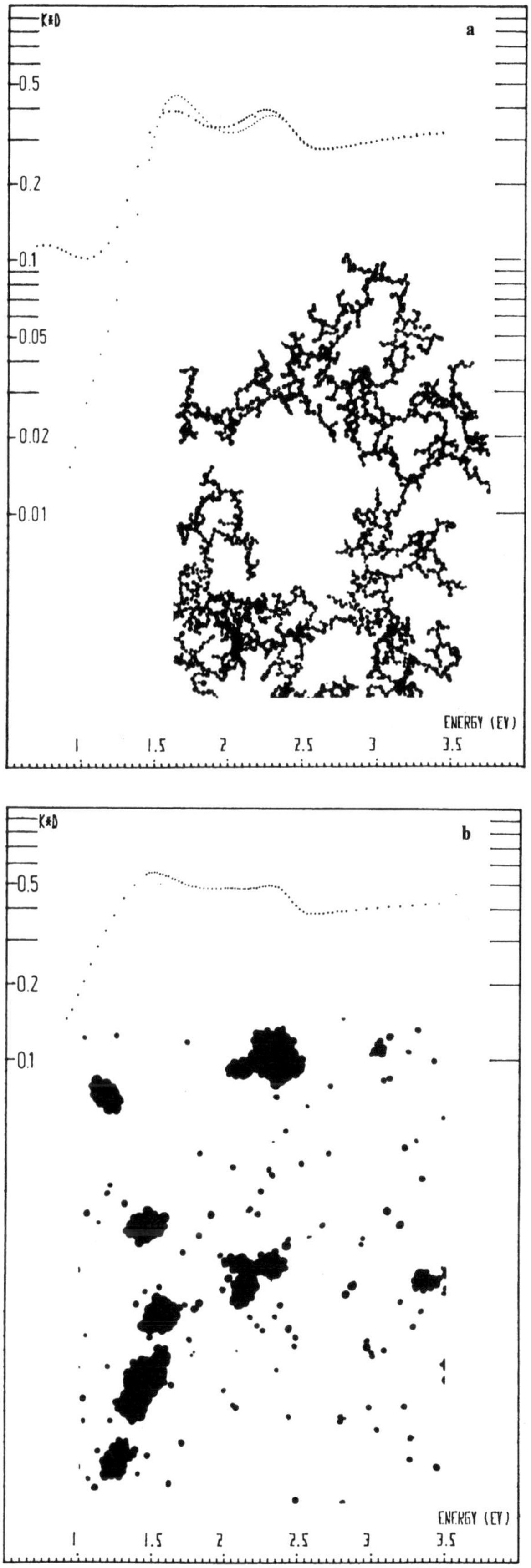

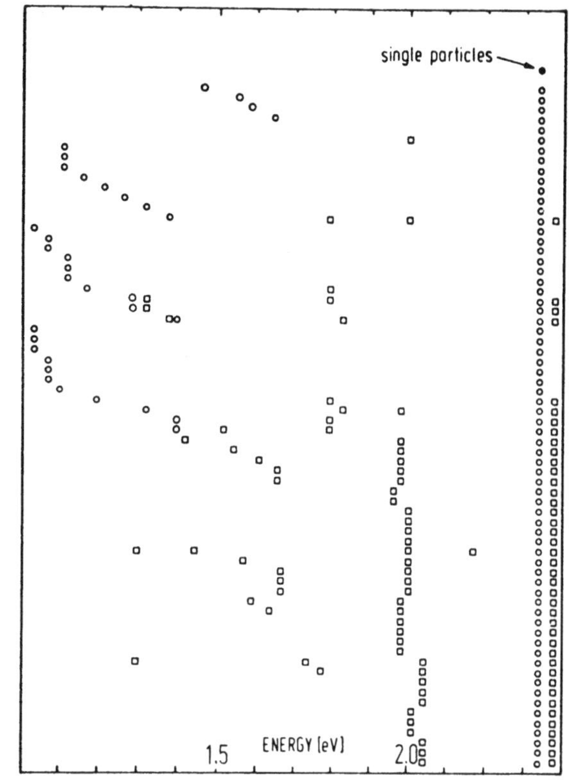

Fig. 10. Compilation of measured resonance peak positions of 69 samples of Au particles ($2R = 56$ nm) [6]. The data of different samples are arranged vertically; the peaks of each sample are plotted along one horizontal line. Circles and squares result from different evaluation methods [6]

the following, of Quinten et al. of explicit extinction spectra of aggregated particles [6] is sketched, enabling quantitative comparisons with our experimental results.

This calculation is based on Mie's general solution [10], but the local field correction term $\Delta' E$ is included, in addition. It is assumed that the particles interact via their electromagnetic scattering fields [14]. Roughly, it is a kind of v. d. Waals-like coupling which causes the plasma resonances of neighbouring particles to split. These scattering fields have been intensively investigated by Ausloos and co-workers [15]. To maintain validity for a wide size range in Mie's results, retardation and higher order electric and magnetic multipoles are considered for arbitrary aggregate sizes, shapes and nearest neighbour distances. Hence, spectra are obtained for all

Fig. 9a and b. TEM micrographs and extinction spectra of case *B* samples of Au particles with quasi-fractal (**a**) and compact (**b**) larger aggregates. Particle sizes: $2R = 36$ nm (a), $2R = 56$ nm (b). Small dots: extinction spectra of 3-dimensional aggregates in the liquid; large dots (Fig. 9a): 2-dimensional projections of the aggregates produced by drying the liquid on a substrate

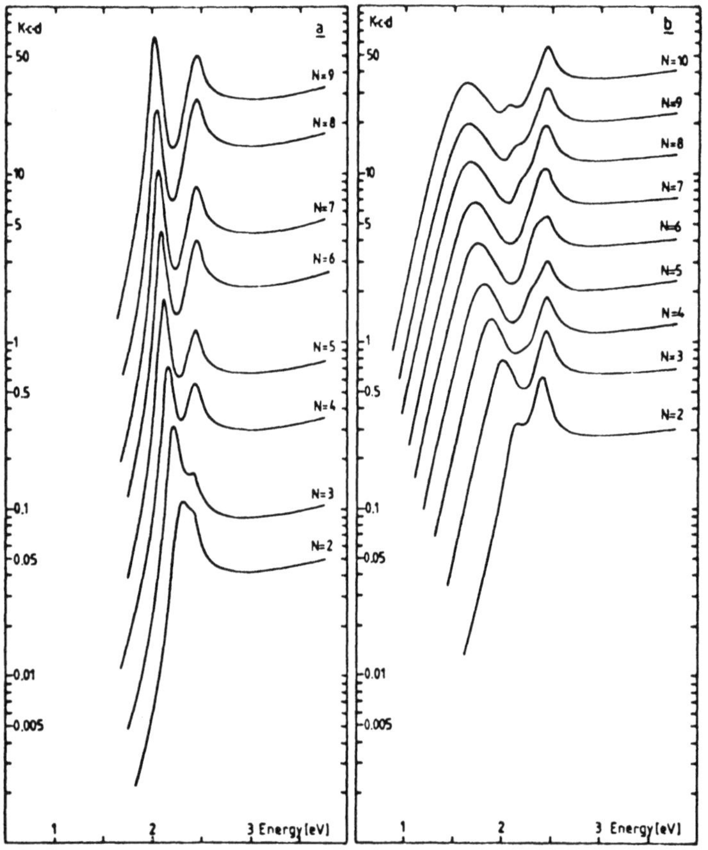

Fig. 11. Calculated extinction spectra of linear aggregates of Au particles [6]. (N: number of particles). Curves a: $2R = 10$ nm (quasistatic conditions); Curves b: $2R = 56$ nm (with retardation) effects)

kinds of aggregates and individual particle sizes from the solid state cluster of Table 1 to particles with diameters of several hundred nanometers. As in (3), the individual particle properties (the dielectric function of the particle material, in this calculation) has to be chosen properly. As a restriction, the particles are approximated by spheres, yet, as performed for the Mie theory, the calculations can, in principle at least, be extended to ellipsoidal particle shapes.

The wave equation

$$(\nabla^2 + k^2)\Pi_i(r, \Theta, \varphi) = 0 \tag{5}$$

is solved by multipole expansion ("partial waves" according to Mie) for the following potentials Π_i at a given particle i in the aggregate:

Π^{inc} of the incident plane wave
Π^{in} of the wave inside the particle
Π^{sca} of the outgoing/scattered wave
Π^{int} of the scattered waves from all particles $j \neq i$ of the aggregate.

The latter potentials can be transformed into *one* potential of an additional incident wave, the *interaction potential* [15]:

$$\Pi^{\mathrm{int}} = \sum_{j \neq i}^{N} \Pi^{\mathrm{sca}}(j) = \frac{1}{k^2 \cdot r_i} \cdot \sum_{l=1}^{\infty} \sum_{m=-l}^{+l} \psi_l(k \cdot r_i) \cdot Y_{l,m}(\Theta_i, \varphi_i) \cdot \sum_{j \neq i}^{N} \sum_{q=1}^{\infty} \sum_{p=-q}^{+q} A_{lm}^{qp} \cdot b_{qp}(j) \tag{6}$$

(k: wave number; Y: spherical harmonics; ψ: spherical Bessel functions; b_{qp}: complex amplitude coefficients of the scattered wave; A_{lm}^{qp}: transformation matrix of the spherical coordinates of particles j into those of particle i. Its explicit form is given in [6].)

These four potentials have to fulfill Maxwell's boundary conditions (additional boundary condition - (ABC-) effects [16] neglected), and thus a system of $N(2 \cdot l + 1)$ equations (N = number of particles in the aggregate) is obtained to calculate the complex amplitude coefficients b_{lm} of the wave scattered from particle i.

One ends up with the extinction constant

$$K(\omega) = 3 \cdot c^2 (2 \cdot \omega^2 \cdot \varepsilon_m \cdot R^3)^{-1} \cdot \mathrm{Im}\left\{ \sum_{i=1}^{N} \sum_{l=1}^{\infty} \sum_{m=-l}^{+l} (-1)^l \cdot b_{lm}(i) \right\} \tag{7}$$

by summing up all particles, all multipoles and the polarization states of the incident light.

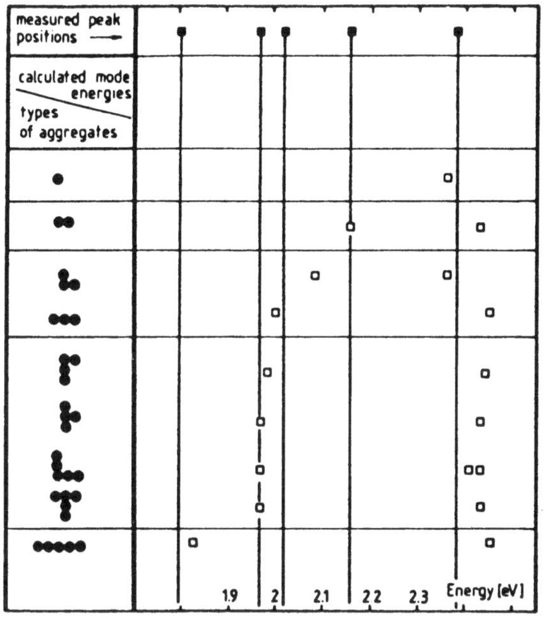

Fig. 12. Numerical comparison of calculated and measured extinction peaks for samples of Au particles ($2R=56$ nm) with small aggregates [6]

If the particle distances $R_{ij} \gg 2R$, then Π^{int} vanishes and (7) equals Mie's result [10].

The numerical evaluation of (7) requires a great deal of computer time. For comparison with the present experimental results, however, it is sufficient to include the full dipolar ($l=p=1$) contributions (i.e. beyond the quasi-static limit). (Computations for quadrupole contributions and for larger aggregates than given here are in progress now.)

Figure 11 gives two series of computed extinction spectra of linear aggregates of (almost) touching Au particles. Obviously, due to the interaction fields, the plasmon polariton peak of the single particle is split into several modes which, in correspondence with our experimental results, cause broad extinction features. This splitting increases strongly with aggregate size, and, beyond the quasistatic limit, with increasing particle size due to inter- and intra-particle retardation effects. These results are discussed in detail in [6]. Here, merely some examples of a quantitative comparison of measured and computed extinction spectra of samples with small aggregates are presented (Fig. 12), which yield assignments for aggregates with N up to five.

The measured values in Fig. 12 are average values taken from several samples compiled in Fig. 10. It is evident that concerning their number and energetic position, the observed peaks are described by the assumed model of plasmon polariton splitting due to electromagnetic interaction fields, i.e. the field correction term $\Delta' E$ of (1) for different aggregates.

It should be pointed out that this collective effect, though strongly influenced by particle size in the size region of larger, bulky particles, does not depend on particle size for small particles ($2R<10$ nm), and hence this also holds for samples of metal clusters, provided that proper individual properties are inserted in (7), and that the clusters can be approximated by spheres.

V. Case C Samples

The systems of this category consist of more or less densely packed single particles plus aggregates with various neighbour distances, coordination numbers, sizes and shapes, which, for their part, again are packed more or less closely together. This is the general - and most realistic - case, as, for instance, shown in Fig. 1.

Now, both field corrections in (1) have to be considered and, hence, case C can be regarded as a combination of cases A and B. (In thick samples, additional collective effects arise from multiple scattering which are disregarded in the present analysis.)

To date no relevant method for separating case A and case B effects in the experimental extinction spectra exists. Consequently, it would not be possible to determine individual particle properties with case C samples, since they may be veiled to a large extent by these collective effects.

Calculations for case C systems have been performed using different approaches [2, 9, 17-19]. A straightforward combination of our cases A and B has been investigated in [2], though only in the quasi-static limit. For this purpose, the effective medium dielectric function of (3) was applied, not using individual particle polarizability but instead by inserting the total polarizability of whole particle aggregates. For example, the polarizability α_{tr} of linear particle triplets, arranged in arbitrary orientation in the sample with filling factor f, was introduced into (3) to obtain the explicit absorption spectra. This α_{tr} is given by

$$\begin{aligned}\alpha_{\text{tr}} = \text{const} \cdot \eta \cdot \Bigg[&\left(1+2\eta\left(\frac{R}{R_w}\right)^3\right) \cdot [((1-Z)(1+\lambda_{+2}\cdot\eta))^{-1} \\ &-((1+Z)(1+\lambda_{-2}\cdot\eta))^{-1}] \\ &+\left(2-2\eta\left(\frac{R}{R_w}\right)^3\right) \cdot [((1-Z)(1+\lambda_{+1}\cdot\eta))^{-1} \\ &-((1+Z)(1+\lambda_{-1}\cdot\eta))^{-1}]\Bigg]\end{aligned}$$

with

$$\eta = (\varepsilon-\varepsilon_m)/(\varepsilon+2\varepsilon_m); \quad Z=\sqrt{257};$$

$$\lambda_{\pm\nu} = \frac{32}{1\mp Z}\cdot F_\nu; \quad F_\nu = \nu\cdot\eta\cdot\left(\frac{R}{R_p}\right)^3 \tag{8}$$

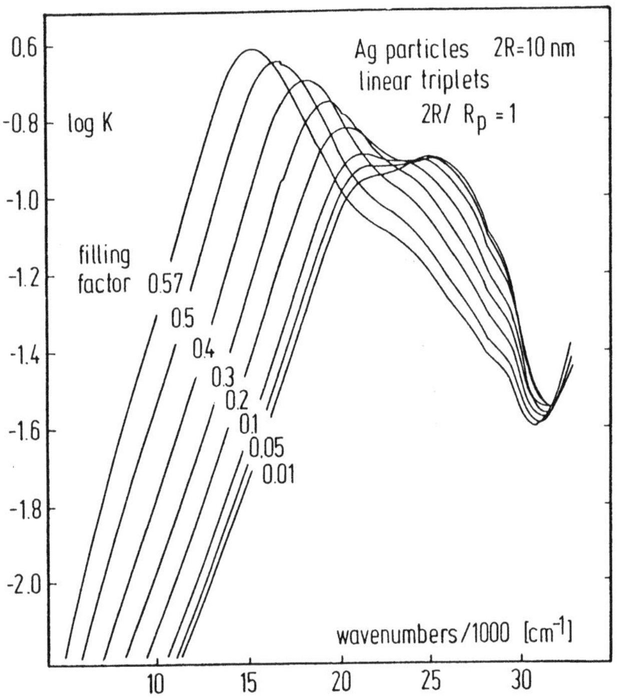

Fig. 13. Calculated absorption spectra of a case C sample (Ag particles, $2R = 10\,\mathrm{nm}$) consisting of linear triplets with varying triplet filling factor f [2]. R_p: particle distance in the triplet

(ε: dielectric function of the particle material; R_p: particle distance).

Some spectra thus computed are shown in Fig. 13. Obviously, again mainly a two-peak spectrum results, as in experimental spectra. Now the relative peak heights and positions are markedly influenced by the packing density of the aggregates, i.e. the triplet filling factor f. In particular, the region of transparency is shifted towards IR when f is increased.

VI. Summary

(1) The optical properties of many-particle systems with dense particle packing and/or aggregation are determined by:

- properties of the individual particle (polarizabilities) which strongly depend on particle size, finite size effects, surface effects and structural "phase" transitions;
- collective properties of the whole particle ensemble which are independent or only weakly dependent on particle sizes but strongly dependent on the details of the local particle arrangement, i.e. the sample topography.

In the framework of Maxwell's theory the collective properties are due to electromagnetic coupling between the particles.

(2) The collective effects can be classified according to effective local field contributions

$$E_{\mathrm{loc}} = E_{\mathrm{Maxwell}} + \Delta E + \Delta' E \qquad (1)$$

- $\Delta E = 0$; $\Delta' E = 0$: Mie case (individual particles)
- $\Delta E \neq 0$; $\Delta' E \approx 0$: effective medium case (plasmon band shifts)
- $\Delta E = 0$; $\Delta' E \neq 0$: well separated particle aggregates (plasmon band splittings)
- $\Delta E \neq 0$; $\Delta' E \neq 0$: general case.

(3) The discussed model calculations for the extinction spectra, (4, 7) and their combination for case C systems, hold for wide ranges of particle size, since, for collective effects, the relevant scale length is the wavelength of light. Therefore, (3) holds for atoms, molecular clusters, solid state clusters and microcrystals as in Table 1. Equation (7) can be applied to solid state clusters, microcrystals and bulk material particles, provided the interaction is restricted to the discussed electrodynamic effects and chemical binding does not occur. (The observed strong size dependences of the optical properties in the small size regions enter via the individual particle properties only.)

(4) To determine optical properties of the single small particle from experimental optical spectra, collective effects have to be detached quantitatively. If, as is most common, the many-particle system topography is ruled by statistical laws, it should be analysed for each sample separately, since the collective effects strongly depend on details of the local particle arrangement which usually cannot be reproduced. Only in the Mie case are collective effects unimportant.

References

1. Kreibig, U., Althoff, A., Pressmann, H.: Surf. Sci. **106**, 308 (1981)
2. Kreibig, U., Genzel, L.: In: Contribution of cluster physics to material science and technology. Proc. ASI Agde/France 1982 Davenas, J., Rabette, R. (eds.), p. 373. Den Haag: Nijhoff 1986
3. Quinten, M., Schönauer, D., Kreibig, U.: Proc. Frühjahrstagung DPG 1986, 988
4. Zsigmondy, R.: Das kolloide Gold. Leipzig: Deuticke 1925
5. Schönauer, D., Kreibig, U.: Surf. Sci. **156**, 100 (1985)
6. Quinten, M., Kreibig, U.: Surf. Sci. (in press)
7. Kreibig, U., Genzel, L.: Surf. Sci. **156**, 678 (1985)
8. Kittel, Ch.: Einführung in die Festkörperphysik. München: Oldenbourg 1973
9. Persson, B., Liebsch, A.: Solid State Commun. **44**, 1637 (1982); Phys. Rev. B**28**, 4247 (1983)
Liebsch, A., Persson, B.: J. Phys. C**16**, 5375 (1983)

10. Mie, G.: Ann. Phys. **25**, 377 (1908)
11. Bruggeman, D.: Ann. Phys. **24**, 636 (1935)
12. Ping Sheng: Phys. Rev. B**22**, 6364 (1980)
13. Niklasson, G., Granqvist, C., Hunderi, O.: Appl. Opt. **20**, 26 (1981)
14. Trinks, W.: Ann. Phys. **22**, 561 (1935)
15. Gerardy, J., Ausloos, M.: Phys. Rev. B**22**, 4950 (1980); B**25**, 4204 (1982); B**27**, 6446 (1983)
Clippe, P., Evrard, E., Lucas, A.: Phys. Rev. B**14**, 1715 (1976)
Ausloos, M., Clippe, P., Lucas, A.: Phys. Rev. B**18**, 1776 (1978)
16. Clanget, R.: Optik **35**, 180 (1972)
Ruppin, R.: Phys. Rev. B**11**, 2871 (1975)
17. Bedeaux, D., Vlieger, J.: Physica **73**, 287 (1974)
18. Felderhof, B., Jones, R.: Z. Phys. B - Condensed Matter **62**, 43 (1985)
19. Geigenmüller, U., Mazur, P.: Physica A (to be published)

U. Kreibig
Fachbereich 11 - Physik
Universität des Saarlandes
D-6600 Saarbrücken
Federal Republic of Germany

Synthesis and Properties of Metal Clusters in Polymeric Matrices

E. Kay
IBM Almaden Research Center San Jose, California, USA

Received July 28, 1986

A one-step plasma deposition process is described which allows the uniform dispersion of small metal clusters throughout a thin film polymer matrix. Plasma parameters and plasma gas phase diagnostics relevant to the control of film composition and structure are discussed. Chemical and structural analytical techniques such as I.R. absorption spectroscopy, E.S.C.A., Auger electron spectroscopy, X-ray fluorescence, X-ray and electron diffraction and microscopy are used to characterize the cluster containing films. Changes in cluster size and shape as a function of volume fraction and as a result of post deposition annealing are described. Optical and electrical properties are presented below and above the onset of percolation and are evaluated in terms of contemporary effective medium theories.

PACS: 78.90+t; 73.90+f; 81.20.Ti; 81.15 Jj

Introduction

The physical and chemical behavior of small metal clusters dispersed in dielectric thin film matrices is of considerable interest both from a fundamental and potential applications point of view. The purpose of this paper is to outline a relatively simple one-step synthetic process whereby a metal can be dispersed uniformly in a polymeric three-dimensional matrix. In addition, optical and electrical properties as a function of metal volume fraction, cluster size and shape will be reviewed and discussed in the context of contemporary effective medium theories. Several applications will be mentioned which take advantage of the fact that the size and shape of the metal clusters can be changed readily by heating the composite thin film after deposition thereby changing, for example, the electrical and optical and thermal properties. The use of a laser beam afford the opportunity of producing these physical property changes in a spatially confined manner which provides possibilities in such varied applications as high density optical storage and circuit writing with predetermined patterns.

Synthesis of Metal Containing Polymer Matrices

There are several examples in the literature which lead to the dispersion of metal clusters in dielectric thin film matrices. Most of these examples resulted in metal containing matrices in which only part, if any, of the metal is present in the metallic, i.e., zero valent, state. In those cases in which metallic cluster dispersions were present, the dielectric matrix was usually a rigid "high temperature" material. Al in Al_2O_3 would be a typical example. The initial purpose of this work was to provide a room temperature synthetic route which produces a chemically inert "low temperature" matrix in which zero valent metal clusters of a given size could be grown from metal atoms during the formation of the composite thin film. This necessitated a choice of materials in which the chemical interaction between the metal and the matrix material is minimal, or alternately stated, a composite system in which the metal-metal interaction is stronger than the metal-matrix interaction. Such a situation is conducive to metal cluster growth provided the metal atoms are sufficiently free to migrate during the synthesis of the composite thin film and can be expected to be self-limiting with respect to cluster size. Choosing a relatively "low temperature" matrix material should also provide the opportunity of allowing the metal clusters formed during film synthesis to migrate after deposition by heating the composite thin film to a higher temperature than that used during synthesis. Such temperature controlled post deposition cluster migration was expected to lead to coalescence of small clusters to form larger clusters at a constant volume fraction. Such a procedure was designed to allow the study of optical and electrical properties of the composite thin film as a function of cluster size and shape and crystallography at a constant volume fraction. Furthermore, a synthetic approach had to be designed in which the volume fraction of metal could be altered at will during synthesis from zero to a hundred percent such that a number of physical properties could be studied below and above the onset of percolation. This approach also affords the study of how the physical properties of the pristine matrix thin film can be modified systematically by the incorporation of metal clusters. Data presented in this paper will demonstrate that depositing gold atoms into a *growing* polymeric thin film fulfills all of the above requirements.

We have considerable experience [1] in the synthesis of uniform polymer thin films by plasma polymerization in an r.f. capacitively coupled diode discharge system. Thin film growth of metals by physical sputtering in a similar discharge configuration is also a well established technique. The procedure outlined here simply provides the conditions under which these two processes can be made compatible with one another such that they can occur simultaneously as a single steady-state deposition process. Sputtered metal atoms arriving at a steady rate at the polymer film surface during polymer deposition assures a uniform composite thin film as a function of thickness.

It is well known that in an r.f. driven capacitively coupled diode plasma system, as schematically shown in Fig. 1, both electrodes are at a negative potential with respect to the plasma and are therefore subject to positive ion bombardment. In the commonly used diode configuration the powered target metal electrode is bombarded by ions of high energy relative to those bombarding the grounded electrode. The substrate is fastened to the grounded, temperature controlled electrode. Alternately the substrate can also be electrically isolated from the grounded electrode and independently powered by an r.f. power source, thereby extracting energetic ions from the plasma over a wide range of energies and bombarding the substrate during film growth.

Injecting either a hydro or a fluorocarbon mono-

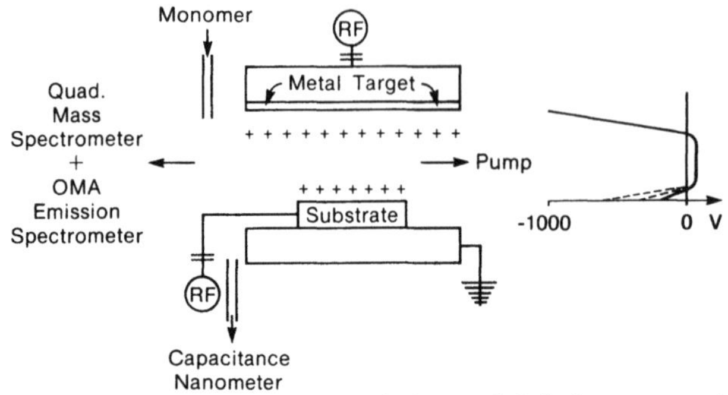

Fig. 1. Schematic of an r.f. capacatively coupled diode system used for metal containing polymer synthesis. Voltage drop between electrodes also indicated

mer into such a diode plasma system will result in electron induced monomer fragmentation and ionization. Different monomers clearly will lead to different fragmentation products [1]. In all cases various ions and radicals will be formed. We have previously shown that, for example, in the case of saturated monomer injection the rate of polymerization on a substrate is proportional to the rate of arrival of unsaturated species of the homologous series: $(CF_2)_n$ [1, 2]. Furthermore, we have shown that the kinetic energy and number of ions bombarding the growing polymer surface also influences the rate of polymerization [3] as well as the structure of the polymer [4].

In a fluorocarbon plasma several competing surface reactions take place simultaneously at all surfaces in contact with the plasma and subject to energetic ion bombardment.

Ion enhanced polymerization:

$$(CF_2)_{n(g)} + \text{Ions} \rightarrow (CF_{2-x})_{n(s)} \downarrow \quad (1)$$

Chemical Sputtering to form volatile species:

$$(CF_{2-x})_{n(s)} + F + \text{Ions} \rightarrow C_nF_{2n+2(g)} \uparrow \quad (2)$$

Physical Sputtering of polymer:

$$(CF_{2-x})_{n(s)} + \text{Ions} \rightarrow C_xF_{y(g)} \uparrow \quad (3)$$

Physical Sputtering of metal:

$$M_{(s)} + \text{Ions} \rightarrow M_{g)} \uparrow \quad (4)$$

If an unsaturated monomer such as tetrafluoroethylene $(CF_2)_2$ is injected into a plasma, then polymerization (reaction 1) dominates because of the abundance of polymer precursors $(CF_2)_2$ and energetic ions. The rate of polymerization increases with ion energy due to the increased production of active surface sites with the result that polymer deposition quickly covers up the powered metal electrode, see Fig. 2. Only be going to very high ion energies can

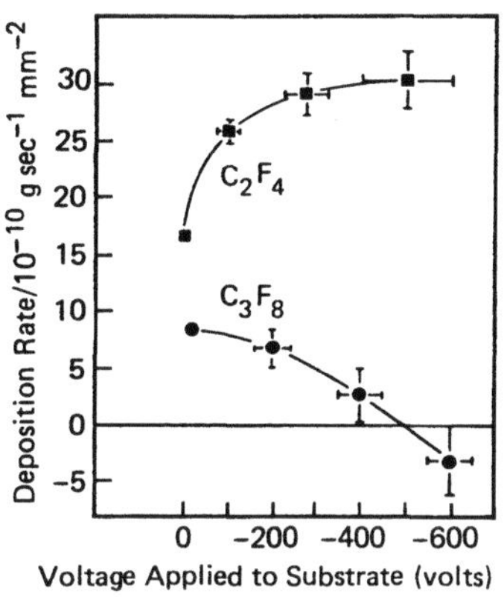

Fig. 2. Polymer Deposition Rate in C_2F_4 and C_3F_8 discharge as a function of ion bombardment energy during film growth

this deposition process (reaction 1) be overcome by physical sputtering (reaction 3).

If a saturated monomer such as C_3F_8 is injected into the plasma electron-impact induced fragmentation leads to: CF_2, and F atoms in about equal amounts and ions (primarily CF_3^+).

Figure 2 clearly indicates that in a C_3F_8 plasma the polymer deposition rate decreases with increasing ion energy and at high enough energies removal of electrode material (metal) takes over. This is a simple reflection of the fact that material removal by reactions 2, 3 and 4 become more and more dominant as the ion energy increases. This is then the condition necessary to keep the powered metal electrode free of polymer deposits, thereby allowing removal of the metal. On the other hand at the grounded substrate electrode surface ($\sim$40 V negative with respect to the plasma potential, see Fig. 1) ion enhanced polymer deposition by reaction 1 dominates.

By adding various partial pressures of an inert gas (e.g. Ar) to the plasma the polymerization reaction can be diluted, whereas the physical sputtering of electrode material due to Ar^+ is enhanced thereby providing systematic control over the ratio of metal sputtered atoms arriving at the substrate from the powered metal target electrode to the polymer precursor species CF_2 arriving at the grounded substrate [5].

Figure 3 clearly demonstrates that monitoring the ratio of metal atoms versus polymer precursors: CF_2 arriving at the substrate surface by fibre optics emission spectroscopy correlates well with the ratio of metal/polymer found in the condensed film as examined by E.S.C.A. Figure 3 further demonstrates that the polymer formation decreases dramatically with increasing substrate temperature which is due to a sharp drop of the sticking probability of the polymer precursor species $(:CF_2)_n$ arriving at the substrate surface as the temperature rises [5].

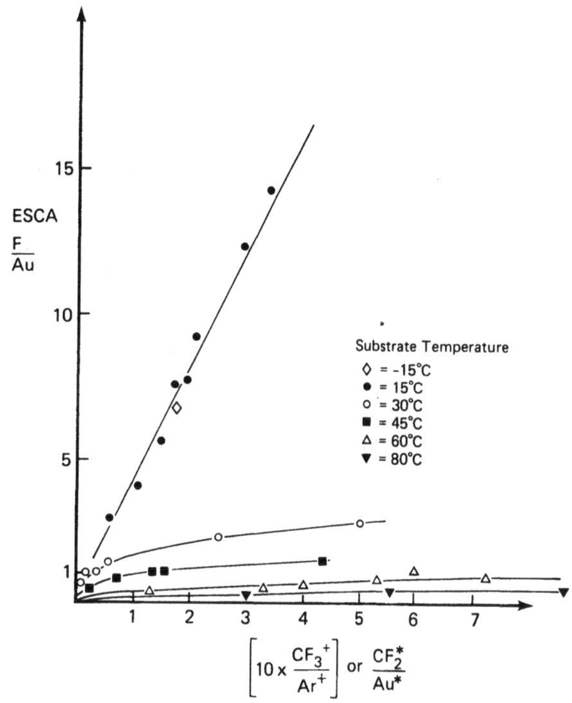

Fig. 3. Key gas species, CF_2 and Au, arriving at substrate surface versus the Au and F found in polymer film at different substrate temperatures

It is therefore clear that either control of the partial pressures of Ar to monomer injected into the plasma or changing the substrate temperature can be used to adjust the metal content of the resultant polymer film.

Structure of Metal Cluster Containing Polymer

Figure 4 demonstrates that the structure and C/F ratio as well as the deposition rate, see Fig. 2, of the polymer matrix can be manipulated by changing the energy of ions bombarding the electrodes. Since the bias voltage imposed on the substrate can in fact be controlled by an r.f. power supply attached to the substrate configuration, it is possible to manipulate the net rate of polymer deposition at the substrate somewhat independently from the reactions going on the higher energy regime at the opposite target electrode. This independent bombardment control also affords the ability to manipulate polymer matrix structure as well as the rate of polymerization of the resultant metal containing films.

The size of the metal clusters can be controlled by one of two procedures. Substrate temperature during deposition or post deposition annealing leading to the formation of larger clusters by cluster coalescence.

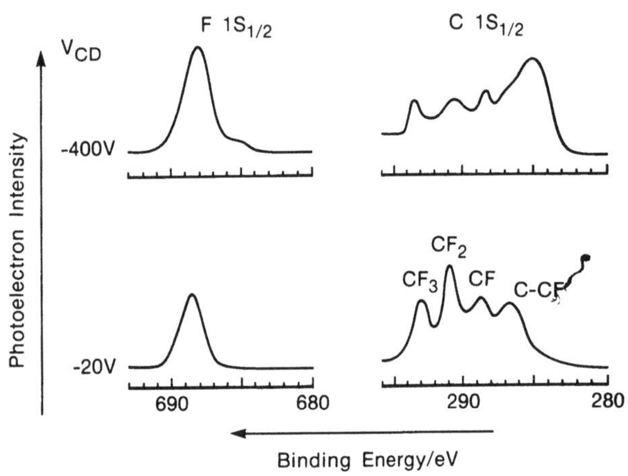

Fig. 4. Effect of ion bombardment on polymer structure during film growth as determined by E.S.C.A.

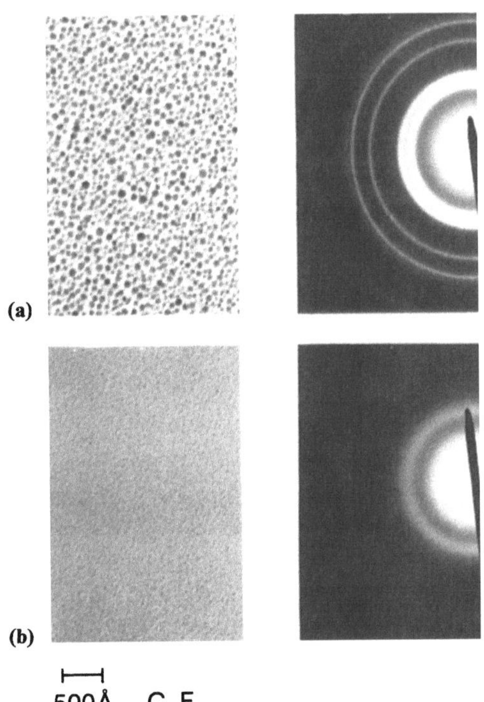

Fig. 5a and b. Transmission electron micrographs of Au clusters embedded in polymer films prepared at different substrate temperatures: **a**: 60° C; **b**: −10° C

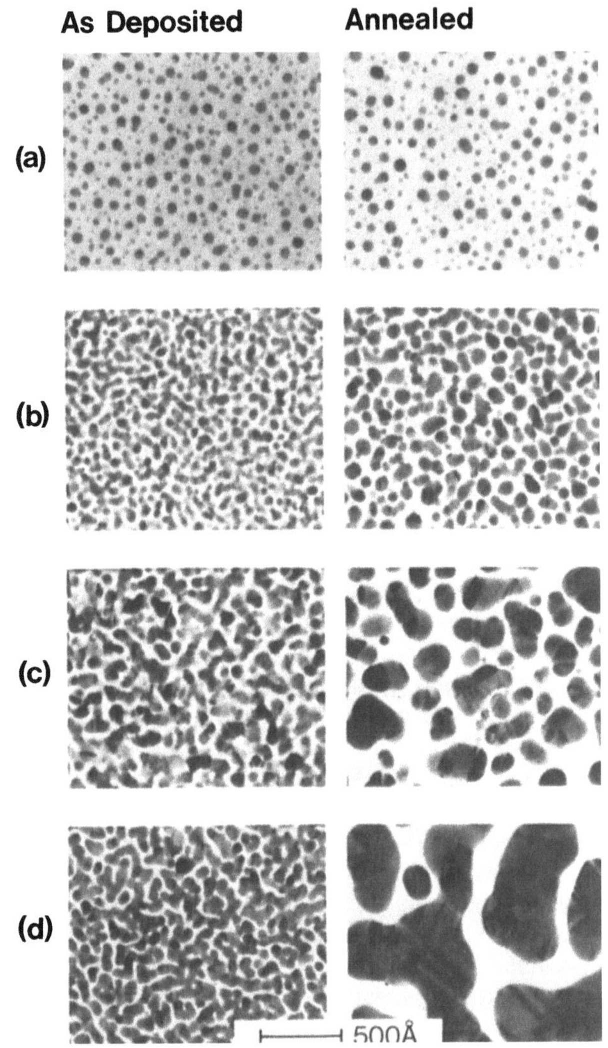

Fig. 6a–d. Transmission electron micrographs of as-deposited and post-deposition annealed Au in polymer matrices for different Au volume fractions p: **a** $p=0.15$, **b** p 0.30, **c** $p=0.55$, **d** $p=0.80$

Comparing Fig. 5 (a with b) shows the difference in cluster size due to different deposition temperatures for approximately similar metal volume fractions. Figure 6 shows the effect of post deposition annealing for different volume fractions of Au in the polymer [6, 7]. The drastic change of cluster size and shape in going from low to high metal volume fractions is interpreted along the following lines:

i. At low metal volume fraction, the initial limiting step is the motion of spherical particles in the matrix before entering into contact. For touching spherical particles of identical size, the time for sintering or neck formation is given by the relation [8, 9]:

$$t=\frac{0.032\kappa T}{\gamma_s \Omega \Delta_s D_s} x \alpha^4$$

where T is the absolute temperature γ_s the free surface tension, Ω the atomic volume Δ_s the thickness of diffusion, D_s the surface self-diffusion coefficient and α the sphere radius. Using the following data: [6] $\gamma_s-1{,}200$ erg.cm^{-2}, [6a] $\Omega=1.7\times10^{-23}$ cm^{-3}, $\Delta_s=3\times10^{-8}$ cm, and $D_s=7.6\exp(-7{,}880/T)$ in cm^2 s^{-1} in this temperature domain, the total time for spheroidization of two touching spheres of radius $\alpha=25$ Å at 180° C is about 1 μs which is much shorter than our annealing time of one hour.

ii. At higher gold volume fraction the sintering mechanism is more complicated due to the non spherical shapes of gold clusters. However, since the clusters are already interconnected, there is no initial limiting step. The average radius of cluster branches can increase by successive sintering processes, but the time of sintering increases rapidly (as α^4). Moreover, the cluster shapes are modified: wormlike clusters close to the percolation threshold will tend to spheroidize as demonstrated for the ideal case of a cylindrical rod [10]. The consequence is an increase of the intercluster distance and further sintering is limited, see Fig. 6b and c.

X-ray and electron diffraction of these Au clusters indicate that Au crystallites are formed at room temperature with pronounced texturing in the 111 direction. Upon annealing up to approximately 200° C the texturing diminishes as these small Au clusters migrate through the polymer matrix and coalesce to larger aggregates of twinned crystal clusters. Electron micrograph of the as-deposited cross sectioned films, see Fig. 7, showed the clusters to be distributed quite uniformly throughout the thickness of the film.

Differential thermal analysis of our highly rosslinked plasma polymerized matrix material showed exotherms around 160° C whose characteristics were consistent with a second order phase transition associated with well known glass to rubber transition in these polymeric materials. The migration of small gold clusters below this transition temperature is minimal. However, just above this glass to rubber transition the Au grains migrate and subsequently coalesce as reflected by dramatic changes in optical properties of the composite thin films. The next section will outline the highlights of our studies on electrical and optical properties as a function of cluster size, shape and volume fraction.

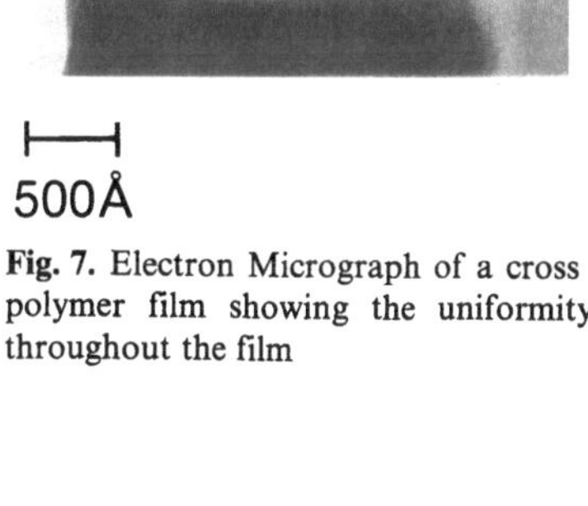

Fig. 7. Electron Micrograph of a cross sectioned metal containing polymer film showing the uniformity of the metal dispersion throughout the film

Transport Properties

A. D.C. *and* A.C. *Electrical Resistivity in the Dielectric Regime.* We consider first the electrical properties and conduction mechanisms in plasma-polymerized films containing a sufficiently low metal volume fraction so that metallic clusters remain isolated, having generally spherical shapes and small diameters (10 to 100 Å). In this composition range, composite films are essentially insulating but the presence of metallic clusters introduces additional conduction mechanisms in comparison to metal-free polymers.

A detailed understanding of the d.c. conduction processes prevailing in metal-free polymers and in metal-containing polymers can be obtained from the analysis of the temperature dependence of the d.c. resistivity ρ. For a wide variety of plasma-polymerized films, some of them synthesized with complex organometallic monomers, the variation of the d.c. resistivity between room temperature and 250° C has been measured [11]. The main conclusion is that ρ is a decreasing function of the temperature and when $\ln \rho$ is plotted versus $1/T$ the data points fall generally on a straight line which corresponds to a thermally activated conduction process. The activation energy E_1 for most metal free plasma polymers is usually greater than 1.1 eV with an average value of 1.36 eV for fifteen different monomers except acetylene [11]. This value corresponds to the energy required to create charge carriers in a structure consisting principally of carbon-carbon bonds. Moreover the observation of a thermally activated photoconductivity in these polymers, with an apparent activation energy of the order of 0.2 eV [11] suggests that the carrier generating mechanism does not require a band transition but rather a molecular excitation to an intermediate non-conducting state from which there is further thermal activation to produce free carriers. In metal-containing polymers where the metal is incorporated as small isolated clusters or possibly as single atoms inside the polymeric structure [11], the thermally activated d.c. conduction process is still observed above room temperature, but E is lower than in metal-free polymers. Different models have been proposed for the temperature dependence of tunneling or hopping conductivity between isolated conducting grains in an insulating matrix [12–17].

Further experimental work is needed to decide

which one of these tunneling mechanisms is the most relevant to interpret d.c. conduction in metal-containing plasma polymers. In addition to the temperature dependence of the resistivity, the study of non-ohmic behavior under high-electric field would be particularly helpful to distinguish between field-induced tunneling [12] for which a $\ln \rho \sim 1/E$ behavior is predicted and the high field limit of fluctuation-induced tunneling [16].

Turning now to a.c. conduction processes, the case of metal-free polymers is first considered. The relaxations observed in dielectric loss measurements as a function of temperature at several frequencies [18, 19] or as a function of frequency at different temperatures [20] in various plasma polymerized hydro- and fluorocarbons, were interpreted as being due to dipole orientation of CO groups, accompanying molecular segment motions. CO moities in these films is the end result of post deposition oxidation of carbon free radicals inevitably built into a plasma polymerized hydro- or fluorocarbon [20].

When metal clusters are incorporated in the polymer, hopping (equivalent to tunneling) between metallic grains rapidly dominates over any other conduction or relaxation processes inherent to the polymeric phase. The frequency dependence of the resistivity is characterized by a plateau at low frequency followed by a decrease which can be approximated by a power-law dependence. The onset of the decrease is shifted to higher frequency as the metal volume fraction increases. Such a behavior can be interpreted along different lines:

i) The tunneling conduction between metallic grains might involve the contribution of a "phonon-assisted" hopping mechanism where the "phonons" would be excited by the a.c. field, this contribution being masked at low frequency depending on the value of the d.c. tunneling conductivity.

ii) The apparent increase of a.c. conduction might arise simply from the contribution of the displacement current due to the polarizability of the insulating phase between metallic grains, even if the tunneling conductivity is frequency independent. We have shown that the second hypothesis which can be tested by an effective medium theory, can be ruled out over the domain of frequency of the available experimental data.

Metallic Conductivity Above the Percolation Threshold

As the metal volume fraction increases, the size and shape of metallic clusters change from isolated spherical units to wormlike isolated units until the point where a percolating "cluster" is established in the film, see Fig. 6. At this critical metal volume fraction p_c, called the percolation threshold, metallic conduction through the percolating cluster competes with tunneling conduction between non-connected branches. In this transition regime, a small increase of p above p_c results in a drastic drop of the resistivity by many orders of magnitude until a pure metallic regime is reached. This behavior is commonly observed in cermets [12], carbon-polymer composites [14], and now in metal polymer films. This is illustrated for example in Fig. 8, where the variation of the d.c. resistivity ρ of gold-plasma polymerized PTFE films, as a function of p is represented for both as-deposited and annealed films [6]. The transition regime occurs at $p_c \approx 0.37 \pm 0.03$ in as-deposited films but is shifted to $p_c = 0.42 \pm 0.03$ after annealing. This increase in p_c is unambiguous since films deposited just at p_c become more resistive after annealing whereas films slightly above p_c become more conducting as indicated by the arrows in Figure 8. Moreover the $\rho(p)$ curve above p_c is also modified by annealing. Both observations show that the electrical properties of such composite materials are determined not only by the metal volume fraction but also by the cluster microstructure which is drastically modified by annealing, see Fig. 6. In fact, the position of the percolation threshold in experimental systems depends strongly on the choice of the metal and the insulator, on the method of synthesis of the composite and on post synthesis treatments. Values of p_c reported in the literature vary from 0.17 for metal-ammonia solutions

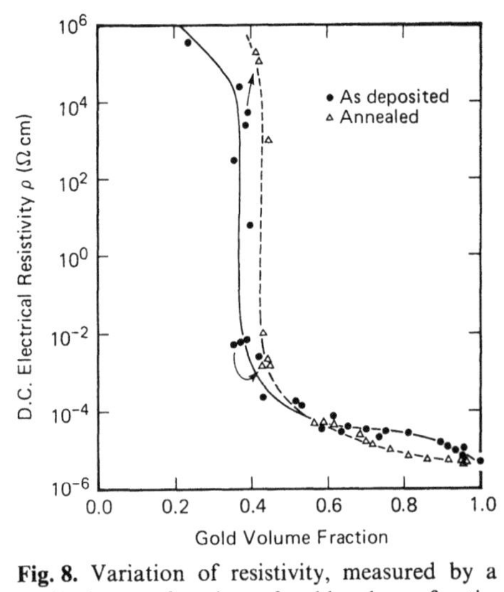

Fig. 8. Variation of resistivity, measured by a four-point probe method, as a function of gold volume fraction for as-deposited and post-deposition annealed films. The arrows indicate how the resistivity of films close to the percolation threshold change upon annealing

[21] up to 0.72 for Al/AlN cermets [21 a] and annealing effects have been observed in W/Al_2O_3 cermets when heated at 1,200° C [12]. To our knowledge no detailed study of percolation conductivity just above p_c has been reported for metal-plasma polymer composites.

We will show that effective medium theories incorporating microstructural information are able to account for the variations in p_c and $\rho(p)$ curves as shown in Fig. 8.

Another question arising when studying thin films is the thickness dependence of electrical conductivity in the planar configuration. Measurements of the planar resistance of gold-plasma polymerized PTFE films have been made *in-situ* during the deposition process [6]. The thickness dependence of the resistivity is represented in Fig. 9 for various metal contents. The initial decrease of ρ is associated with the growth of the first layer of gold clusters via nucleation and coalescence. This situation, which is analogous to the early growth stages of a pure metallic film, has received full treatment based on a computer model of the evolution of the film resistance as a function of coverage [22]. However, in the present case, the metal cluster growth is constrained by the codeposition of the polymer. The lateral extension of the first clusters determines the 2D resistivity of the percolating films, but further film growth progressively opens 3D conducting channels. The transition from 2D to 3D percolation has been analyzed both theoretically and experimentally in the case of a mixture of conducting and insulating spheres [23]. Very thin Pb–Ge composite films have been shown to exhibit a 2D percolation behavior characterized by a conductivity critical exponent t_c of about 1.1 instead of 1.7 for 3D [24]. In our films, the 3D behavior is reached asymptotically and ρ is almost constant above 2,000 Å within experimental precision.

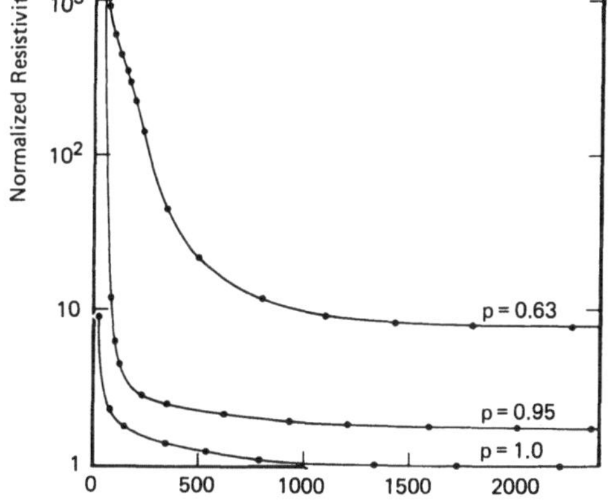

Fig. 9. Thickness dependence of the resistivity during film growth. The results have been obtained by multiplying the in-situ resistance measurements by the measured thickness. The normalization constant is the value obtained for a thick sputtered gold film

Optical Properties

The optical properties of discontinuous metallic films or granular composite films, consisting of metal clusters embedded in a dielectric matrix, have been of interest since the beginning of the century when Maxwell-Garnett [25] gave the first theoretical framework to explain the resonant absorption or "dieletric anomaly" which characterize such systems. When observed in transmission, these films have a different color than thin semitransparent continuous metallic films. For example, gold composite or discontinuous films at relatively low metal volume fraction have a pink color, whereas semi-transparent continuous gold films are green. Also in contrast to continuous films, granular metal films have poor reflectivity.

There have been numerous investigations on Cu, Ag and Au discontinuous films [26], Au or Ag colloidal particles [27, 28] or cermet films such as $Ag-SiO_2$ or $Au-SiO_2$ [29] to verify the applicability of various theoretical models. Concerning metal plasma polymer composites, several authors have reported optical transmittance measurements [20, 31, 5, 32]. The most thorough study, up to now, is our work [7] for gold-fluorocarbon polymer films. The transmittance curves for three samples, having as deposited gold volume fractions $p = 0.17$, 0.52 and 0.91 are shown in Fig. 10a, b, c, respectively. At low p, see Fig. 10a, the spectrum exhibits an absorption peak, the so-called "dielectric anomaly" characteristic of isolated metallic particles embedded in a dielectric and no infrared absorption. In high p, see Figure 10c, as-deposited films, this resonance is still present. Although shifted to the red, where the IR absorption tends to reach pure gold behavior. A systematic red shift of the absorption maximum of the dielectric anomaly for similar films was reported earlier by Kay and Hecq [5]. After annealing, several effects take place: *i.* a strong enhancement of the resonance at low p, see Fig. 10a, characterized by an increased ab-

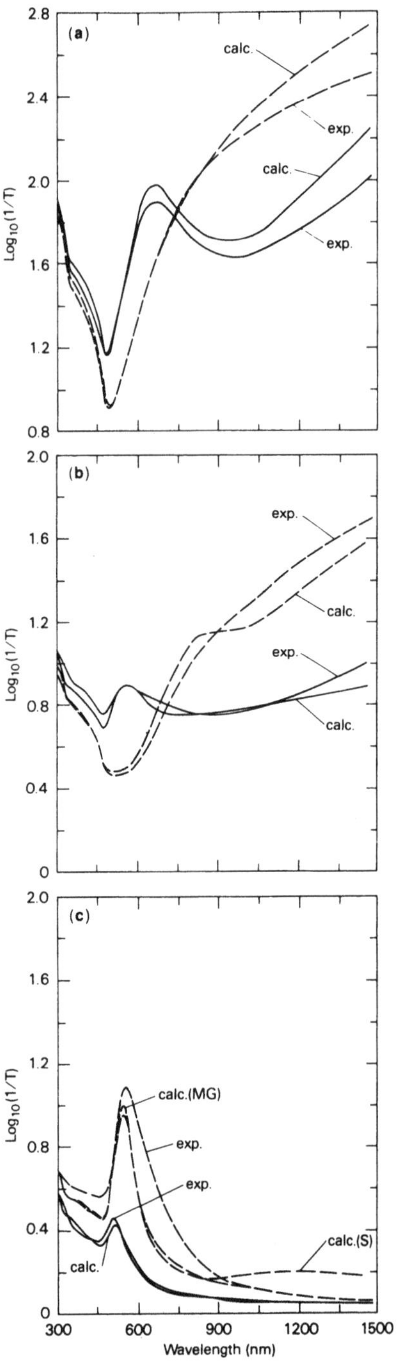

Fig. 10a–c. Experimental and calculated optical transmittance spectra of as-deposited (solid curve) and annealed (dashed curve) films at different volume fractions: **a** $p=0.17$, **b** $p=0.52$, **c** $p=0.91$

sorption ratio between the maximum and the UV region; *ii.* the disappearance of the resonance at high p, see Fig. 10c; *iii.* an increase of IR absorption at intermediate p, see Fig. 10b.

These effects have to take into account the microstructural modification of the films, both in cluster

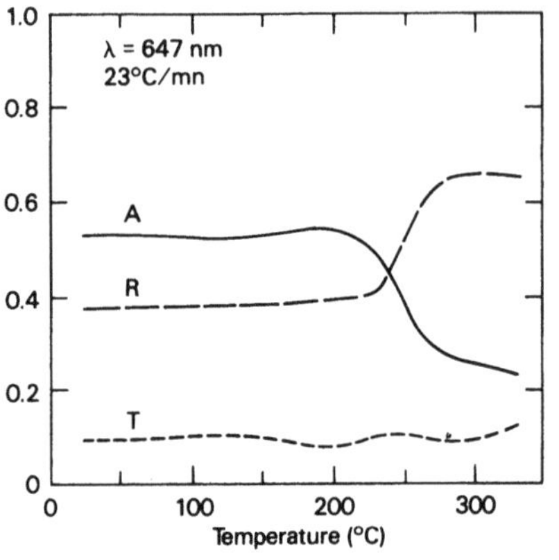

Fig. 11. Variation of transmittance (T), reflectance (R) and (word) $A=1-T-R$ of a gold cluster containing polymer films as a function of annealing temperature at wavelength λ-647 nm and $p=0.56$

shape and crystal grain size, as observed by TEM analysis, see Fig. 6. In this respect, the use of effective medium theories provides an enlightening interpretation as shown in a later section. To illustrate the correlation between change in optical properties and structural modification in the film, we show in Fig. 11 the variations in transmittance, reflectance and absorbance at 647 nm of a film of intermediate p when submitted to a fast heating ramp. The drastic inversion between reflectance and absorbance, starting at about 200° C, reflects the migration and coalescence of gold clusters. Due to the fast heating rate, this critical event is delayed to a higher temperature than the glass transition temperature, $T_g \simeq 160°$ C.

Interpretation by Effective Medium Theories

The optical properties of discontinuous metallic films or granular composite films, consisting of metal clusters embedded in a dielectric, have been of interest since the beginning of the century when Maxwell-Garnett [25] described the first theoretical framework to explain the resonant absorption of dielectric anomaly which characterizes such systems. In this theory (referred to hereafter as MG theory), the dielectric function of the composite medium is related to the polarizability of a single metallic particle surrounded by the dielectric via the Clausius-Mossotti (CM) equation. Although this theory correctly predicts the position of the anomalous absorption peak in the visible region at low metal volume fraction and for small spherical particles, it fails to describe the evolution of the optical-absorption spectrum at high metal volume fraction when the metal clusters are interconnected and no longer spherical. This limitation comes essentially from the fact that the MG theory does not treat the metal and the dielectric phase on a symmetrical basis as a function of volume fractional changes. For this reason, no percolation threshold is predicted in the electrical conductivity. In spite of these intrinsic limitations, the MG theory has been refined to take into account the effect of particle shape in the form of ellipsoids, [27, 29] the effect of multipole interactions between spherical particles [33–36] and more recently the effect of topological disorder [37]. The particular case of the two-dimensional island-film structure has also received attention by solving the Clausius-Mossotti problem for a two-dimensional lattice [38, 39]. The effect of oxide coatings or adsorbed gases on the metallic particles [40] has also been considered, as well as modified substrate dielectric properties for discontinuous metallic films.

Instead of using the Clausius-Mossotti approach to compute the dielectric constant of a composite medium, Bruggeman [41] initiated the effective-medium-theory (EMT) formalism. The EMT approach, which has been generalized by Stroud [42] involves two steps:

i. The definition of microscopic structural units, composed of the known reference materials involved in the composite medium. These structural units must reflect the microstructure of the medium by their shapes. They are supposed to be embedded in a host material which is the so-called effective medium (EM). The expressions of the depolarization electric field induced in the EM by each unit are then derived by solving the Poisson equation with adequate boundary conditions.

ii. The derivation of the EM dielectric constant includes the self-consistency conditions which require that the macroscopically averaged depolarization field for randomly distributed structural units vanish.

In the Bruggeman model, the structural units are spheres, or more generally, ellipsoids of each component and their probability of presence in the EM is simply their respective volume fractions. In the case of a metal-dielectric composite medium, this model does predict a percolation threshold for the electrical conductivity, but does not describe correctly the optical dielectric anomaly predicted by the MG theory. Even with some refinements such as the introduction of effective depolarization factors [43] or retardation effects, [44] the Bruggeman model remains inadequate for the description of optical properties.

A better understanding of the reasons which do make the MG theory successful in predicting the dielectric anomaly has come from the realization that the MG equation can be derived from an EMT model in which one structural unit is made to consist of metal coated with dielectric and embedded in the EM [45, 46]. Sheng showed, by using a pair-cluster EMT model, [46] that the dielectric anomaly is due to the short-range interaction between nearest-neighbor grains of metal and dielectric. Both the pair cluster and the MG models take into account explicitly the metal-dielectric interface at a microstructural level, whereas the Bruggeman model considers only metal-EM and dielectric-EM interfaces.

Another EMT model has been proposed by Sheng [47] to incorporate the advantages of both the MG and Bruggeman models. Two MG types of structural units consisting, respectively, of metal coated with dielectric and dielectric coated with metal are considered. In each unit, the volumes occupied by each component are proportional to their volume fractions. The probabilities of occurrence of these units have been related to a single growth model of cermet films where diffusion and coalescence of each phase take place.

a. Metallic Conductivity. The Sheng model is applied to the computation of the d.c. conductivity σ as a function of gold volume fraction p, normalized to the pure gold conductivity, by setting $\sigma(p=0)=0$ for the polymer (tunneling conductivity is neglected) and $\sigma(p=1)=1$ for gold. Experimental data of resistivities given in Fig. 8 have been converted into normalized conductivities by using the resistivity of pure sputtered gold films as a reference. The results for as-deposited and annealed films are displayed in Fig. 12. In the computation of $\sigma(p)$ the values of the structural parameters: γ_m for polymer coated gold unit, and γ_i for gold coated polymer units have to be related to the actual film microstructure. TEM pictures of polymer rich films reflect the shape of polymer coated gold units whereas pictures of gold rich films reflect the shape of gold coated polymer units. From Fig. 6a, it can be deduced that γ_m has to be 1 for both as-deposited and annealed films. The choice of γ_1 from Fig. 6d is not straightforward, but the image of a polymer phase wrapping the branches of gold cluster branches can be approximated by considering flat platelets of polymer randomly oriented. Consequently, γ_1 should be $\ll 1$. Indeed, the better agreements between experimental and computed ($\sigma(p)$) curves are obtained with $\gamma_i=0.02$ for as-deposited film and $\gamma_i=0.08$ for annealed films as shown in Fig. 12, whereas $\gamma_i=1$ (spherical or prolate shapes) do not give satisfactory results. Cross sections of the structural units deduced from this model are shown in Fig. 13 for $p=0.5$. The reduction of oblateness of the gold coated polymer unit upon annealing is con-

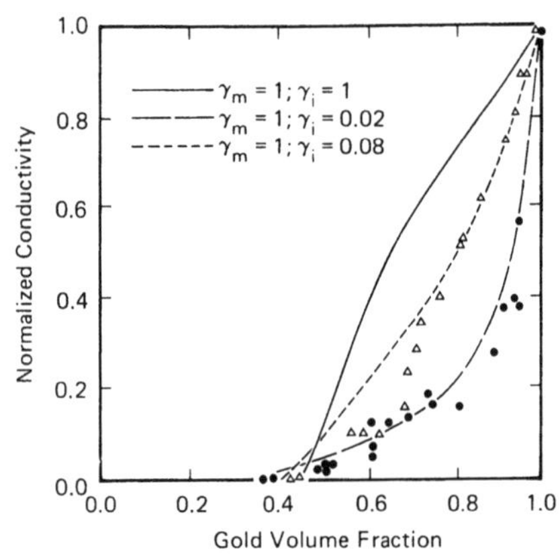

Fig. 12. Normalized conductivity as a function of gold volume fraction for as deposited and annealed δ samples. Solid curves are computed according to Sheng EMT and are labeled by the values of the two parameters γ_m and γ_i indicating the oblateness of polymer coated gold and gold coated polymer inclusions respectively

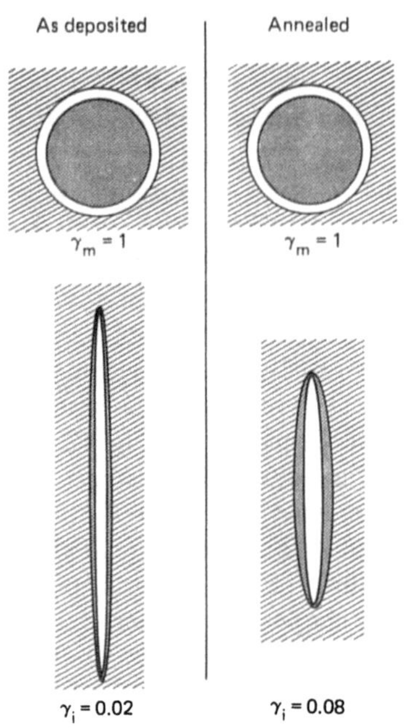

Fig. 13. Cross sections of structural units embedded in the effective medium considered in the application of the Sheng EMT to gold/polymer films at $p=0.5$. The polymer coated gold particles are spherical for as-desposited and annealed films. The gold coated polymer units are on the other hand very oblate spheriods assumed to be randomly oriented. Annealing corresponds to a reduction of the oblateness of these units

sistent with the TEM observations and account for the upward shift of the percolation threshold.

By applying his theory to the case of W/Al_2O_3 cermet films, Sheng also concludes that a reduction of oblateness of metal coated insulator units takes place upon annealing [47]. The corresponding γ_1 values of 0.035 and 1 are, however, higher than for our films which indicates that Al_2O_3 inclusions in tungsten tend to spheroidize more easily than polymer in gold. This observation can be related to a difference of metal/insulator interfacial surface tension between the two phases [6]. In the gold/polymer system the free surface tensions of the two phases are very different: $\gamma_s(Au)=1{,}200$ and $\gamma_s(CF_x)=18$ erg. cm^{-2}. Consequently, a gold particle in polymer will spheroidize to minimize its energy, but on the contrary an inclusion of polymer in gold will "wet" the maximum area. In the W/Al_2O_3 system the free surface tensions of the two phases measured at their respective melting points are almost identical: $\gamma_s(W)=550$ and $\gamma_s(Al_2O_3)=570$. It is therefore expected that in annealed films the two phases have symmetrical behavior as far as the microstructure is concerned. Indeed the application of the Sheng model concludes that $\gamma_i=\gamma_m=1$ in this case.

b. Optical Properties. The application of EMT to the propagation of an electromagnetic wave in a composite medium is legitimate provided that the typical scale of heterogeneities of radius r, is smaller than the probing wavelength λ. Niklasson et al. [48] have shown that the Maxwell-Garnett and Bruggeman models, which retain only the first term in the Mie expansion series of the scattered depolarization field are correct to a precision better than 10% if $(r/\lambda)<0.03$. Such a condition is largely fulfilled in as-deposited gold-fluorocarbon films over the entire range of p, where the scale of inhomogeneities is less than 100 Å, as shown in the TEM pictures of Fig. 6. In the case of annealed films, this condition is more critical at high p. Another requirement is that the complex dielectric constants of the reference phases (gold and polymer) are correctly defined, which implies taking into account the possible effect of tunneling through polymer and the effect of finite metallic grain size. The former effect should be negligible at optical frequencies, following the same arguments as Cohen et al. [29] used for the case of Ag- or $Au-SiO_2$ cermets. The latter effect is, however, of considerable importance and has been taken into account [7].

The agreement between simulated and experimental spectra is in general satisfactory but some limits of the model have to be pointed out:

i. The Sheng model at $p\sim0.3$ for the annealed sample optical spectrum in Fig. 10a tends to give too much emphasis to the gold-coated polymer units which contribute to the IR absorption. This relates to the choice of the probability law, $l(p)$ used in the model. For comparison the computed spectrum according to the Maxwell-Garnet model at the same volume fraction is also represented in Fig. 10a.

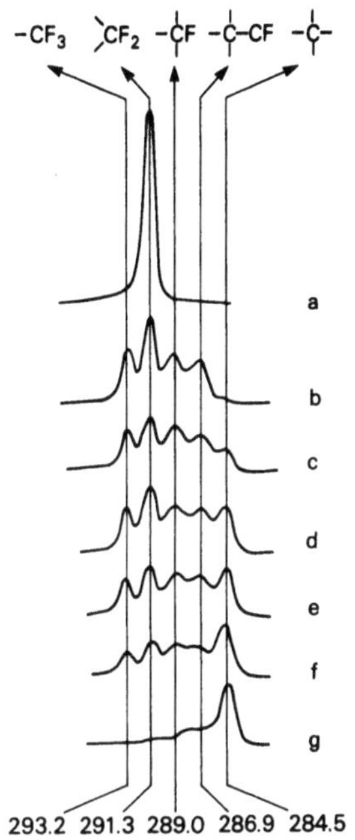

Fig. 14. Surface photo electron E.S.C.A. spectra of $C(ls)$: **a**: conventional polytetrafluorethylene, **b**: plasma polymerized PTFE, **c**: gold containing polymer $p=0.1$, **d**: $p=0.2$, **e**: $p=0.4$; **f**: $p=0.5$ and **g**: $p=0.75$

ii. The experimental resonance peak at low p is much more broadened than the Calculated one and no parameter adjustment can reduce this discrepancy. However, we have shown that the model of Liebsch and Persson [37] which treats the problem of a topological disorder of metal clusters along a Clausius-Mossotti approach may partially account for the observed resonance broadening [7].

We have also tested a three-phase Sheng model assuming the presence of a carbon interface between gold and polymer in order to check whether the evolution of the ESCA C_{1s}, as the gold volume fraction incrases in the film, see Fig. 14, might be attributed to a bulk effect instead of a simple surface contamination. Clearly the bulk hypothesis can be ruled out since no agreement is obtained between experimental and computed spectra using the three-phase model [7].

In order to simulate the experimental optical transmittance spectra shown in Figure 10, a computer program was written [7]. We have shown [7] that the drastic inversion between reflectance and absorbance at $\lambda=647$ nm observed on annealing a film at intermediate p, see Fig. 11, is reasonably described by the Sheng model assuming that p increases due to film contraction, $\varkappa$ increases from 30 to 110 Å and γ_i increases from 0.045 to 0.35.

References

1. Kay, E., Dilks, A.: Critical Review, J. Vac. Sci. Technol. **18(1)**, 1 (1981)
2. Millard, M.M., Kay, E.: J. Electrochem. Soc. **129**, 160 (1982)
3. Kay, E., Dilks, A.: Thin Solid Films **78**, 309 (1981)
4. Kay, E.: Erosion and Growth of Solids Stimulated by Atom and Ion Beams. Kiriakidis, G., Carter, G., Whitton, J.L.: (eds.) NATO ASI Series, Series E: Appl. Sci **112**, 247–274 (1986), Martinus Nijhoff Publishers, Dordrecht, Boston, Lancaster, (published in cooperation with Nato Scientific Affairs Division)
5. Kay, E., Hecq, M.: J. Appl. Phys. **55**, 370 (1984)
6. Perrin, J., Despax, B., Hanchett, V., Kay, E.: J. Vac. Sci. Technol. **A4(1)**, 46 (1986)

6a. Dilks, A., Kay, E.: Macromolecules **14**, 855 (1981)

7. Perrin, J., Despax, B., Kay, E.: Phys. Rev. **B32:2**, 719 (1985)
8. Nichols, F.A.: J. Appl. Phys. **37**, 2805 (1966)
9. Blachere, J.R., Sedeki, A., Meiksin, Z.H.: J Mater. Sci. **19**, 1202 (1984)
10. Nichols, F.A., Mullins, W.W.: J. Appl. Phys. **36**, 1826 (1965)
11. Bradley, A., Hammes, J.P.: J. Electrochem. Soc. **110**, 15 (1963) **110**, 543 (1963)
12. Abeles, B., Sheng, P., Coutts, M.D., Aire, Y.: Adv. Phys. **24**, 407 (1975); Sheng, P., Abeles, B., Arie, Y.: Phys. Rev. Lett. **31**, 44 (1973)
13. Simanek, E.: Solid State Commun. **40**, 1021 (1981)
14. Efros, A.L., Schlovskii, B.I.: J. Phys. C: Solid State Phys. **8**, L49 (1975)
15. Entin-Wohlman, O., Geten, Y., Shapira, Y.: J. Phys. C: Solid State Phys. **16**, 116 (1983)
16. Sheng, P.: Phys. Rev. **B21**, 2180 (1980); Silhel, E., Gittleman, J.I., Sheng, P.: J. Electronic Mat. **11**, 4 (1982)
17. Devenyi, A., Manaik-Devenyi, R., Hill, R.M.: Phys. Rev. Lett. **29**, 1738 (1972)
18. Tibbitt, J.M., Shen, M., Bell, A.T.: J. Macromal. Sci. Chem. **A10**, 519 (1976)
19. Perrin, J., Kay, E., Siemens, R.: J. Macromol. Sci. (in press)
20. Hetzler, U., Kay, E.: J. Appl. Phys. **49**, 5617 (1978)
21. Jortner, J., Cohen, M.H.: Phys. Rev. **B13**, 1548 (1976)

21a. Affinito, J., Fortier, N., Parsons, R.R.: J. Vac. Sci. Technol. **A2**, 316 (1984)

22. Meiksin, Z.H., Sedehi, A.,Blachere, J.R.: Thin Solid Films **115**, 33 (1984)
23. Clerc, J.P., Giraud, G., Alexander, S., Guyon, E.: Phys. Rev. **B22**, 2489 (1980)
24. Deutscher, G., Rappaport, M.L.: J. Phys. Lett. (Paris) **40**, L219 (1979)
25. Maxwell-Garnett, J.C.: Philos. Trans. R. Soc. London **203**, 385 (1904); 205, 237 (1906)
26. Norrmann, S., Anderson, T., Granquist, C.G., Hunderi, O.: Phys. Rev. **B18** (1978)
27. Doremus, R.H.: J. Chem. Phys. **42**, 414 (1965)
28. Kreibig, U., Althoff, A., Pressman, H.: Surf. Sci. **106**, 308 (1981)
29. Cohen, R.W., Cody, G.D., Coutts, M.D., Abeles, B.: Phys. Rev. **B8**, 3689 (1973)
30. Wielanski, R.F., Beale, H.A.: Thin Solid Films **84**, 425 (1981)
31. Asano, Y.: Thin Solid Films **105**, 1 (1983)
32. Biedermann, H.: Vacuum (GB) **34**, 405 (1984); Biedermann, H., Holland, L.: Nucl. Instrum. Methods **212**, 497 (1983)
33. Doyle, W.T.: J. Appl. Phys. **49**, 795 (1978)

34. McPhedran, R.C., McKenzie, D.R.: Proc. R. Soc. London Ser. **A 359**, 45 (1978)
35. Sangani, A.S., Acrivos, A.: Proc. R. Soc. London Ser. **A 386**, 263 (1978)
36. Lamb, W., Wood, M., Ashcroft, N.W.: Phys. Rev. **B 21**, 2248 (1980)
37. Liebsch, A., Perrson, B.N.J.: J. Phys. C. **16** 5375 (1983)
38. Dignam, M.J., Moskovitz, M.: J. Chem. Soc. Faraday Trans. **2**, **69**, 56 (1973)
39. Bedaux, D., Viegler, J.: Physica (Utrecht) **67**, 55 (1973); **73**, 287 (1974)
40. Donnadieu, A.: Thin Solid Films **6**, 249 (1970)
41. Bruggeman, A.G.: Ann. Phys. (Leipzig) **24**, 636 (1935); **24**, 665 (1935)
42. Stroud, D.: Phys. Rev. **B 12**, 3368 (1975)
43. Granqvist, C.G., Hunderi, O.: Phys. Rev. **B 18**, 2397 (1978)
44. Granqvist, C.G., Hunderi, O.: Phys. Rev. **B 16**, 1353 (1977)
45. Smith, G.B.: J. Phys. **D 10**, L 37 (1977); Appl. Phys. Lett. **35**, 668 (1979)
46. Sheng, P.: Phys. Rev. **B 22**, 6364 (1980)
47. Sheng, P.: Phys. Rev. Lett. **45**, 60 (1980); Opt. Laser Technol. **13**, 253 (1981)
48. Niklasson, G.A., Granquist, C.G., Humderi, O.: Appl. Opt. **20**, 26 (1981)

E. Kay
IBM Almaden Research Center
650 Harry Road
San Jose, CA 95190 – 6099
USA

Guest-Host Interaction and Photochemical Transformation of Silver Particles Isolated in Rare Gas Matrices

P.S. Bechthold*, U. Kettler**, H.R. Schober, and W. Krasser
Institut für Festkörperforschung der Kernforschungsanlage Jülich, Federal Republic of Germany

Received April 7, 1986; final version May 15, 1986

Silver dimers and small clusters isolated in rare gas matrices are studied by UV-VIS absorption, emission, and resonance Raman spectroscopy. One, two, and three dimer trapping sites can be identified in Xe, Kr, and Ar matrices, respectively. The sites are identified by computer simulation to be either of D_{4h} symmetry, with the molecules aligned in the ⟨100⟩-direction of the *fcc*-host lattice, or of D_{2h} symmetry, with the molecules occupying single or double vacancies and aligned in the ⟨110⟩-direction. Low energy external modes of the dimers are observed in the resonance Raman spectra. They probe the guest-host interaction of the molecules and are assigned to librational modes.

Trimers and larger clusters are found to be very photosensitive. Three different trimers can be observed in Xe and Kr matrices, respectively. They can be transformed reversibly into each other by laser irradiation. This indicates that probably three corresponding isomers can get stabilized in the two matrices. Further evidence for this interpretation is obtained from emission spectroscopy. Despite the photosensitivity of the trimers it is possible to stabilize the concentration of a particular species by a dual beam technique. Thus we obtained a resonance Raman spectrum of one species in a Xe matrix and a preresonance spectrum of the corresponding species in a Kr matrix.

PACS: 33.20, 35.40, 53.20P, 78.30

1. Introduction

Despite the recent progress in cluster spectroscopy obtained with free molecuar beams and documented in other contributions to this issue much of our present knowledge emerges from studies of particles which were isolated in cryogenic matrices, particularly rare gas matrices. UV-VIS absorption- and emission-, IR absorption-, magnetic circular dichroism-, Raman and resonance Raman-, photoemission-, electron spin resonance, and Mössbauer-spectroscopies as well as EXAFS (extended x-ray absorption fine structure) studies have successfully been applied to matrix-isolated small metal particles [1, 2]. The particle density in a beam is still too small for most of these techniques. In the matrices the clustering occurs on condensation of a sufficient metal flux together with the matrix gas on a cold target and can be enhanced even by thermal annealing or photoaggregation. At a metal concentration of a few percent one can achieve particle densities which are appropriate for the spectroscopic techniques mentioned above. Thus, the matrices serve to enrich and stabilize the particles for spectroscopic investigations but they also perturb their spectral properties.

In order to understand the nature of the metal-metal bond it is therefore necessary to study the matrix effects in some more detail.

Small silver particles are of particular interest for such an investigation, first because they are among the most intensively studied systems experimentally [3–17] and theoretically [18–23]. Secondly they constitute the catalytically active centers in silver halide photography [24], one of the most effective catalytic

* Current address: Chemistry Division, Argonne National Laboratory Argonne, Illinois 60439, USA
** Present address: Department of Physics, University of California, Santa Barbara, CA 93 106, USA

processes known, and thirdly, silver is the substrate most often used for surface enhanced Raman scattering [25] which is induced by aggregates on the bulk metal surface. Details of the latter two processes are still under debate.

Here we present optical absorption, emission, and resonance Raman measurements of silver dimers, trimers and larger clusters in Xe, Kr and Ar matrices. Special emphasis is put on the interpretation of guest-host interactions and photochemical transformations. For dimers a computer simulation allows us to identify trapping sites and librational modes of the molecules. For excitation we use the various lines of argon and krypton ion lasers and a *cw*-dye laser tunable in the 416–480 nm range. The experimental arrangement has been described elsewhere [15, 26]. The silver concentrations in the matrices were in the range of 0.5–2%. Matrix temperatures were chosen between 8 K and 25 K depending on the rare gas involved.

2. Dimers

The observed low energy absorption bands due to the ${}^1\Sigma_u^+ \leftarrow {}^1\Sigma_g^+$ transition of disilver are summarized in Table 1 [10, 12, 15]. In a xenon matrix a single peak with a half width of about 25 nm, is found at 393 nm. In a krypton matrix a slightly asymmetric absorption profile indicates the overlap of two different bands at about 389 and 398 nm, respectively. This is due to two different Ag_2 trapping sites in the krypton matrix, as proved by the resonance Raman measurements. The corresponding two trapping sites in argon matrices can be clearly distinguished by their absorption peaks at 385 and 410 nm, respectively. A third Ag_2 absorption line was identified in argon matrices at 443 nm [3, 15]. It is the only ${}^1\Sigma_u^+ \leftarrow {}^1\Sigma_g^+$ absorption line of Ag_2 in rare gas matrices which is red shifted with respect to the gas phase value of 435 nm [27]. This documents already the unusual character of the associated trapping site.

Table 1 gives the peak positions of laser-induced fluorescence bands which are correlated with the absorption bands described above [10, 15, 26]. The emission bands are broad and do not show any vibrational fine structure. Fortunately, the resonance Raman effect can supply the desired vibrational data because it correlates observed vibrational frequencies with a specific absorption band and thus a particular species in the matrix. This correlation facilitates the assignment of the vibrations to a cluster of a certain size, but may serve also to study vibrational frequencies of isomers or – as in this case – to identify and distinguish multiple site occupation of the same particle. To achieve resonance Raman scattering the exciting laser wavelength has to be tuned to the absorption band considered. Vibrational bands of the associated species are then selectively enhanced with respect to others and can be assigned unambiguously to the absorption band. The resonance Raman spectra are excited with the 406.74 nm line of the krypton ion laser except for the species with the 443 nm absorption line in argon matrices which is excited by the dye laser at 445 nm. Figure 1 shows the resulting spectrum for a xenon matrix. It is dominated by two intense lines at 47.5 and 198 cm^{-1}. The weaker sharp structures are due to overtones (at 95 cm^{-1} and 395 cm^{-1}) and combination bands (spaced by 47.5 cm^{-1}) of these lines. The broad side bands are due to phonons of the host lattice [15]. Since the translational invariance of the host lattice is removed by the trapping of the dimers, phonons throughout the Brillouin zone are activated in the spectrum and couple to the sharp impurity modes. The line at 198 cm^{-1} is close to the internal stretching mode of gaseous Ag_2 ($\omega_e = 192.4$ cm^{-1} [27]). Therefore it can be assigned to the stretching vibration of Ag_2 in xenon. The matrix shift of 6 cm^{-1} is due to the repulsive cage forces of the lattice. The low energy mode at 47.5 cm^{-1} is due to the same molecule because it forms the combination bands with the internal stretching vibration. It is an external mode of the molecule because the dimer has only one internal frequency. Therefore, this mode directly probes the interaction with the xenon host lattice at the trapping site of the molecule. Its energy is just above the maximum phonon energy of crystalline xenon (43 cm^{-1} [28]). The mode is therefore weakly localized and considerable motion of the surrounding xenon atoms must be involved [29]. The occurrence of such sharp and reproducible lines suggests that all the dimer mol-

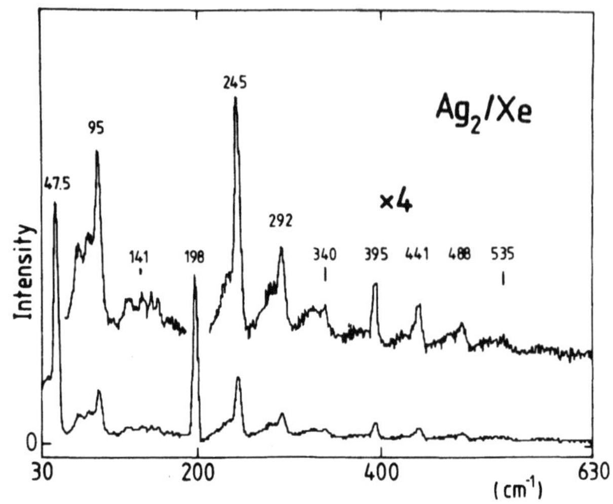

Fig. 1. Resonance Raman spectrum of Ag_2 molecules isolated at 25 K in a xenon matrix. Excitation wavelength 406.74 nm

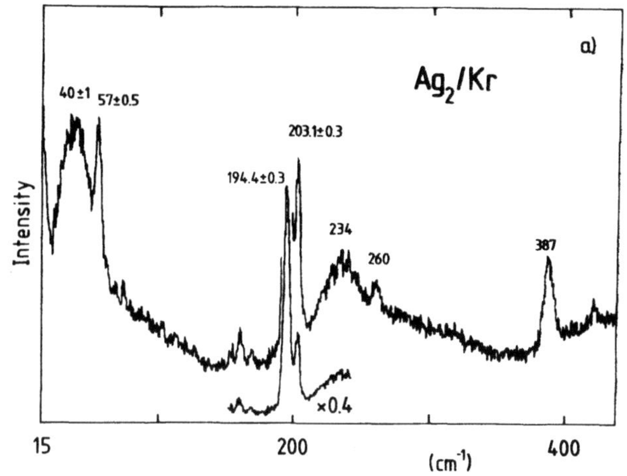

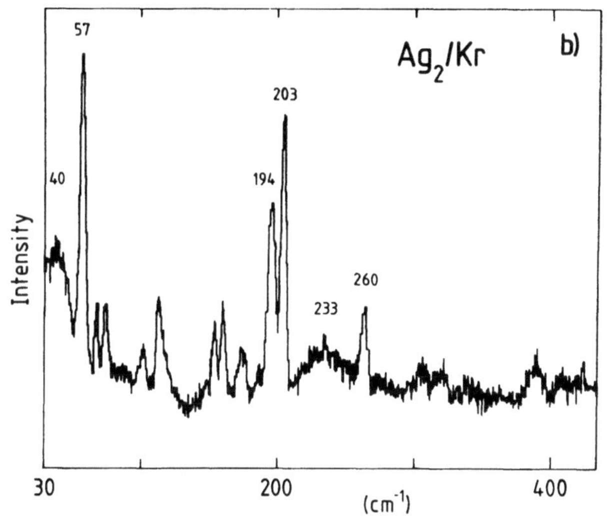

Fig. 2a and b. Resonance Raman spectra of Ag_2/Kr at 20 K, demonstrating the simultaneous occupation of two trapping sites and the correlation of modes as described in the text. **a** Freshly prepared matrix. **b** After a short annealing at 50 K. The new modes coming up during the annealing process may be due to larger clusters. (The small peak at 418 cm^{-1} is due to the sapphire substrate [39])

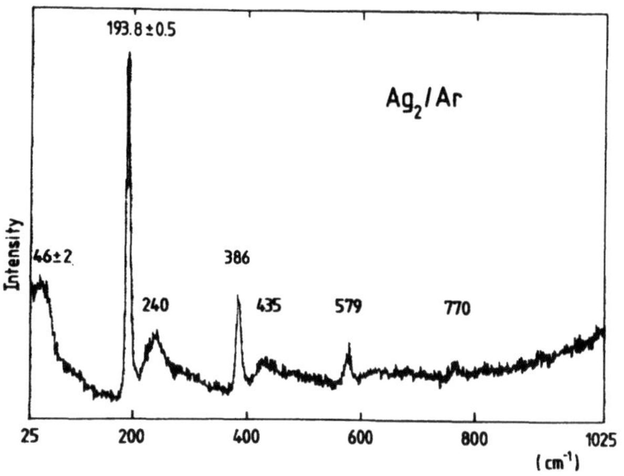

Fig. 3. Resonance Raman spectrum of Ag_2 in an argon matrix recorded at 12 K with 406.74 nm excitation

ecules occupy the same trapping site. Different sites would exert different perturbations and thus generate various vibrational modes.

These arguments are supported by the resonance Raman spectrum obtained with a krypton matrix. It shows two separated lines at 194.4 and 203.1 cm^{-1} in the region of the disilver stretching frequency (Fig. 2a). Obviously, the molecules in both trapping sites of the krypton matrix get simultaneously excited. In addition we see a low energy localized mode at 57 cm^{-1} which together with the line at 203 cm^{-1} forms a combination band at 260 cm^{-1}. The broad resonant structure peaking at 40 cm^{-1}forms a combination band with the line at 194.4 cm^{-1}. It lies within the region of the krypton lattice modes which have their maximum energy at 50 cm^{-1} [28]. Therefore it is no longer localized but couples to a multitude of lattice modes. The envelope of all these modes gives the broad resonant mode observed. The strong background above the maximum lattice mode of 50 cm^{-1} may be due to anharmonic effects or second order processes. The interrelation of the various bands is easily verified if the matrix is annealed for a short time at 50 K. One observes a simultaneous decrease of the bands at 40, 194 and 234 cm^{-1} with respect to the lines at 57, 203 and 260 cm^{-1} (Fig. 2b). Since simultaneously the low energy side of the absorption band decreases gradually we correlate the vibrational modes to the absorption and emission lines as given in Table 1. In argon matrices we get an internal vibration at 193.8 cm^{-1} and a broad resonant mode at 46 cm^{-1} when exciting with the 406.74 nm laser line (Fig. 3). Now, the laser line lies within the absorption band at 410 nm so that the resonance condition is better realized. Therefore we can identify three overtones and combination bands.

The species with the 385 nm absorption band cannot be simultaneously excited and its resonance Raman spectrum could not yet be obtained. With an excitation wavelength of 445 nm we get the resonance Raman spectrum of the species with the 443 nm absorption band (Fig. 4). It shows the internal vibration at 190.5 cm^{-1} and an overtone at the slope of a strong emission line. The weak structures within and close to the band of the host lattice phonons (maximum vibrational energy 67.5 cm^{-1} [28]) are also seen in a combination band with the internal mode.

In summary, the experimental data prove the existence of one, two, and three well defined dimer trapping sites in xenon, krypton, and argon matrices, re-

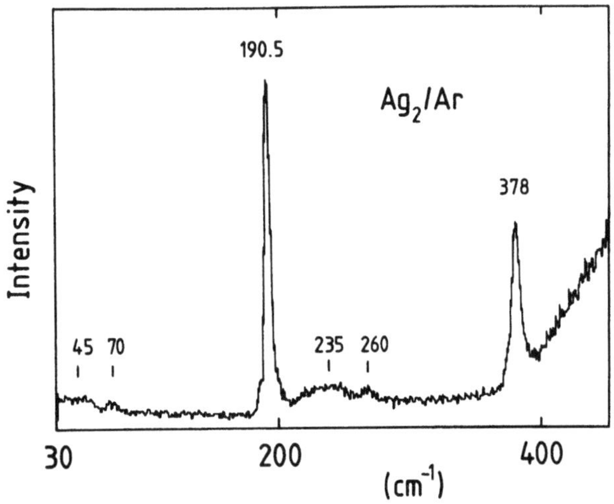

Fig. 4. Resonance Raman spectrum of Ag_2 in an argon matrix excited with 443 nm radiation

Fig. 5. Schematic representation of the computational model. Dark dots represent the Ag_2 molecule. Atomic positions in regions I and II were relaxed in the energy calculations whereas the atoms in region III were held at their ideal lattice positions. The vibrations were calculated for the sub-crystallite I with the atoms in II held at their relaxed positions. Region I contains up to 400 atoms, regions I+II up to 1,500 atoms

spectively. The sites are characterized by slight shifts of the Ag_2 stretching frequency and by external modes within or slightly above the phonon frequency range of the host lattice. The results are summarized in Table 1.

To identify the geometry of the trapping sites and the external modes we performed a computer simulation of disilver in the three rare gas matrices. The computations were done in two steps: In the first step we calculated the relaxed (meta-) stable configurations of the dimer in the matrices. In the second step we calculated the vibrational spectra for these configurations. Details of the calculations will be reported elsewhere [30]. Briefly, we assume the potential energy to be composed of centrosymmetric pair potentials and determine the local minima of the potential energy of a crystallite of up to 1,500 moveable atoms which is embedded in an infinite undistorted *fcc*-host lattice and includes the silver dimer at its center (Fig. 5). The calculations were done with a modified version of the program DEVIL originally developed at AERE Harwell [31]. The pair potential for the Ag_2 ground state is constructed from gas phase spectroscopic data using the Rydberg-Klein-Rees (RKR) method [32]. For the rare gas-rare gas interaction we took a Morse potential because it best reproduced the phonon dispersion curves of the host lattice, but use of Lennard-Jones (6–12) potentials led to essentially, the same results. The silver-rare gas interaction potential was of a purely repulsive Born Mayer form $V(R)=Ae^{-BR}$, where the parameters A and B were obtained from fits to Thomas-Fermi-Dirac interaction energy calculations [33]. This is justified because defect properties are often mainly determined by the repulsive part of the potential.

In the second step we calculated eigenfrequencies and eigenvectors of the dynamical matrix for a crystallite of up to 400 atoms (region I in Fig. 5), centered around the Ag_2 and embedded in the relaxed host crystal as calculated in the energy minimization procedure. We find in all three matrices a stable site with D_{4h} symmetry where the molecule occupies a single vacancy and is aligned in the $\langle 100 \rangle$-direction. The most stable site in Kr and Ar matrices is the double vacancy site with D_{2h} symmetry and the molecule aligned along $\langle 110 \rangle$. The third stable configuration in Ar has D_{2h} symmetry with the molecule aligned along $\langle 110 \rangle$ in a single vacancy site. Other sites are unstable. The calculated eigenfrequencies and assignments of those Raman active modes in which the Ag_2 molecule significantly participates are compared with the experimental values in Table 1. The external modes are identified to be librational modes. In the case of the resonant (in band) modes we give only the frequencies corresponding to maximal disilver amplitudes, i.e. the maxima of the local vibrational spectral functions. In general, we find good agreement between experimental and calculated frequencies, except for the internal mode frequency of the single vacancy D_{2h} species in argon.

In our rigid potential model one expects an increase of the dimer frequency with respect to the gas

Table 1. Spectroscopic data and mode assignments of Ag_2 molecules isolated in rare gas matrices and of gaseous Ag_2

Matrix gas		Xe	Kr		Ar			Gas phase
Site symmetry		$D_{4h}(1V)$[a]	$D_{4h}(1V)$	$D_{2h}(2V)$	$D_{4h}(1V)$	$D_{2h}(2V)$	$D_{2h}(1V)$	
Absorption band (nm)		393	389	398	385	410	443	435
Fluorescence band (nm)		480	454	505	–	473	480	
Stretching vibration[b] (cm^{-1})	exp.	198	203.1	194.4	–	193.8	190.5	$\omega_e = 192.4$
	calc.	A_{1g} 201	A_{1g} 202	A_g 192	A_{1g} 200	A_g 193	A_g 201	
External localized mode[b] (cm^{-1})	exp.	47.5	57	–	–		70	
	calc.	E_g 49.7	E_g 63		E_g 79.8		B_{1g} 70 B_{3g} 70.9	
Resonant mode (cm^{-1})	exp.	–	–	40	–	46		
	calc.	E_g <8.5	E_g <7.5	B_{1g} 30[c] B_{3g} 31[c]	E_g <6	B_{1g} 28[c] B_{3g} 28[c]	B_{1g} <12 B_{3g} <8	

[a] $1V$, $2V$ refer to single and double vacancy sites, respectively
[b] Resolution $\pm 0.5\ cm^{-1}$
[c] Only the calculated mean frequency of the Ag_2 in band vibrational spectrum is given

phase, because in this configuration the Ag_2 molecule is compressed. However, the experimental frequency is decreased. It seems that the rigid potential concept no longer holds. The distance of the Ag_2 to its nearest neighbour argon atoms (lying in the direction of the molecular axis) is minimal. Therefore, one might assume a small hybridisation of the electrons of the Ag_2 and these argon atoms, thus reducing the Ag-Ag bond strength. Such an electronic relaxation should also cause an even stronger reduction of the electronic energy in the excited state (exciplex formation). Consequently, the energy of the transition to the first excited state should be reduced in agreement with the observation. The librational mode should be less affected. The calculation gives a B_{1g} mode at 70 cm^{-1} and a B_{3g} mode at 70.9 cm^{-1} whereas experimentally a weak mode is observed at 70 cm^{-1} (Fig. 4).

3. Trimers

With silver concentrations of about 2% two new absorption bands are observed at 425 nm and 442 nm in freshly prepared xenon matrices (Fig. 6). They were attributed to trisilver from concentration-dependent measurements [4, 9]. Their strength can be enhanced by a short annealing of the matrix.

A similar behaviour is observed in krypton matrices, however, the corresponding bands peaked at 417 nm and 423 nm overlap so that an asymmetric structure results (Fig. 7). In argon matrices the respective trimer bands overlap with the dimer absorption and cannot be identified unambiguously [26].

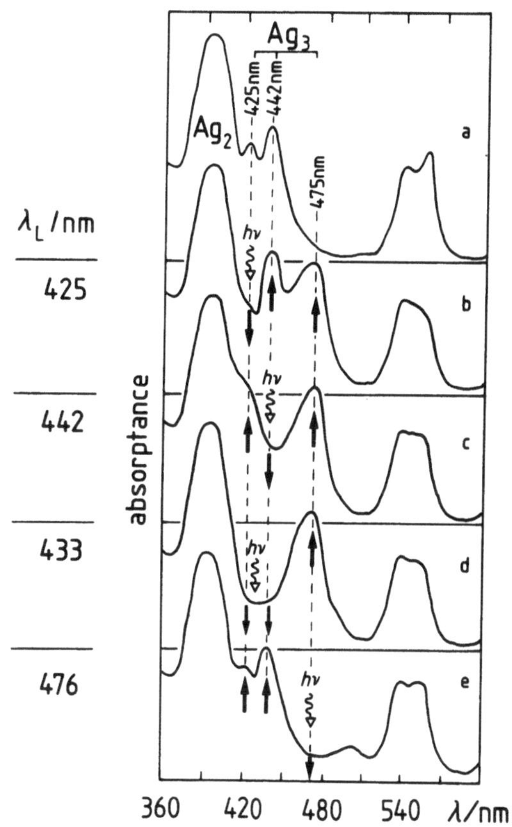

Fig. 6. Absorption spectra of silver particles isolated in solid xenon at 25 K demonstrating the photoinduced transformations of trisilver. Sinuous arrows mark the wavelengths of laser irradiation which are also numerically given at the left. Dark arrows indicate the decrease or increase of the marked absorption band. The strong band at 393 nm is due to dimers. The bands at 535 and 557 nm are possibly due to Ag_4 [26]. The light source of the absorption spectrometer does not measurably change the shape of the spectra

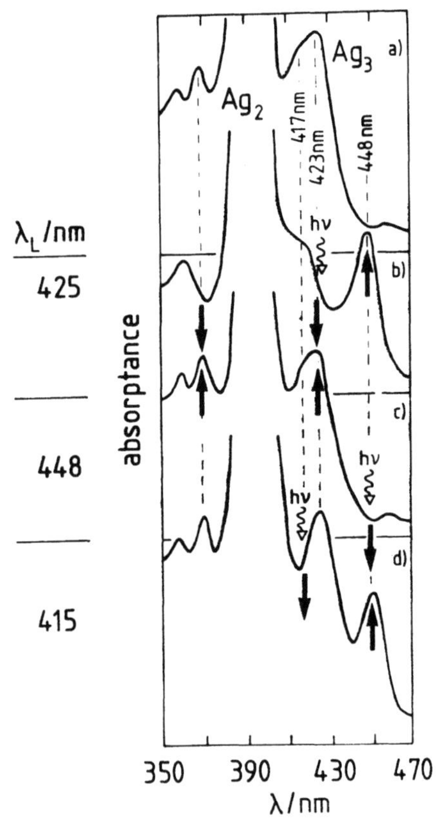

Fig. 7. Absorption spectra showing the phototransformation of Ag_3 in solid krypton at 20 K. The meaning of the arrows is the same as in Fig. 6

The trimers are found to be particularly photosensitive. Upon weak laser irradiation into one of the absorption lines the corresponding band disappears and a new third absorption band is created at 475 nm in xenon and at 448 nm in krypton matrices, respectively [9, 16]. This is demonstrated in Figs. 6 and 7 with various excitation lines. Irradiation into the newly created bands almost re-establishes the original spectra. Due to this reversibility the phototransformations can be cycled several times.

Table 2. Spectroscopic data of Ag_3 in xenon and krypton matrices

Matrix gas	Xe			Kr		
Absorption bands (nm)	425	442	475	417	423	448
Emission bands (nm)	555	624/659	760	445	478/490	563
Localized mode (cm^{-1})		111			115	
Resonant mode (cm^{-1})		38				

The absorption lines are correlated to emission bands as given in Table 2. The identification of these bands is complicated by the photosensitivity of the corresponding trimers, by reabsorption due to larger clusters, and in krypton matrices by partial overlap with dimer emission bands [16, 26]. The emission line at 555 nm in xenon is partly reabsorbed by the very photosensitive absorption bands at 535 and 557 nm (Fig. 6) and causes the changes in these bands observed during the trisilver phototransformation.* The fluorescence line at 659 nm has a side band which may be due to a second trapping site of the same trimer species. A broad emission at 485 nm in freshly prepared krypton matrices can be split by a weak annealing of the matrix into two bands peaked at 478 nm and 490 nm, respectively. These two bands could be associated with two trapping sites in the krypton matrix. If an emission is directly excited by irradiation into the corresponding absorption band one also observes weak emissions at the other two lines due to the phototransformation processes.

Since the phototransformation is completed in a few seconds, resonance Raman spectra cannot be generated in the conventional way. But with an auxiliary laser beam driving the backtransformation process we are able to stabilize the concentration of the species with the 442 and 423 nm absorption bands in xenon and krypton matrices, respectively. For the xenon matrix we get a resonance Raman spectrum with an intense line at 111 cm^{-1} and two overtones (Fig. 8). If we tune the exciting dye laser across the corresponding trimer absorption band at 442 nm we see in a very narrow range of excitation wavelengths a broad structure in the spectrum of the scattered light with the 111 cm^{-1} line superimposed at the slope (Fig. 9). The peak of this broad structure at 38 ± 2 cm^{-1} is within the band of host lattice phonons. Thus, in analogy with our observations for dimers, we interpret it as a resonant mode of the corresponding trimer. With the krypton matrix we can only produce a preresonant spectrum because the auxiliary laser line is only at the edge of the 448 nm absorption band (Fig. 10). We see a trimer mode at 115 cm^{-1}** and simultaneously the dimer mode at 194 cm^{-1}.

* The absorption bands at 535 and 557 nm were tentatively assigned to Ag_4 [26]. For the band at 535 nm we observed similar photoreversible reactions as for the trimers. Upon irradiation into this band a new line at 485 nm is created, while the original band disappears completely. It reappears during the back transformation, but a third line at 499 nm is also generated

** This line has to be distinguished from the Raman line at 120 cm^{-1} reported for silver particles in krypton matrices by Schulze et al. [5]. We have observed that line, too, but only at 514.5 nm excitation

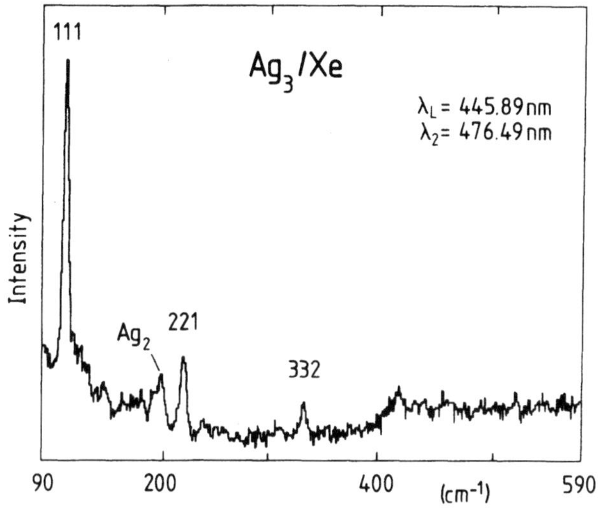

Fig. 8. Dual-beam resonance Raman spectrum of the silver trimer with the 442 nm absorption band in solid xenon (excitation wavelength $\lambda_L = 445.89$ nm, wavelengths of the auxiliary argon laser beam $\lambda_2 = 476.49$ nm). The fundamental vibration of the trimer is at $111 \pm 2\ \mathrm{cm}^{-1}$, also clearly visible are two overtones and the nonresonant dimer mode at $198\ \mathrm{cm}^{-1}$. The peak at $418\ \mathrm{cm}^{-1}$ is due to the sapphire substrate [39]

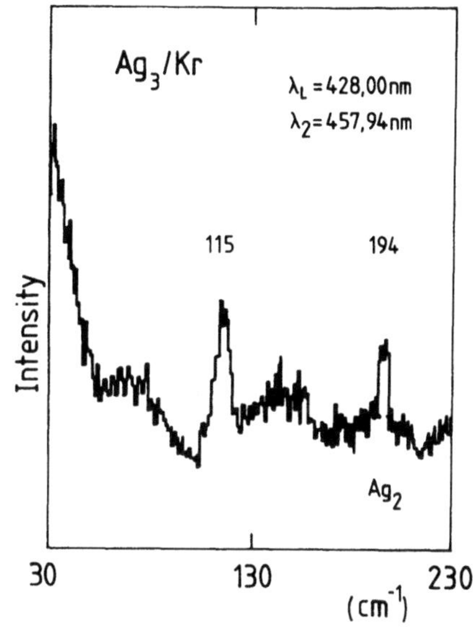

Fig. 10. Preresonant Raman spectrum of Ag_3/Kr obtained with the dual-beam technique (excitation wavelength $\lambda_L = 428$ nm, wavelength of the auxiliary argon laser $\lambda_2 = 457.94$ nm). The line at $115\ \mathrm{cm}^{-1}$ is due to the trimer with the 423 nm absorption band. The line at $194\ \mathrm{cm}^{-1}$ is due to a dimer

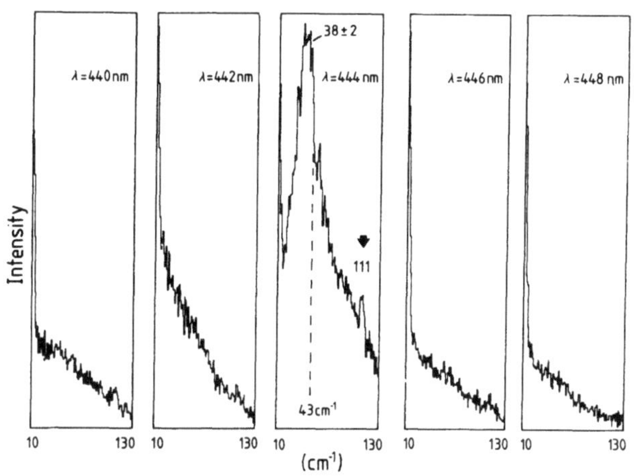

Fig. 9. Dependence on the excitation wavelength of the strong resonant mode at $38\ \mathrm{cm}^{-1}$ (single scans). The maximal host lattice phonon frequency is at $43\ \mathrm{cm}^{-1}$. The arrow marks the internal mode of the trimer at $111\ \mathrm{cm}^{-1}$

The rapid reversible phototransformation of the trimers can most easily be explained by photoisomerization processes of the trisilver/matrix cage units [9, 16]. Photoaggregation processes take place to a small extent only. Although we cannot exclude with certainty that the effects are caused by trimers of more or less the same form trapped in different lattice sites there is some evidence that different geometrical isomers of Ag_3 might be involved. The large relative Stokes shifts of the emission bands indicate a "true" isomerization rather than a trapping site effect. On the other hand the side band of the 659 nm line in xenon and the small splitting of the 485 nm line in krypton matrices might be due to trapping sites.

The D_{3h} equilateral triangular geometry of trisilver is unstable due to the Jahn Teller effect. The symmetric degenerate $^2E'$ electronic ground state will split by a distortion to C_{2v} symmetry into $^2A_1$ and $^2B_2$ states with configurations $6a_1^2\,7a_1^1$ and $6a_1^2\,5b_2^1$, respectively. Ab initio calculations [21–23] show that the $^2B_2$ state (apex angle $>60°$) is lower in energy and has a very shallow bending mode potential leading in the linear limit (apex angle 180°) to a $^2\Sigma_u^+$ state only slightly above the $^2B_2$ potential minimum. At apex angles below 60° the $^2A_1$ state becomes the ground state. Its potential minimum is less then 0.14 eV above the $^2B_2$ minimum. Such an energy difference can be supplied by the matrix cage, so that the matrix can invert the sequence of the $^2B_2$ and $^2A_1$ states. This was verified in ESR measurements which indicate a change of the Ag_3 electronic ground state from $^2B_2$ in a C_6D_6 matrix [13] to $^2A_1$ in a N_2-matrix [17].

This proves that different isomers can get stabilized in different environments. It seems that after photoexcitation of Ag_3 the cluster/cage unit can relax to a new geometry which is associated with an isomerization of the molecule. Moreover, matrix effects may decrease the energy separation of the $^2A_1$ and $^2B_2$ potential minima and, in addition, reduce the heights of the potential barrier between them. In such a case a pseudorotation of the cluster may result as a consequence of the dynamic Jahn Teller effect [34, 35]. Pseudorotation was observed for Li_3 in argon [36] and adamantane matrices [37] while for Na_3 and K_3 in argon matrices a pseudorotating form and a rigid obtuse triangular form ($^2B_2$ ground state) can exist [34, 38]. Recently also Cu_3 was proposed to be pseudorotating [35]. For trisilver, however, no indication of a pseudorotation has so far been found. To make a decision whether different isomers of Ag_3 exist in noble gas matrices ESR measurements and a more complete vibrational analysis are needed.

4. Conclusion

The influence excerted on silver dimers and trimers by xenon, krypton, and argon rare gas matrices has been studied in detail. For the dimers it is found that different trapping sites exist and that the number of absorption, emission and vibrational bands is increased correspondingly. By a computer simulation we have been able to identify for the first time the various trapping sites and assign them by the corresponding vibrational modes. For the trimers and larger clusters the situation is even more complex. The observation of photoreversible reactions indicates that not only trapping site effects but also isomerization processes have to be considered. Despite these complications matrices are and will be an indispensible tool to gain fundamental and useful information on metal molecules and clusters. Furthermore, studies as the one reported here may in the end contribute to a better understanding of such complex matrices as e.g. laser glasses and crystals, where clusters are responsible for concentration quenching effects, and phototropic glasses where clusters provide the reversible light induced changes of optical absorption. To extend the investigations to larger clusters, we would propose to combine advantages of cluster beams, i.e. mass separation, and of matrices, stabilization and accumulation, to trap mass selected clusters in matrices.

References

1. Ozin, G.A., Mitchell, S.A.: Angew. Chem. Int. Ed. Engl. **22**, 674 (1983)
2. Weltner, W., Van Zee, R.J.: Annu. Rev. Phys. Chem.**35**, 291 (1984)
3. Gruen, D.M., Bates, J.K.: Inorg. Chem. **16**, 2450 (1977)
4. Schulze, W., Becker, H.-U., Abe, H.: Chem. Phys. **35**, 177 (1978); Ber. Bunsenges. Phys. Chem. **82**, 138 (1978)
5. Schulze, W., Becker, H.-U., Minkwitz, R., Manzel, K.: Chem. Phys. Lett. **55**, 59 (1978)
6. Ozin, G.A., Huber, H.: Inorg. Chem. **17**, 155 (1978)
7. Mitchell, S.A., Ozin, G.A.: J. Am. Chem. Soc. **100**, 6776 (1978)
8. Welker, T., Martin, T.P.: J. Chem. Phys. **70**, 5683 (1979)
9. Ozin, G.A., Huber, H., Mitchell, S.A.: Inorg. Chem. **18**, 2932 (1979)
10. Leutloff, D., Kolb, D.M.: Phys. Chem. **83**, 666 (1979)
11. Grinter, R., Armstrong, S., Jayasooria, U.A., McCombie, J., Norris, D., Springall, J.P.: Faraday Symp. Chem. Soc. **14**, 94 (1980)
12. Mitchell, S.A., Kenney-Wallace, G.A., Ozin, G.A.: J. Am. Chem. Soc. **103**, 6030 (1981)
13. Howard, J.A., Preston, K.F., Mile, B.: J. Am. Chem. Soc. **103**, 6226 (1981)
14. Mitchell, S.A., Ozin, G.A.: J. Phys. Chem. **88**, 1425 (1984)
15. Bechthold, P.S., Kettler, U., Krasser, W.: Solid State Commun. **52**, 347 (1984); Surf. Sci. **156**, 875 (1985)
16. Kettler, U., Bechthold, P.S., Krasser, W.: Surf. Sci. **156**, 867 (1985)
17. Kernisant, K., Thompson, G.A., Lindsay, D.M.: J. Chem. Phys. **82**, 4739 (1985)
18. Shim, I., Gingerich, K.A.: J Chem. Phys. **79**, 2903 (1983)
19. Martins, J.L., Andreoni, W.: Phys. Rev. **A 28**, 3637 (1983)
20. Rabii, S., Yang, C.Y.: Chem. Phys. Lett. **105**, 480 (1984)
21. Flad, J., Igel-Mann, G., Preuss, H., Stoll, H.: Chem. Phys. **90**, 257 (1984)
22. Basch, H.: J. Am. Chem. Soc. **103**, 4657 (1981)
23. Andreoni, W., Martins, J.L.: Surf. Sci. **156**, 635 (1985); (to be published)
24. Boudon, J.: Growth and properties of metal clusters. Amsterdam: Elsevier 1980
25. Otto, A.: In: Light scattering in solids IV: In: Topics in Applied Physics. Cardona, M., Güntherodt, G. (eds.), Vol. 54, p. 289. Berlin, Heidelberg, New York, Tokyo: Springer 1984
26. Kettler, U.: Ph.D. Thesis, Universität zu Köln, (1984) (Berichte der Kernforschungsanlage Jülich No. 1980 (1985))
27. Ruamps, J.: Ann. Phys. (Paris) **4**, 1111 (1959); Brown, C.M., Ginter, M.L.: J. Mol. Spectrosc. **69**, 25 (1978); Srdanov, V.I., Pešić, D.S.: J. Mol. Spectrosc. **90**, 27 (1981)
28. Endoh, Y., Shirane, G., Skalyo J., Jr.: Phys. Rev. **B 11**, 1681 (1975)
29. Barker, A.S., Jr., Sievers, A.J.: Rev. Mod. Phys. **47**, Suppl. 2, (1975)
30. Bechthold, P.S., Schober, H.R.: (to be published)
31. Schober, H.R.: J. Phys. **F 7**, 1127 (1977)
32. Telle, H., Telle, U.: Comp. Phys. Comm. **28**, 1 (1982)
33. Abrahamson, A.A.: Phys. Rev. **178**, 76 (1969); see also: Torrens, I.M.: Interatomic potentials, New York: Academic Press 1972
34. Thompson, G.A., Lindsay, D.M.: J. Chem. Phys. **74**, 959 (1981)
35. Moskovits, M.: Chem. Phys. Lett. **118**, 111 (1985)
36. Garland, D.A., Lindsay, D.M.: J. Chem. Phys. **78**, 2813 (1983)
37. Howard, J.A., Sutcliffe, R., Mile, B.: Chem. Phys. Lett. **112**, 84 (1984)
38. Thompson, G.A., Tischler, F., Garland, D., Lindsay, D.M.: Surf. Sci. **106**, 408 (1981)
39. Porto, S.P.S., Krishnan, R.S.: J. Chem. Phys. **47**, 1009 (1967)

P.S. Bechthold
U. Kettler
H.R. Schober
W. Krasser
Institut für Festkörperforschung
Kernforschungsanalge Jülich GmbH
Postfach 1913
D-5170 Jülich 1
Federal Republic of Germany

Ionized Cluster Beam Technique for Thin Film Deposition

T. Takagi
Ion Beam Engineering Experimental Laboratory, Kyoto University, Sakyo, Kyoto, Japan

Received April 7, 1986; final version June 4, 1986

The ionized cluster beam (ICB) technique can be classified as an ion-assisted technique for film formation, and it has the feature of transferring low energy and equivalently high current beams. The clusters can be created by condensation of supersaturated vapour atoms produced by an adiabatic expansion process. These clusters are large size macro-aggregates of 100–2,000 atoms formed by pure expansion of vapourized solid state materials. The clusters are first partially ionized by an electron impact, then the kinetic energy can be added to the ionized clusters. The ICB has unique capabilities of film deposition due to cluster properties and its low energy ion beam transport in a range from thermal energy to a few hundred eV, with the ability to use the effective influence from the ions without space charge problems. This allows high quality deposition and epitaxy of materials at low temperature onto a wide variety of substrates and even permits the formation of thin film materials not previously possible.

PACS: 81.15J; 68.55; 36.40

I. Introduction

The ionized cluster beam (ICB) deposition technique was developed at Kyoto University in 1972 [1]. In ICB deposition, solid state material is vapourized in a heated crucible and the vapour is ejected through the nozzle of the crucible into a high vacuum chamber. A suitable cluster source for ICB deposition is a cylindrical nozzle. Operating conditions are selected so that vapour emerging through the crucible nozzle undergoes adiabatic expansion and subsequent cooling to a supersaturated condition [2]. The clusters thus formed consist of 100–2,000 atoms. These clusters, therefore, may be called "large size vapourized solid state material clusters". They are ionized in a singly charged state. These positively charged clusters can then be accelerated towards the deposition surface of the substrate by an applied potential.

The unique capabilities of ICB deposition are attributed to the properties of the cluster state. Enhancement of adatom migration is one of the most significant properties of cluster beam deposition. Other important characteristics of ICB are extremely low charge-to-mass ratios, preventing space-charge problems and to transport equivalently high-current and low-energy ion beams in a range just adequate for film formation, that is, from thermal energy to a few hundred eV. Also, self-cleaning of the substrate surface by incident kinetic energy of ionized clusters can be expected.

By using these unique properties of the clusters, the ICB technique could be applied in preparing high quality films and functional devices [3, 4]. In this chapter, the operation mechanism of the equipment, film formation mechanisms and characteristics of films prepared by the ICB are reviewed.

II. ICB Sources and Deposition Equipment

In ICB deposition, the source of atoms which form the clusters can be solid state materials heated to sufficient temperature to cause adiabatic expansion through the nozzle (pure expansion method). It was commonly believed that metals would have little tendency to condense, because they are characterized by high surface tension coefficients [5]. However, our calculations of nucleation rates for metals have shown that they have values comparable with those of alkali metals and water.

Numerical calculations of nucleation rates have shown that for metals the rate is strongly proportional to a function of $(\sigma/kT)^3$ instead of σ^3 only. Since metal has to be heated to high vapourization temperature T, (σ/kT) does not change much even when σ is high. Consequently, the nucleation barrier height and the critical cluster size become comparable to those of gases. We have made several calculations for metals and gases which have shown that the energy barriers for nucleation of metals are indeed comparable to, or even lower than, those of gas and water [6, 7]. The results are also confirmed by Yang and Lu for many different metals and semiconductors [8]. Theoretical discussions and computer simulation of cluster formation using the experimental results are described by us in another chapter of this volume [9].

The construction and operation conditions of the ICB source have been described elsewhere [2]. The cluster size was estimated by retarding field and static energy analysis, time-of-flight (TOF) and TEM observation methods [6]. These results showed that the cluster size ranged from hundreds to two thousand atoms per cluster. Cluster structure has also been analysed by electron beam diffraction [10]. The cluster beam can be formed so as to be observable in an electron diffraction device. The cluster beam was collimated with the aperture and crossed by a 45 keV electron beam. The diffraction pattern was recorded on the photographic plate. The structure and radial distribution function were analysed by the densitometer trace of the diffraction. The structure is amorphous and the interatomic distances between atoms are much longer than those of bulk crystal.

At the source, crucible heating, ionization and acceleration parts are controlled independently. Typical construction of the source with single crucible and nozzle is shown in Fig. 1. The crucible can be heated to higher than 2,000 °C. The ratio of ionized clusters to total clusters can be adjusted by changing the ionization electron current. Typical degrees of ionization are 5%–7% at ionization current $Ie = 100$ mA, 7%–15% at $Ie = 150$ mA and 30%–35% at $Ie = 300$ mA [11]. A dual cluster beam system having two groups of crucibles has also been developed. A single crucible system with multiple nozzles which forms a highly uniform beam with rectangular cross section is shown in Fig. 2 [12]. To obtain a spatially uniform distribution of the ionized cluster beam over the large substrate area at a wide range of acceleration voltages, an ion extraction and acceleration electrode system with three electrodes can be used, as shown in Fig. 1. Different combinations of applied voltages for these electrodes can be chosen by a system computer to keep the unformity of ionized clusters in total flux.

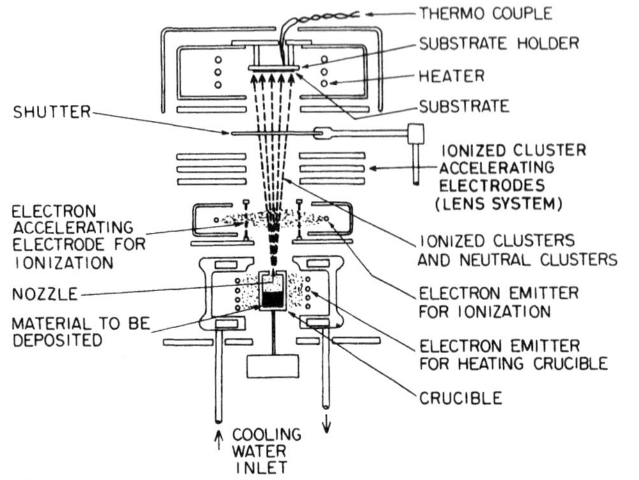

Fig. 1. Ionized cluster beam source with single crucible and single nozzle

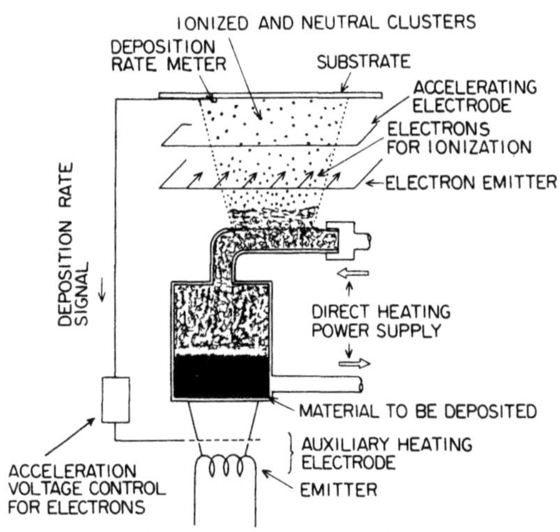

Fig. 2. Ionized cluster beam source with single crucible and multiple nozzles (curtain beam)

III. Film Formation Kinetics by ICB

The ICB deposition system utilizes a new ion transport method where a high current and low energy ion beam can be transported in the energy range of the order of a few eV to tens of eV, which is suitable for film formation. For example, a cluster accelerated by an acceleration voltage of 5 kV has an energy of 5 keV. However, the energy per cluster atom is only from 2.5 to 50 eV, provided that the cluster size is from 100 to 2,000. In addition, a cluster ion beam of 1 mA has an equivalent flux of

0.1 to 2 A of an atomic ion beam. It has already been revealed that the ionic charge has a notable effect on enhancing film formation activity and chemical reaction even without cluster acceleration or even if the beam contains only a few per cent of ions in its total flux [3]. The ICB technique enables efficient utilization of the effect of ions without problems of space charge limitation or charge accumulation on insulators. Some of the remarkable properties of this technique are the increase of surface diffusion energy of adatoms on the substrate, self-cleaning of the substrate surface, and control of the probability of the depositing material sticking to the substrate and to the film itself. These fundamental effects depend on the kinetic energy of the incident ions and materials. Some of the pronounced effects were experimentally observed [13].

In ICB deposition, the following processes in film formation are considered. When the clusters bombard the substrate surface, both ionized and neutral clusters are broken up into atoms which are then scattered over the surface with high surface diffusion energy. A scattered atom is physically attracted to the substrate surface but it may move over the surface (adatom migration) because of high kinetic energy parallel to the surface arising through conversion of incident kinetic energy. The adatom interacts with other adatoms and/or substrate atoms to form a stable nucleus, finally stopping its movement to become a chemically adsorbed atom. So, enhanced adatom migration results in the increased nucleation rate and increased growth rate of the island.

Unique effects of kinetic energy in film formation by ICB can be elucidated by measuring the deposited mass M on the substrate as a function of substrate temperature. The deposited mass M is expressed by [14]

$$M = \dot{M}t - \frac{N_0 \dot{M}}{I^*} \exp\left(-\frac{U}{kT}\right),$$

where $U = \phi_{ad} - \phi_d$, ϕ_{ad} is the activation energy for desorption and ϕ_d is the activation energy for surface diffusion, $\dot{M}$ is the mass impingement rate on the substrate, I^* is the rate of formation of critical nuclei, and N_0 is the density of adsorption sites on the substrate surface. Since ϕ_{ad} is generally larger than ϕ_d, the curve of mass versus the reciprocal of the temperature has a positive slope. Figure 3 shows the deposited mass versus the reciprocal of the substrate temperature. Crystalline states of the films deposited at different acceleration voltages are also shown. For the non-ionized clusters corresponding to conventional deposition methods, the mass de-

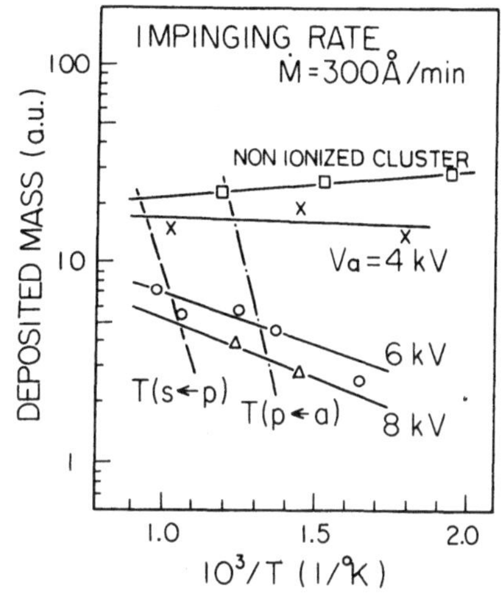

Fig. 3. Deposited mass versus reciprocal substrate temperature. $T(s \leftarrow p)$ and $T(p \leftarrow a)$ indicate polycrystalline to single crystal and amorphous to polycrystalline transition temperatures, respectively

posited increases with decreasing substrate temperature. The mass deposited at a given substrate temperature decreases with increasing acceleration voltage and the slope changes with acceleration voltage. The change of the slope from positive to negative sign with increasing acceleration voltage is considered to be due to the change in values of the parameters N_0, I^*, and U. This is one of the important features of the ICB deposition process which differentiates it from conventional deposition processes using atomic, neutral or ionized atomic-state particles, where the energy U does not change [15].

Migration and surface cleaning effects can be clearly seen in actual film deposition by comparing the Si epitaxial growth in normal high vacuum (HV) and ultra-high vacuum (UHV) chambers. Figure 4 shows the diffraction patterns of films deposited under different deposition conditions. For deposition in the HV chamber, which was evacuated by an oil diffusion pump to a base pressure of 10^{-7}–10^{-6} Torr, silicon epitaxial films can be obtained above 6 kV or higher acceleration voltages with 10% ionized clusters in total flux. An amorphous or polycrystalline structure was formed in a range of 0–4 kV. In this deposition no special cleaning except for chemical processing was used prior to the deposition. On the other hand, in the epitaxial growth of Si on an atomically clean and well ordered silicon surface in the UHV chamber, only 200 V acceleration, equivalent to approximately 0.2 eV/atom if cluster size is assumed to be 1,000 atoms, was enough to obtain epitaxial films at a substrate temperature (Ts) of 500 °C, because the surface cleaning effect was not necessary. Conventional deposition usually requires a substrate temperature of higher than 800 °C. By

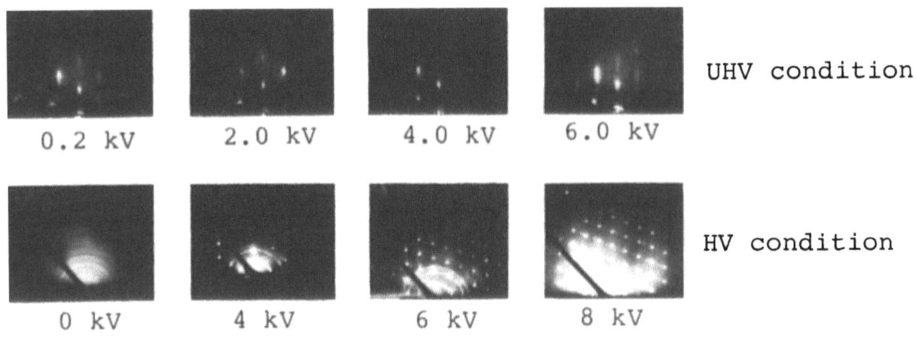

Fig. 4. Electron diffraction patterns of epitaxial silicon films by ICB at different acceleration voltages

increasing the acceleration voltage, the crystalline quality could be improved. The difference between the operating conditions in HV and UHV shows that additional incident energy supplied by higher acceleration voltage to Si reaching the substrate surface in a conventional HV served to remove both the native oxide from the substrate surface and impinging residual gas atoms during deposition. From the Rutherford backscattering spectra using 185 keV H^+ and SIMS analysis, no peak of oxygen at the interface between the deposit and the substrate could be detected [16].

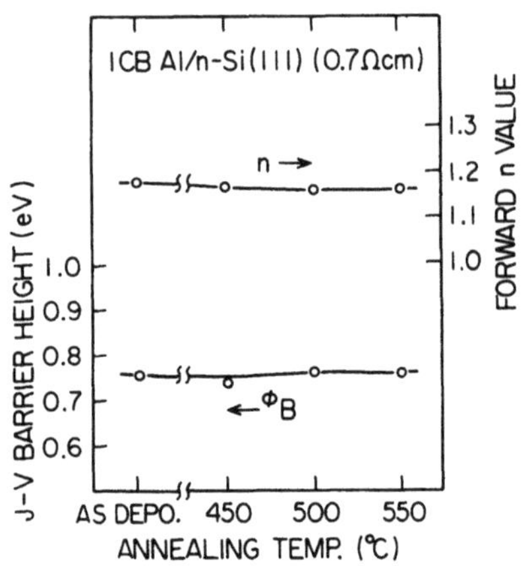

Fig. 5. Dependences of the forward *n* value and the $J-V$ barrier height of the Al-Si junction on annealing temperature

IV. Examples of Film Applications

Here we describe only a few areas of recent important progress in ICB applications.

i. Semiconductor Devices

a. Metallization and Interconnects. Metallization and interconnect technologies are important key factors in advancing VLSI technology further. Although aluminium is widely used for contact electrodes and interconnects in silicon semiconductor devices, the quantitative limits of the Al/Si system imposed by such problems as electromigration and interface stability have not been clearly determined. We have shown these limits by using epitaxial Al films on silicon substrate [17]. The thermal stability of the epitaxial Al/Si system has been studied in the 450–550 °C temperature range [18]. No hillocks and valleys as normally observed in the Al films prepared by conventional vacuum deposition could be seen; no degradation of film crystallinity was observed by Rutherford backscattering spectroscopy (RBS). Auger electron spectroscopy (AES) measurement showed that the interface remained abrupt even after annealing at 550 °C. Electrical stability was evaluated using the Schottky barrier diodes. The barrier height and the *n* value remain stable after annealing. Figure 5 shows thermal stability of electrical characteristics of Schottky barriers. The barrier height and the *n* value are 0.75 eV and 1.17, respectively and the value showed little change after anneals to temperatures up to 550 °C. This behaviour is in stark contrast to that of the films made by sputtering or evaporation. An electromigration test was made for 10 μm wide, 1,000 μm long strips of 400 nm thick films. After flowing current at a density of 10^6 A/cm^2 at 250 °C, there was no change in resistance after 400 h of operation as opposed to sputtered Al films that normally fail after 20 h. The results show that many difficulties encountered with Al metallization for VLSI application are thus apparently not intrinsic to the Al/Si system itself, but are consequences of particular deposition techniques. The reason is simply due to the single crystalline structure of the deposited film, which so far could not be formed using the conventional method.

Epitaxial deposition of metals on single crystalline dielectric materials becomes important for three-dimensional VLSI devices. As a trial of Al epitaxial growth on sapphire and CaF_2 substrates the ICB method has been used. Epitaxial Al films on these substrates could be obtained. Thermal stability of the film has also been shown to be stable [4].

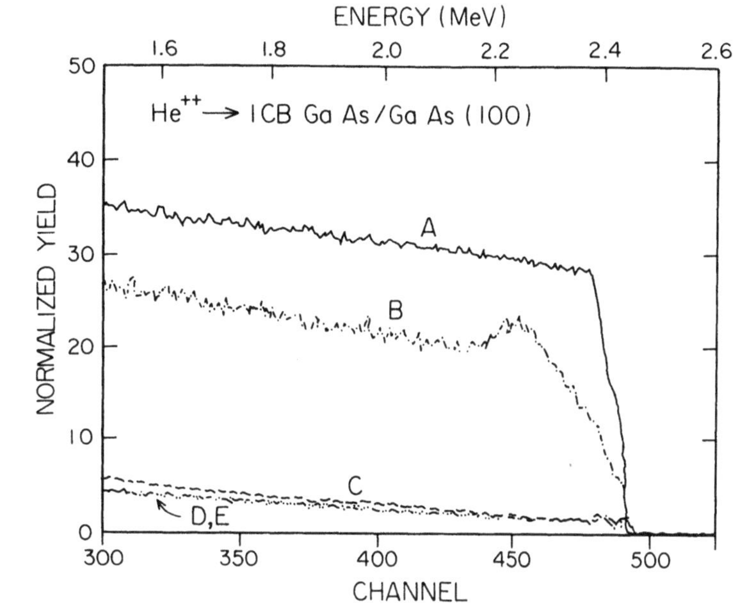

Fig. 6a–e. 3.0 MeV He^{++} channelling spectra for GaAs films deposited under different conditions: **A** random spectrum, **B** *Va* = 500 V, *Ts* = 500 °C, **C** *Va* = 1 kV, *Ts* = 500 °C, **D** *Va* = 1 kV, *Ts* = 600 °C and **E** substrate

b. Compound Semiconductors. Some GaAs thin films with high crystalline quality were recently prepared. Compared with MBE, ICB could be more suitable for industrial production because of much higher allowable deposition rates for crystal growth, and the possibilities of using a much larger substrate area, lower substrate temperature and more versatile doping methods. The films were deposited on (100) LEC GaAs substrates using dual Ga and As cluster beams, both partially ionized and accelerated to the substrate which was held at 500–600 °C under a background pressure of only 10^{-6} Torr. It is expected that the sticking coefficient of ICB deposited Ga is enhanced and less arsenic flux is required compared to that used in MBE. Reflection high energy electron diffraction (RHEED) patterns of the film deposited at an acceleration voltage of 1 kV showed a streaked pattern and Kikuchi lines, indicating surface smoothness in addition to the excellent crystallinity obtained. The RBS channelling spectra shown in Fig. 6 demonstrate that there is no measurable difference in the spectrum of the film compared to that of the substrate [19, 20].

The study of thin multilayered films in the 1–10 nm range has become very important in both fundamental physics and advanced device application. A CdTe-PbTe multilayered structure has been constructed on an InSb substrate at 200 °C by means of a dual ICB source using CdTe and PbTe ingots as source materials.

This CdTe/PbTe multilayered structure was evaluated using X-ray diffraction, which showed satellite peaks around the main peaks corresponding to the CdTe(111) and PbTe(111) planes. The differences in

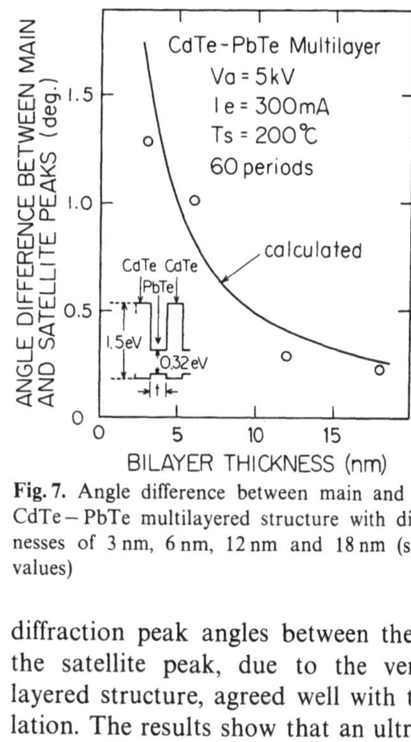

Fig. 7. Angle difference between main and satellite peaks for a CdTe – PbTe multilayered structure with different bilayer thicknesses of 3 nm, 6 nm, 12 nm and 18 nm (solid line: theoretical values)

diffraction peak angles between the main peak and the satellite peak, due to the very thin multiple layered structure, agreed well with theoretical calculation. The results show that an ultra-thin layer with good crystallinity and good abrupt junction could be formed even with a thickness of 1.5 nm. Figure 7 shows the peak dependences as a function of deposited film thickness.

The existence of an $n = 1$ miniband in the potential well structure of 60 period films with a layer thickness of 6.5 and 8 nm respectively was also confirmed by optical absorption spectra [21].

c. Insulating Films and Single Crystalline MIS *Devices.* Insulating films such as oxides, nitrides and

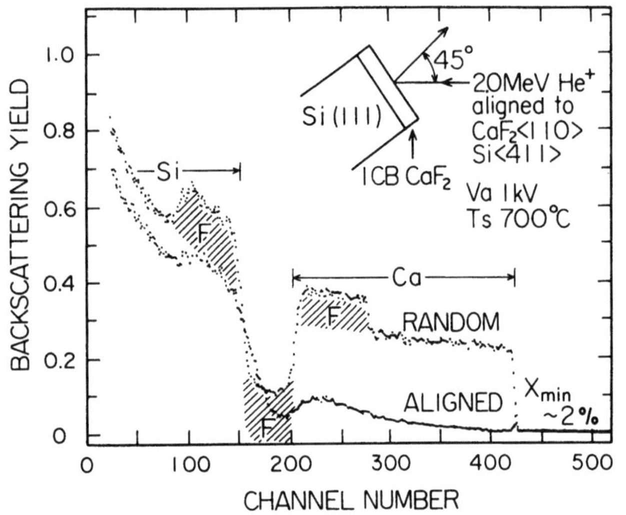

Fig. 8. RBS spectra CaF_2 film deposited on Si(111) substrate by ICB at $Va = 1$ kV, $Ts = 700$ °C

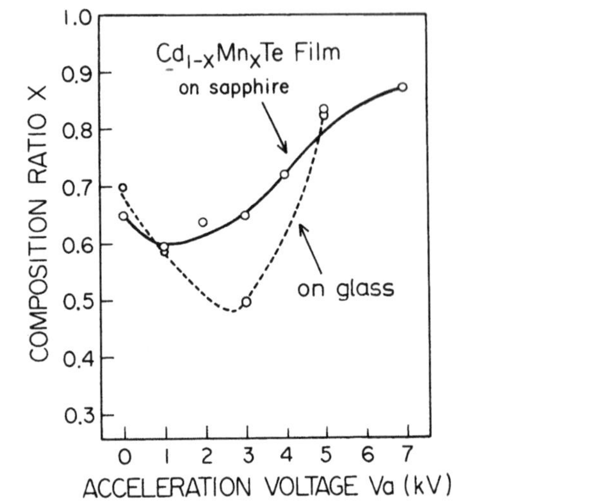

Fig. 9. Change of composition ratios of $Cd_{1-x}Mn_xTe$ films by controlling acceleration voltage on sapphire and glass substrates

carbides can be formed by ICB deposition in reactive gases. In the reactive ICB deposition the reactive gas pressure is maintained at about 10^{-5}–10^{-4} Torr, so that plasma is not produced in the chamber. Development of process technology regarding formation of a very thin dielectric film at low substrate temperature is very important in device fabrication. The highest quality conventional SiO_2 produced to date has been a high temperature (800–1,200 °C) thermally grown oxide. The necessity of high temperature restricts the device processing that can be done before thermal oxidation. Encouraging ICB SiO_2 deposition at a substrate temperature of 200–300 °C has been obtained by Minowa et al. [22]. Wong et al. deposited SiO_2 films which displayed high dielectric strength (60% yield exceeding 8×10^6 (V/cm) [23]. Recently, stoichiometric and dense Al_2O_3 has been deposited using ICB at a substrate temperature of 100 °C in our laboratory.

We are currently studying deposition of various kinds of single crystalline dielectric films on which epitaxial metal could be deposited. We have already deposited single crystalline CaF_2 with a smooth surface on Si(111) and Si(100) substrates. Figure 8 shows preliminary results of RBS channelling analysis of CaF_2 film on a Si substrate. The minimum yield is 2% for ⟨110⟩ incidence. The film had a smooth surface, good adhesion and no cracks [24].

Epitaxial Al film has also been grown on CaF_2 by ICB. These results show that a single crystalline MIS device can be possible. This type of monolithic metal/insulator/semiconductor layer structure will bring about construction of higher density VLSI and high performance three-dimensional devices [4].

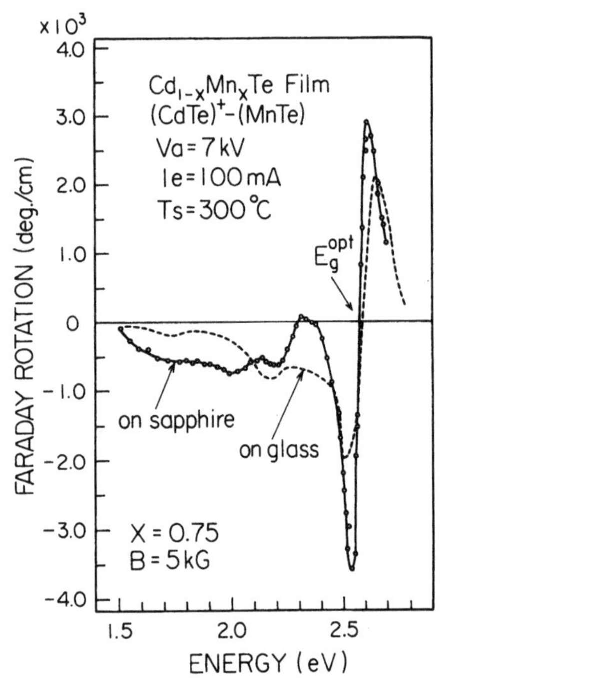

Fig. 10. Faraday rotation spectra of $Cd_{1-x}Mn_xTe$ films grown on sapphire and glass substrates ($Va = 7$ kV, $Vo = 300$ V, $Ie = 100$ mA and $Ts = 300$ °C)

ii. Application to Magneto-Optical Devices

Currently, bulk material is used for optical isolator circulator for high density operations. A $Cd_{1-x}Mn_xTe$ film, which shows a large Faraday rotation, could be one of the candidates in fabricating a highly sensitive thin film-type device. Epitaxial $Cd_{1-x}Mn_xTe$ films with large Faraday rotation were deposited on 300 °C sapphire (0001) and on a glass substrate by simultaneous deposition of ionized

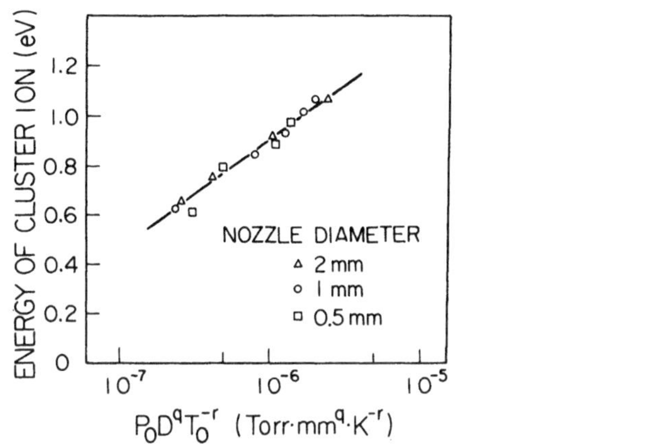

Fig. 11. Energy of anthracene cluster beam plotted against scaling parameter

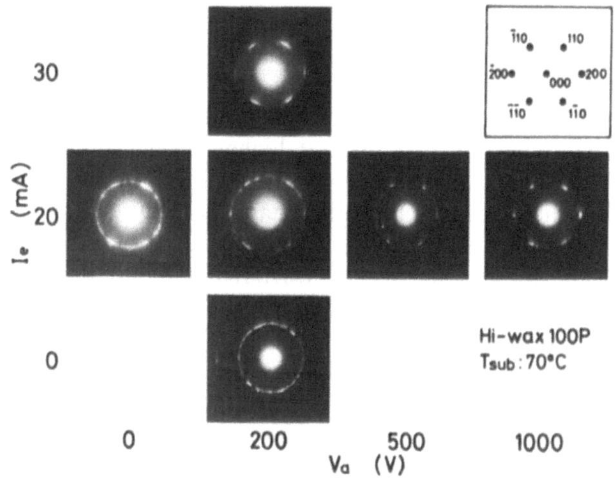

Fig. 12. HEED patterns of polyethylene films deposited at different ion acceleration voltages and ionization currents

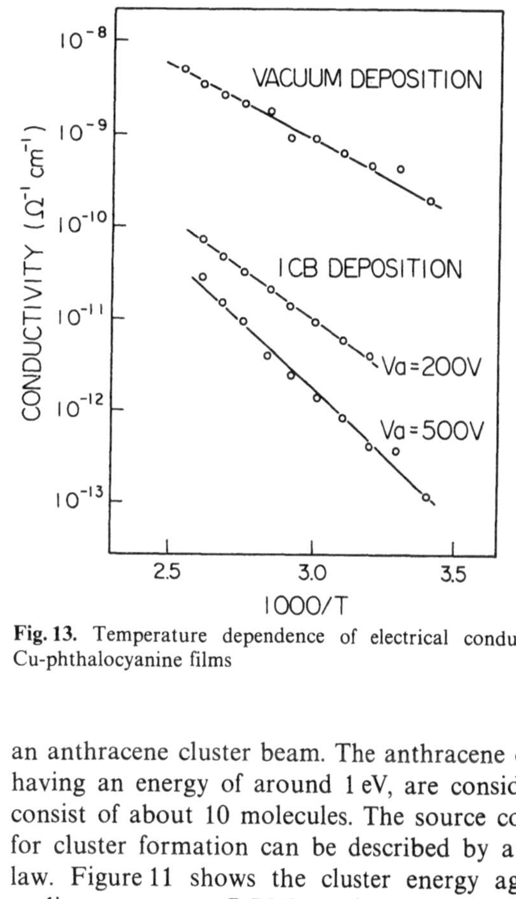

Fig. 13. Temperature dependence of electrical conductivity of Cu-phthalocyanine films

CdTe and neutral MnTe clusters. Through ICB deposition of $Cd_{1-x}Nn_xTe$, we found that the film composition, that is x, can be changed by controlling the ion acceleration voltage [25]. Figure 9 exemplifies this for $Cd_{1-x}Mn_xTe$ film deposition using a dual source ICB system. When ionized CdTe cluster beams and neutral MnTe cluster beams were used, film composition x varied for 0.5 to 0.9. When the MnTe cluster beam was ionized, crystalline quality could be controlled without any change in x. Since the optical bandgap of this material varies linearly with composition x, the photon energy where the Faraday rotation changes from negative to positive can also be varied by depositing the film at different acceleration voltages. Figure 10 shows one example of the dispersion of Faraday rotation for film deposited on sapphire and glass substrates. The results show that the dispersion became much sharper and larger when the crystalline state of the film could be improved. The cross point of the dispersion curve at 2.6 eV can be changed by composition x, consequently also by the acceleration voltage because the cross point corresponds to the optical bandgap Eg^{opt}. This versatility in the ICB method could extend development of functional devices with special optical characteristics.

iii. Organic Materials

Just as ICB can form highly oriented inorganic films, the fundamental effects associated with ICB could be exploited to produce high quality organic film. Recent research on anthracene, polyethylene and Cu-phthalocyanine film depositions have indicated such possibilities in organic film formation.

The formation process of organic clusters has been studied by an energy analysis experiment using an anthracene cluster beam. The anthracene clusters, having an energy of around 1 eV, are considered to consist of about 10 molecules. The source condition for cluster formation can be described by a scaling law. Figure 11 shows the cluster energy against a scaling parameter $P_0D^qT_0^{-r}$, where q and r are constants whose best fit values were 1 and 2.25, respectively. All the points in the figure representing different nozzle diameter fall in one straight line. This result experimentally verifies that the organic clusters are formed by homogeneous nucleation processes during the supersonic expansion of the vapour through the nozzle.

An evaluation of the crystallite characteristics of the polyethylene film has been made. Figure 12 shows the HEED patterns of the films deposited at different acceleration voltages and ionization current. The film deposited at higher acceleration voltage indicated high crystal orientation. The molecular chains in the crystallites stand perpendicularly to the substrate. Recently, we investigated the ICB effects on the electrical characteristics of an organic semiconductor [26]. Figure 13 shows the temperature dependence of the electrical conductivity of Cu-phthalocyanine films. The results indicate that the activation energy of the electrical conductivity can be increased by applying the appropriate acceleration voltage during deposition. Investigations have been carried out regarding effective doping methods for oxygen or Mg during Cu-phthalocyanine deposition.

Insulating-gate FETs were prepared by depositing phthalocyanine films on silicon substrates covered with SiO_2 films. The substrate was used as a gate electrode. Typical field effect characteristics, indicating that the phthalocyanine films have *p*-type conduction, were observed. In addition, an appropriate Mg doping increased the transconductance g_m by a factor of more than 10. Detailed investigation is now in progress.

5. Conclusions

Thus far most of the ICB applications have been directed towards electronic device fabrication. Other applications exist where there is a crucial need for dense, smooth, adherent films with well-controlled microstructure. One example is the optical coatings industry, where multilayer laser mirrors and antireflection coatings frequently display variations in scattering, sensitivity to moisture, and (for laser mirrors) uncontrolled damage threshold. Further ICB techniques for the deposition of TiO_2, SiO_2, Al_2O_3 and MgF_2 for these applications are being currently developed. Yet another application where no work has been done using ICB is surface modification of thin film coatings, which are hard and corrosion resistant and have a low coefficient of friction. Again, ICB offers control over those deposition parameters precisely which may make these coatings useful.

References

1. Takagi, T., Yamada, I., Kunori, M., Kobiyama, S.: Proceedings of the 2nd International Conference on Ion Sources. pp. 790–795. Vienna: Österreichische Studiengesellschaft für Atomenergie GmbH 1972
2. Yamada, I., Takaoka, H., Inokawa, H., Usui, H., Cheng, S.C., Takagi, T.: Thin Solid Films **92**, 137 (1982)
3. Takagi, T.: Material Research Society Symposium Proceedings. Vol. 27, pp. 501–511. Pittsburg, Pennsylvania: Materials Research Society 1984
4. Yamada, I., Takaoka, H., Usui, H., Takagi, T.: J. Vac. Sci. Technol. A**4**, 722 (1986)
5. Stein, G.D.: In: Proceedings of the International Ion Engineering Congress – ISIAT '83 & IPAT '83 –, Kyoto. Takagi, T. (ed.), pp. 1165–1176B. Tokyo: Inst. Elec. Eng. of Japan 1983
6. Yamada, I.: ibid. pp. 1177–1192A.
7. Yamada, I., Takagi, T.: In: Proceedings of the Xth International Symposium on Molecular Beams, pp.VII-B1-B5. Cannes: DRET & CEA, France 1985
8. Yang, S.-N., Lu, T.-M.: J. Appl. Phys. **58**, 541 (1985)
9. Yamada, I., Usui, H., Takagi, T.: Z. Phys. D – Atoms, Molecules and Clusters **3**, 137 (1986)
10. Yamada, I., Stein, G.D., Usui, H., Takagi, T.: In: Proceedings of the 6th Symposium on Ion Sources & Ion Assisted Technology. Takagi, T. (ed.), pp. 47–52. Kyoto: Research Group of Ion Engineering 1982
11. Takagi, T., Yamada, I., Sasaki, A.: Institute of Physics Conference Series 38, pp. 142–150. Bristol, London: Institute of Physics 1978
12. Takagi, T., Yamada, I., Takaoka, H.: Surf. Sci. **106**, 544 (1981)
13. Takagi, T.: J. Vac. Sci. Technol. A**2**, 382 (1984)
14. Neugebauer, C.A.: Handbook of Thin Film Technology. Maissel, L.I., Gland (eds.), Chap. 8, pp. (8-3)-(8-44). New York: McGraw-Hill Book Company 1970
15. Babaev, V.O., Bybov, Ju.V., Guseva, M.B.: Thin Solid Films **38**, 1 (1976)
16. Yamada, I., Saris, F.W., Takagi, T., Matsubara, K., Takaoka, H., Ishiyama, S.: Jpn. J. Appl. Phys. **19**(4), L181 (1980)
17. Yamada, I., Inokawa, H., Takagi, T.: J. Appl. Phys. **56**, 2746 (1984)
18. Yamada, I., Palstrøm, C.J., Kennedy, C., Mayer, J.W.: Material Research Society Symposia Proceedings, Vol. 37, pp. 401–406. Pittsburg, Pennsylvania: Material Research Society 1984
19. Yamada, I., Takagi, T., Younger, P.R., Blake, J.: SPIE. Advanced Applications of Ion Implantation. Vol. 530, pp. 75–83. Los Angeles: Intern. Soc. Opt. Eng. 1985
20. Younger, P.: J. Vac. Sci. Technol. A **3** (3) Pt. 1, 588 (1985)
21. Takagi, T., Takaoka, H., Kuriyama, Y., Matsubara, K.: Thin Solid Films **126**, 149 (1985)
22. Minowa, Y., Yamahoshi, K.: J. Vac. Sci. Technol. B**1**, 1148 (1983)
23. Wong, J., Lu, T.M., Metha, S.: SPIE. Advanced Applications of Ion Implantation. Vol. 530, pp. 84–86. Los Angeles: Intern. Soc. Opt. Eng. 1985
24. Yamada, I., Usui, H., Inokawa, H., Takagi, T.: Extended Abstract of the 17th Conference on Solid State Devices and Materials. pp. 313–316. Tokyo: Japan Sosiety of Applied Physics 1985
25. Koyanagi, T., Obata, Y., Matsubara, K., Takaoka, H., Takagi, T.: In: Proceedings of the 8th Symposium on Ion Sources & Ion Assisted Technology. Takagi, T. (ed.), pp. 285–288. Kyoto: Research group of Ion Engineering 1984
26. Usui, H., Yamada, I., Takagi, T.: J. Vac. Sci. Technol. A**4** (1), 52 (1986)

T. Takagi
Ion Beam Engineering
Experimental Laboratory
Kyoto University
Sakyo, Kyoto 606
Japan

Growth and Properties of Particulate Fe Films Vapor Deposited in UHV on Planar Alumina Substrates

H. Poppa[1], **C.A. Papageorgopoulos**[2], **F. Marks**[3], **and E. Bauer**[4]
Stanford/NASA Joint Institute for Surface & Microstructure Research, Department of Chemical Engineering, Stanford University, Stanford, California, USA
IBM Almaden Research Center, San Jose, California, USA

Received April 7, 1986; final version June 30, 1986

An integrated experimental approach was used to prepare and characterize well-defined particulate deposits of Fe on a variety of planar Al_2O_3 supports. The Fe particles were vapor deposited in UHV on in-situ cleaned and characterized amorphous and single crystal alumina supports at controlled substrate temperatures and deposition rates. Integrated support and deposit properties were monitored in-situ by AES, XPS, and LEED; particle number densities and sizes were monitored by standard TEM. It was found that *a)* metal exposure is an insufficient and often misleading measure of particle size when the support surface properties are unknown or poorly controlled, *b)* judicious combination of depostion temperature and impingement flux lead to size- and number density-controlled particulate deposits of good thermal stability, which can be improved by annealing. Preliminary results of XPS and of CO-TPD and H_2-TPD measurements exhibiting particle size effects are also reported.

PACS: 82.65 − i

Introduction

Small metal particles and clusters ranging in size from a few atoms to thousand atoms and eventually on to tens of thousands of atoms are often found to exhibit physical and chemical properties that are very different from the respective bulk material properties. When these particles and clusters are placed on solid supports, which in the case of chemical catalysts are usually refractory oxides, the physical and chemical properties can again be strongly modified. The corresponding effects are known as Metal Support Interactions (MSI) and have received extraordinary attention in the recent past in the areas of catalysis and surface science. This strong interest is due to the important practical implications of MSI in the field of catalysis and due to fundamental questions regarding the solid state transition from a collection of a few atoms to the bulk state [1].

Metal support interactions were first systematically addressed by the work of Schwab et al. [2] and Solymosi et al. [3] and the principles and further work in this area were expanded and reviewed more recently by Boudart [4], Ponec [5], Tauster et al. [6] and by Bond [7].

The area of MSI is still actively investigated today because of the extreme complexity of the subject. It is often very difficult to distinguish between the interrelated effects of particle size, particle shape and support material, between electronic and chemical reaction effects. Detailed systematic studies are, therefore, indispensable. Modern surface analytical and microstructural tools have played an increasingly important role in the search for answers to the small particle/MSI problem [8]. These studies can be roughly divided into two groups. The first applies advanced

[1] IBM Almaden Research Center, San Jose, CA 95120-6099, USA
[2] Department of Physics, University of Ioannina, Ioannina, Greece
[3] Physikalisches Institut, Technische Universität, D-3392 Clausthal, Zellerfeld, FRG
[4] Department of Physics, Stanford University, Stanford, CA 94305, USA

techniques such as IR spectroscopy [9, 10] and XPS (X-ray photoelectron spectroscopy) [11] to systematically varied support oxides and metal particle sizes while preparing the particles by traditional wet chemical means in porous oxide powders.

Since a number of powerful modern analysis techniques are very surface specific, the second group of investigators prepares metal particles by evaporation onto the surface of single crystal, polycrystalline or amorphous planar oxide supports [12]. This group is further subdivided by the use of standard vacuum or UHV techniques. Most popular, because less involved experimentally, is the practice of evaporating relatively thick (mostly continuous) metal layers onto thin (electron transparent for subsequent microscopy) oxide films at room temperature and in conventional vacuum. These deposits are then annealed (sometimes in H_2) at increasingly higher temperatures to form particles of increasing size [13, 14, 15, 16]. The limitations of this approach are insufficient particle size control (in particular for very small sizes) and compromised particle and support cleanliness. This shortcoming is addressed by UHV evaporation of the metal onto in-situ cleaned planar supports (often of single crystal nature) [17, 18, 19]. The main problem with this approach is usually the total lack of reliable information of the nucleation and growth conditions of the metal on the substrate in question, of any direct measurements of respective particle sizes (by electron microscopy or by diffraction techniques) and the lack of systematic studies of elevated substrate temperatures during vapor deposition. These shortcomings can be fatal when systematic particle size effects in the size range from 0.5 nm to 5 nm are the main object of the study and when additional investigations of size dependent particle properties by, for instance, TPD (temperature programmed desorption) or by chemical reaction measurements are to be pursued.

An interesting study that combines both the above-mentioned approaches (model system Rh on (0001) rutile and Rh on traditional titania powders) has also been reported [20].

In the present study we report on the preparation and preliminary use of a model catalytic system according to an integrated experimental approach recently proposed [8, 21]. The particular model system chosen was Fe on a variety of differently prepared planar alumina substrates. Fe was selected because of its catalytic importance for the synthesis of ammonia and because of the existence of well-controlled basic studies [22]. We also wanted to establish whether the expected difficulties of TEM imaging of such a relatively light metal and the strong affintiy of Fe to oxygen would be prohibitive for the study of such systems. In contrast to another recent investigation where all the vapor depositions were performed at room temperature (RT) [23], we now systematically employ elevated substrate temperatures during particle growth and, in addition to a similar study with the same model system [24], we now study the influence of much higher metal deposition fluxes. Also, with the help of more TEM studies, we take a closer look at the influence of substrate surface structure on particle sizes and number densities for otherwise identical metal deposition conditions.

Experimental

The basic experimental UHV system has been described before [21] and includes standard facilities for Auger and X-ray photoelectron spectroscopy, TPD (temperature programmed desorption) and for LEED.

A variety of thin film and bulk Al_2O_3 substrates were prepared outside the UHV system (see Table 1).

These specimens could be resistively heated and cleaned in-situ before metal deposition and included: *a)* Mechanically polished and chemically etched bulk sapphire slabs of $(1\bar{1}02)$ orientation and with reasonable degree of surface order (specimen type I a and type I b; see the LEED pattern of Fig. 1 which shows fairly well developed reflections on a high background), *b)* ion-beam [25] or rf-sputter deposited amorphous alumina films a-Al_2O_3 on polycrystalline Ta ribbon carriers (specimen type II a), and *c)* similarly sputter deposited a-Al_2O_3 thin films of approximately 10 nm thickness which were supported on TEM grids covered with ultra-thin carbon films (specimen type II b). Three of the TEM control specimens of type II b were mounted in good thermal contact behind 2 mm holes in the Ta ribbon samples (specimen type II a). The temperature of all resistively heated samples was measured with a spot IR pyrometer (which was also used to calibrate some spot

Table 1. Summary of specimen/support types

Specimen Type	I α-Al_2O_3	II a-Al_2O_3	III Mo (110)
Crystallinity	single crystal $(1\bar{1}02)$	amorphous	single crystal (110)
Preparation	polishing & etching	ion beam or rf sputter deposition	thinned & etched
Support	a) thin bulk slab b) e-transp. thinned disk	a) Ta-ribbon b) ultra-thin C-film	Mo-ribbon

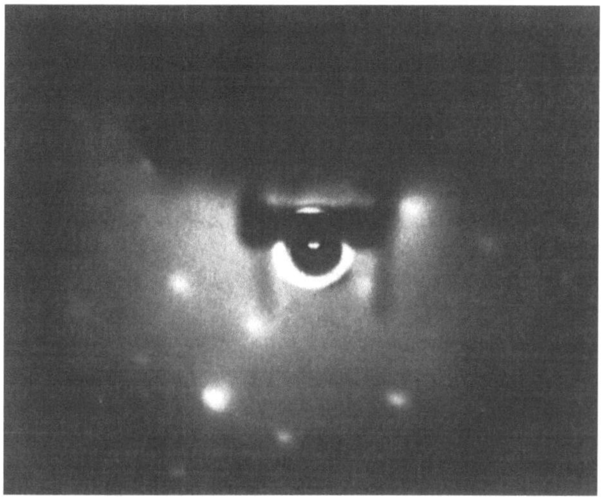

Fig. 1. LEED pattern (117 eV) of α-Al_2O_3 (1$\bar{1}$02) recrystallized after ion bombardment cleaning at 950 K

welded thermo couples). Usually three of these alumina samples were mounted in various combinations on a multiple specimen manipulator; for some tests, a single crystal Mo (110) ribbon was substituted for one of the alumina samples (specimen type III).

Two ways of in-situ specimen surface cleaning were employed. The principal means of cleaning of as-received oxide surfaces was exposure to a RT high pressure (0.03 mbar) rf oxygen discharge [26], which removed all surface carbon and supplied saturation oxygen coverages to the usually oxygen deficient oxide surfaces. Argon ion bombardment with 0.5–2.0 kV beam energy (while simultaneously flooding the specimen with low energy electrons to reduce charging) was also used, particularly when previously deposited metal had to be removed in order to re-use the same support. In all cases of sputter cleaning, the support surfaces became slightly oxygen deficient and the single crystal sapphire surface was "amorphized". Annealing at elevated temperatures was usually sufficient for restoring the surface oxygen concentrations by diffusion from the bulk. The necessary annealing temperatures depended on the type of alumina substrate. The amorphous film supports recovered by extended annealing at 700 K, the maximum temperature allowed before crystallization starts [12]. The sapphire surfaces were annealed at about 800 K, when the amorphized surface state was to be retained, or heated to 1,000 K, when restoration of the previous single crystalline surface state (characterized by a 1×1 LEED pattern) was desired. (This recrystallization temperature corresponds to the recovery of the Al sublattice in sapphire while temperatures of the order of 1,300 K are needed for the full recovery of the oxygen sublattice [27]; higher annealing temperatures led to surface reconstructions [28, 29] which will be discussed in a future publication [30]).

Two basically different types of metal evaporation sources were used. Sublimation sources were employed for the low-rate Fe depositions previously discussed [24] but a miniaturized work-accelerated e-beam evaporator was constructed for deposition rates which were up to $100\times$ higher (about 1 equivalent monolayer per second, eML/s). By evaporating through a mask opening, up to three deposits of different thickness could be placed on a single oxide substrate. The evaporation rate was monitored with a quartz crystal micro-balance and deposition times as low as $2s$ could be realized by utilizing a hand-operated shutter. Whith well outgassed Fe sources, the pressure during high rate evaporations was maintained between 5×10^{-10} and 2×10^{-9} mbar.

When using Mo (110) substrates as supports, the surface preparation procedures were as previously reported (see, for instance [31]).

Results and Discussions

The results at the present date are divided into particle preparation/charaterization procedures and some – still preliminary – particle property measurements. The latter are meant to demonstrate the usefulness of well-controlled preparation conditions for meaningful measurement with model catalytic systems. Fe is the only metal discussed here and no support materials other than aluminas are mentioned, but the preparation procedures and the general experimental approach are applicable to a wide variety of other metal/support systems as well and should, therefore have more general usefulness.

Condensation, Nucleation and Growth

During the early phases of our experiments leading to the preparation of model systems of supported metal particles, low metal deposition rates were usually employed and the oxide supports were held at RT (although a few studies at elevated temperatures were also reported [32, 33]. In this way, metal particles smaller than 1 nm could easily be prepared. However, not varying the two most important experimental parameters that influence the nucleation and growth behavior of the particles – deposition temperature and rate – was not very satisfactory, and the controlled sintering of these particles when exposed to reactive gases at elevated temperatures – needed for many catalysis oriented studies – limited further applications.

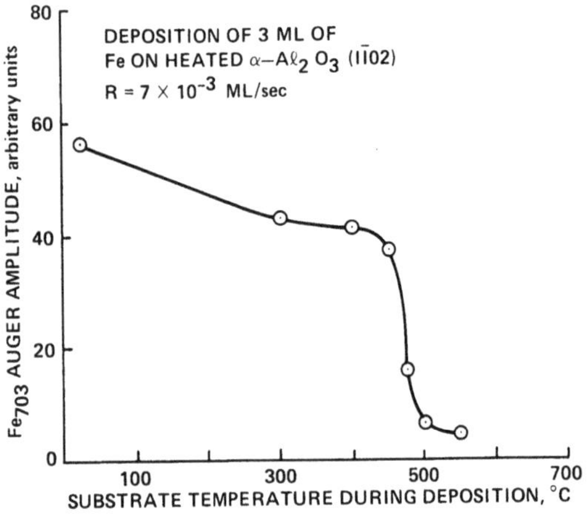

Fig. 2. Auger peak-to-peak (p-p) amplitude of 3 equivalent monolayers of Fe deposited at substrate temperatures ranging from room temperature of 850 K

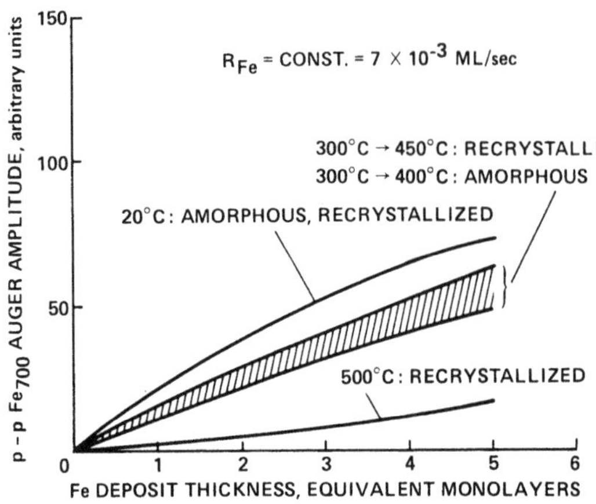

Fig. 3. Change of Fe Auger (p-p) amplitude during cumulative deposition at different substrate temperatures on amorphous and on recrystallized α-Al_2O_3 (1$\bar{1}$02) supports

We, therefore, started to systematically investigate the influence of substrate temperature during deposition, T_s, and impinging vapor flux, R_{Fe}. The effect of both experimental parameters is only known in very general terms for such "unusual" deposition systems as Fe/alumina [34], although they certainly influence many of the essential deposit properties. Some of these properties are size, habit, crystallographic structure, size and habit distributions, number densities, growth behavior, and the general nucleation mechanism on planar alumina supports. The latter is of particular importance because it determines most other deposit properties and because there is a growing conviction among nucleatin researchers [35] that most heterogeneous nucleation situations are governed by nucleation at preferred defect sites, as was shown to be the case for mica supports many years ago [36].

Differentiated AES (and sometimes XPS) measurements of the Fe (47 eV, 703 eV), Al (51 eV), and O (502 eV) transitions were used routinely for monitoring the Fe deposits [24]. When substrate charging problems were encountered at higher Fe coverages (θ_{Fe}), the standard primary electron beam energy of 2.5 kV had to be lowered to 1 kV. Figure 2 shows the effect of increasing T_s on the condensation behavior of 3 equivalent monolayers of Fe on a α-Al_2O_3 (1$\bar{1}$02) surface when deposited with a small impingement rate. A major change in the condensation mechanism occurs between 700 K and 800 K. As a consequence, $T_s = 600$ K was chosen as the standard elevated deposition temperature because it is more than 100 K lower than the temperature at which the condensation behavior changes drastically and because one can expect that particles grown at this temperature should be thermally stable at later treatment temperatures of at least that magnitude. The influence of T_s and support crystallinity is explored in Fig. 3. This figure also shows a drop in the condensation rate of Fe with deposition temperature and displays the general condensation behavior as a function of deposition time for amorphous and recrystallized surfaces (a-Al_2O_3 recrystallizes above 700 K). From the last plot (and from similar AES measurements on a larger variety of differently prepared alumina surfaces) it appears as if at $T_s = 600$ K only minor differences in condensation behavior exist for the different types of alumina supports [24]. This, however, can be quite misleading since AES and XPS do NOT measure the actual degree of Fe dispersion on the support surface (see later).

The Fe surface coverage itself can be quickly assessed by AES [37, 38, 39] when a simplified model of film growth is assumed. Auger peak ratio plots [40] for the growth of Fe on sapphire [41] are shown in Fig. 4 which also displays some of our experimental data. (The following electron inelastic mean free path lengths were used [42]: Fe (1.29 nm), Al (0.477 nm), O (1.498 nm), Mo (0.596 nm); the monolayer thickness for Fe and alumina was 0.205 nm [43], and the respective Auger sensitivity factors for 3 kV primary electrons were: Fe (0.2), Al (0.24), O (0.5), and Mo (0.33). The uppermost curve refers to layer by layer growth whereas the set of solid curves below corresponds to particulate Vollmer-Weber growth for varying Fe coverages. A check of the general applicability of such a simplified theoretical growth model is provided by the data for the known layer by layer growth of Fe/Mo (110), which fit Fig. 4 reasonably well. We can, therefore, have some confidence in the interpretation of our experimental Fe on sapphire data shown for low and

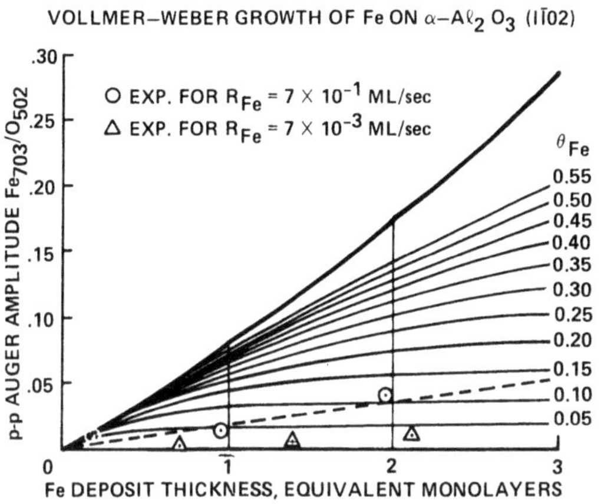

Fig. 4. Calculated dependence of the deposit/support Auger (p-p) ratio on the mode of growth and experimental Auger data for high and low impingement fluxes of Fe (see text). The Fe deposits are made at a substrate temperature of 600 K

high Fe deposition rates in Fig. 4. (A sticking coefficient of 0.7 is assumed, which is based on the comparison of comparable AES measurements on Mo (110) at RT and on sapphire at 600 K; Figure 4 demonstrates also that the sticking coefficient estimate is not critical because of the flatness of the low coverage curves). The Fe coverages deduced from Fig. 4 are still too low by about a factor of 5 when compared to coverages measured directly by post-deposition TEM. This indicates that a more sophisticated growth model should be used in future AES peak ratio analyses. It is also obvious from Fig. 4 that the experimental Fe coverages for the low R_{Fe} deposits are roughly 10 × lower than for the deposits made at 100 × higher rate. Therefore, correspondingly noisy Auger (and XPS) signals resulted for the deposits made with a low deposition rate.

AES ratio plots like Fig. 4 can in principle also provide information on the average thickness of the growing 3-dimensional Fe islands when combined with reliable sticking coefficient measurements. In this case, the island thickness in ML can be obtained by $n_{ML} = d/(d_{ML} \times \theta_{Fe})$ where d is the actual deposit thickness, d_{ML} the monolayer thickness, and θ_{Fe} is inferred from the AES plot.

Although the preceding AES measurements clearly established the fact that Fe grows on alumina in the form of 3-dimensional islands or particles (and all other previous metal/oxide systems investigated in this laboratory have always shown particulate growth), it is reassuring to demonstrate this directly because of recent reports of UHV evaporated Fe growing in the layer by layer mode on another refractory oxide substrate (single crystal rutile [44]). Figure 5 shows normalized Auger amplitude plots of Fe grown in subsequent depositions in the same deposition system on Mo (110) and on α-Al_2O_3 (1Ī02). In spite of a small difference in backscattering of the two substrate types [45], the figure clearly demonstrated that all Fe Auger signals on sapphire (type I specimen) are much smaller than the corresponding Fe amplitudes on Mo (110) (type III specimen) which is known to induce layer growth. (The much reduced Auger signal strength for the low rate deposition of Fe is also obvious again.)

The confirmation of particulate growth in general and of the actual degree of dispersion of the Fe deposit in particular is only possible by direct imaging microstructural techniques. Therefore, good resolution and image contrast enhancing techniques of TEM are indispensable. This is especially true in the case of Fe deposits because of the lower mass density and the attendant imaging difficulties which were demon-

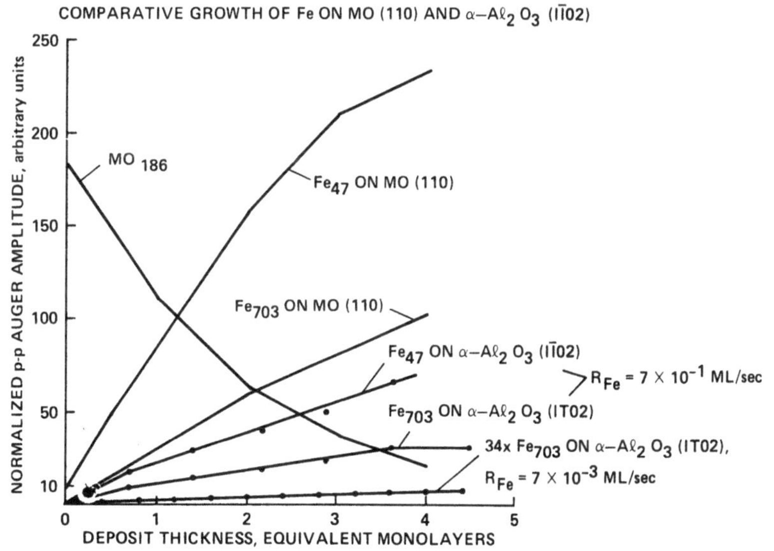

Fig. 5. Auger measurements of the comparative growth of Fe on Mo (110) and on α-Al_2O_3 (1Ī02); the high impingement rate results refer only to the insulating substrate; the lower impingement rate was also used for the deposition on Mo (110). The substrate temperature for the Mo (110) deposition was 300 K whereas 600 K was used for the depositions on sapphire

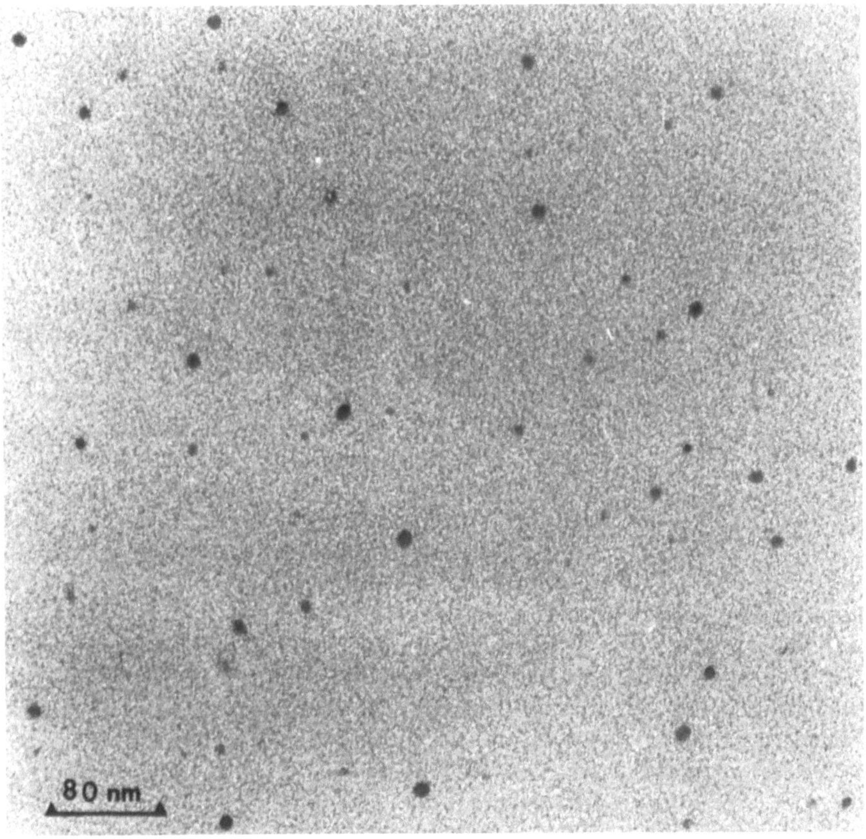

Fig. 6. Electron micrograph of a $2s$ deposit of Fe on a-Al_2O_3 at a substrate temperature of 600 K and with a rate of 0.7 ML/s

strated in a previous report [24] dealing with low-rate Fe deposits.

In addition to the interest in much higher rate Fe depositions (because of the poor AES and XPS signal/noise ratios obtained for small number densities), it was clearly essential to establish also what effect higher rates would have on actual particle number densities and size distributions, and to what degree these deposit properties could be controlled. Figure 6 shows an electron micrograph of a $2s$ Fe deposit on a-Al_2O_3 made at $T_s = 600$ K with a deposition rate of 0.7 ML/s. As previously discussed [24], all post-deposition TEM in this report was also accomplished by using standard TEM techniques (after the thin film Fe/alumina specimens had been removed from the UHV deposition system). Therefore, only the most basic particle properties such as size and average particle to particle distances will be evaluated here because of the known effects of non in-situ TEM techniques on more subtle particle properties such as their shape (see, for instance, [46]). Since Fig. 6 exhibits a wider size distribution than previously seen with Fe and a variety of other metals when deposited at much lower rates (see, for instance [23]), Fe deposits at 0.35 and 0.17 ML/s were also prepared. The results of size distribution measurements on all three

Table 2. Size distribution half widths (in mm) for Fe deposits on α-Al_2O_3 supports at $T_s = 600$ as a function of Fe impingement rate (ML/s)

R_{Fe} (ML/s)	0.7	0.35	0.17
FWHM (nm)	1.8	0.8	0.5

deposits in terms of their FWHM are listed in Table 2.

It can be seen that the distributions are much narrower for the lowest rate. Since the number densities of particles, N_{max}, for all three rates were also found to be approximately the same (1.5×10^{10} cm^{-2}), the 0.17 ML/s rate would seem to be preferable, especially since the associated longer exposure times improve the experimental reproducibility of results. Figure 7 shows a $2s$ Fe deposit on a-Al_2O_3 achieved with such an intermediate rate of at $T_s = 600$ K. The average particle size in this deposit is about 3.0 nm.

In the case of defect controlled preferred nucleation which seems to prevail here ($N_{max} = $ const), differences in the "structure" of the particles with deposition rate, which lead to varying growth rates of individual particles, must be the reason for the observed differences in size distribution. The nature of such structural differences is, however, not clear at all at present.

Clearly the most revealing results of this study were obtained when comparing the "states of dispersion" of Fe deposits made under identical conditions but on different types of alumina surfaces. Both resulted in practically the same Auger measurements within our limits of experimental error. Figure 8 com-

Fig. 7. TEM of a 2*s* deposit of Fe on *a*-Al_2O_3 at 600 K with a rate of 0.175 eML/s

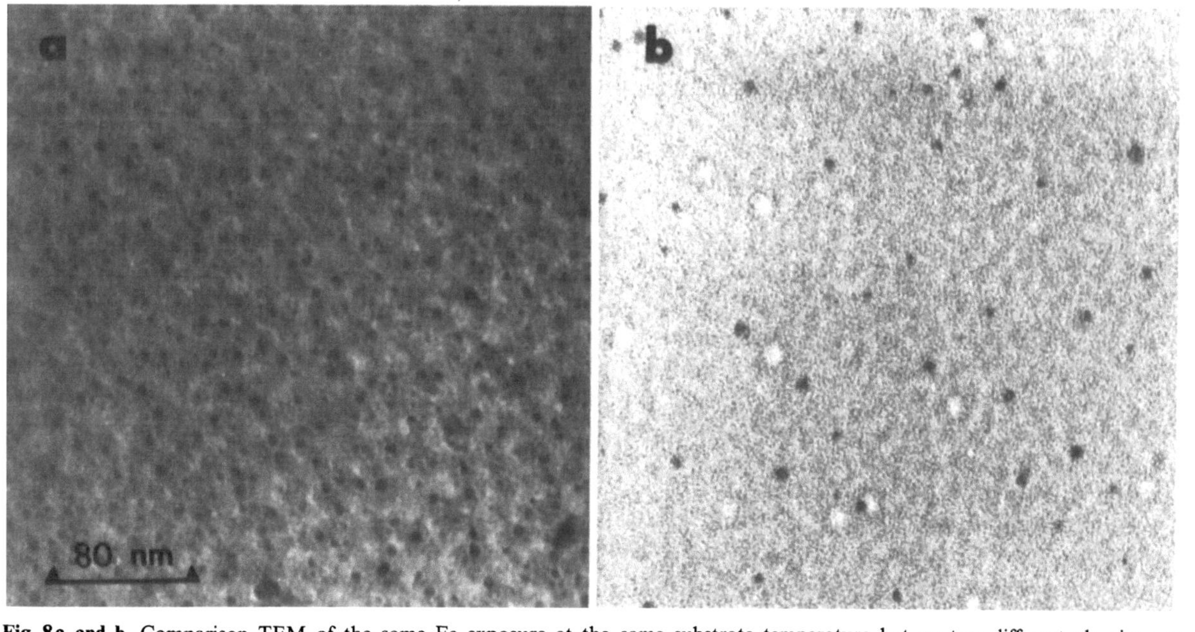

Fig. 8a and b. Comparison TEM of the same Fe exposure at the same substrate temperature but on two different alumina supports: (**a**) 4*s* of Fe at 0.35 ML/s on α-Al_2O_3 $(1\bar{1}02)$; (**b**) 8*s* of Fe at 0.175 ML/s on a-Al_2O_3

pares a 4*s* Fe exposure with a deposition rate of 0.35 ML/s on a α-Al_2O_3 (1$\bar{1}$02) (type I(b) specimen) with a 8*s* Fe exposure of 0.175 ML/s rate on an amorphous alumina specimen (type II (b)); the deposition temperature was 600 K in both cases. The dispersions of the two Fe deposits of identical metal exposure on the two different alumina supports are absolutely different, although the respective Auger signals (Fe_{703}, O_{502}, Al_{51}) are practically the same. (One has to remember that the metal coverage is of the order of about 1% only and even low-energy Auger signals from the support, like Al_{51}, are not very useful for characterizing metal dispersions that differ in island thickness.) This result is a perfect example of the limited value of integrating surface measurements when applied to particulate deposits, especially when higher energy spectroscopic transitions are utilized. Our interpretation of the results of Fig. 8 assumes that the particles on the sample with the noisier background (*b*) are larger and more 3-dimensional than the probably much thinner particles on the low background single crystal support (which somewhat improves the image contrast). The larger and "thicker" particles can then be thought to be easily subdivided into approximately the same number as found on (*a*); the same particle mass will then lead to comparable AES results, especially for the case of relatively large electron escape depths.

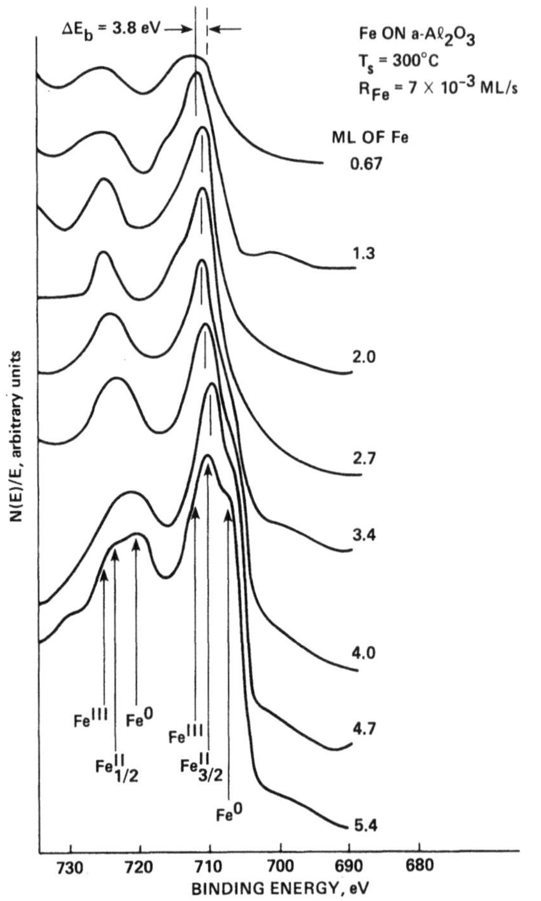

Fig. 9. Smoothed XPS spectra of cumulative Fe deposits on α-Al_2O_3; the metal exposures range from 0.7 to 5.4 eML

Initial XPS Measurements

Because of experimental problems, XPS data are available only from low rate Fe deposits and the spectra evaluated at present were, therefore, quite noisy, especially for the small deposits; they need to be repeated. Figure 9 shows a series of Fe deposits ranging in metal exposure from about 0.7 to 5.4 eML. After correcting for sample charging (by using the substrate peaks), the composite peaks are identified as shown, where the low thickness peaks seem to be mainly of Fe^{II} character with possibly a minor component of Fe^{III} mixed in. Later a $Fe^{II}_{1/2}$ shake-up peak [47] may be contributing and the first sign of metallic Fe^0 is seen at about 3 eML exposure. At present it is impossible to say whether the small amount of residual oxygen in our Fe source [24] influenced the all-oxide character of the small particles. (In the future, attempts will be made to clean up the Fe particles by H_2 reduction at 650 K [48], a treatment which these particles should easily survive without too much sintering, see later.) The main question, however, that needs to be clarified in future tests in the large shift in binding energy of about 3.8 eV, which is much higher than what is usually measured for small particles (see, for instance, [1, 17]) and is interpreted variously in terms of relaxation shifts, initial state shifts, or by a Coulomb effect [49]. If recently expressed doubts about the Coulomb effect [50] can be overcome, it would seem likely that small and highly oxidized metal particles on insulating substrates should be prime candidates for charging effects, and this would then provide a possible explanation for our results.

Annealing of Particles and Desorption of H_2 and CO

As mentioned above, the thermal stability of particulate deposits is of major concern because of their potential as model catalysts. Systematic annealing tests are, therefore, desirable and the annealing results of Fig. 10, which are presented in terms of Fe Auger signal changes as a function of subsequent 3 min anneals, give some indication at what support temperatures major particle rearrangements and/or interactions with the support occur. Our deposition temperature T_s of 600 K was partly chosen because it was

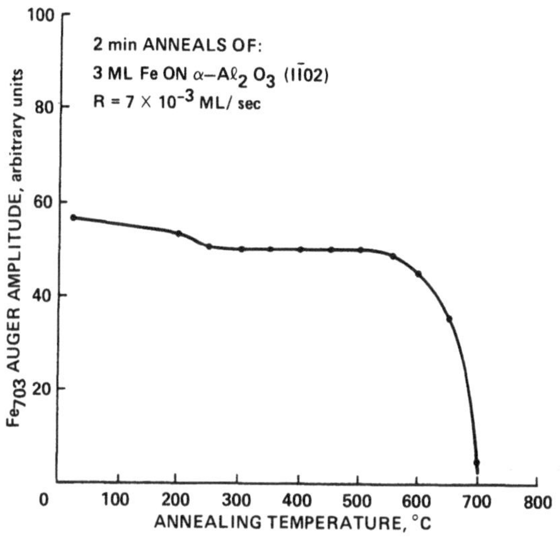

Fig. 10. Auger measurements of a 3 eML deposit of Fe subjected to subsequent 3 min annealing treatments at increasing temperatures

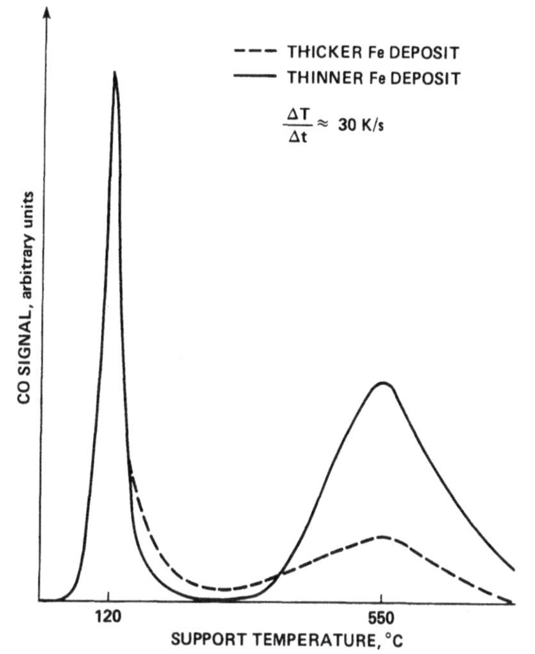

Fig. 11. Temperature programmed desorption spectra of 1 L of CO absorbed at RT on two different particulate deposits of Fe on α-Al_2O_3 (1$\bar{1}$02)

well above 500–550 K where the first major change is seen in Fig. 10, and well beyond the α-CO desorption peak temperature on Fe (110), which is 400 K. The next drastic change in Fig. 10 starts around 850 K where appreciable interdiffusion and reaction with the sapphire substrate takes place resulting probably in spinel formation. (See also [14] where a similar reaction between heavily oxidized Fe deposits and an amorphous alumina support film beginning at about 700 K was detected, whereas Tatarchuk et al. [48], found a strong reaction between Fe and TiO_2 only above 770 K, which is again different from results for Pt/TiO_2 [51]). The relatively good high temperature stability of our particulate deposits is advantageous for two important applications: It seems possible from Fig. 4 that we might be able to record the CO thermal desorption/recombination peak at about 850 K (see Fig. 11 and the discussion below) and it will definitely be feasible to treat particulate Fe deposits at 650 K in H_2 for in-situ reduction/cleaning.

One of the main objectives of working with well defined particulate model catalysts is to look for particle size effects. It was, therefore, attempted to study the decomposition of CO on Fe and see if particle size effects could be found that were similar in nature to the ion bombardment enchanced CO dissociation results reported recently [52]. Figure 11 shows TPD spectra obtained after 1 Langmuir (L) adsorption of CO at RT from two Fe deposits of differing thickness. Since the total Fe surface area is also different in the two cases, the spectra of Fig. 11 were normalized to the low-T CO peak. The thinner deposit exhibits an appreciably higher recombination peak possibly demonstrating what Gonzalez et al. [52] consider the effect of special surface sites on the smaller particles which enhance CO dissociation. Care is indicated, however, with such a simplified interpretation because an appreciable degree of sintering (and support reaction) was present here which caused a slow decay of all spectra upon TPD repetition. The actual Fe particle sizes corresponding to the two deposits are also not known at present since control experiments with TEM monitor specimens have not been performed.

In contrast to CO adsorption, H_2 is expected to constitute a much "gentler" particle surface probe because of its much lower adsorption energy. Figure 12 shows a series of desorptions from a fresh Fe deposit (T_s = 600 K, 9.4 eML metal exposure) exposed to 2.4 L of H_2. The initial peak (trace (1) in Fig. 12) decays rapidly within 5 desorption cycles to (3) although the particles never experienced desorption temperatures higher than 625 K. Residual CO during the higher pressure H_2 exposures is assumed to be causing this decay. Since C and/or O was suspected as surface contaminants, high-T CO recombination removal was attempted (see above), which first led to some recovery of the original β_1 desorption peak (4) and later to the development of a new higher

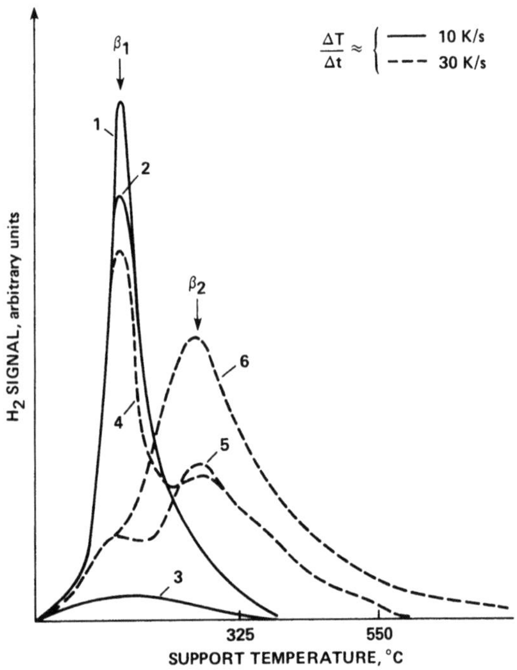

Fig. 12. Temperature programmed desorption spectra of 2.4 L of H_2 on a fresh Fe deposit on α-Al_2O_3 ((1$\bar{1}$02); for details see text

temperature peak β_2. With further high-T desorption cycling, the β_2 peak, which we interpret as representing H_2 desorption from the from the particle interior, dominates the spectrum. The disappearance of the β_1 peak is accompanied by a gradual decrease of the Fe Auger signal, which is an indication of appreciable sintering.

More systematic desorption studies with both gases are certainly desirable and – as is obvious from all the data presented above – conscientious control and constant monitoring of deposit particle sizes is absolutely indispensable.

Conclusions

The preparation and microstructural properties of particulate UHV deposits of Fe on a variety of planar, in-situ cleaned and characterized alumina substrates were studied. The intention of the study was to demonstrate with this particular metal/support system how well-defined and thermally stable model deposits can be prepared and to give an example of some subsequent in-situ measurements that can be performed in a standard surface analysis facility. Using this integrated surface analysis/TEM experimental approach should enhance the prospects of untangling the complex and interrelated problems of supported metal systems, with particle size and support effects being of foremost interest.

Specifically, the following results were obtained:

1. The surface compositional and microstructural nature of the planar support surface can be controlled experimentally which is essential for particle nucleation and growth and, therefore, size. Metal exposure alone is clearly an insufficient measure of particle size.

2. Integrating surface measurements such as AES/XPS are valuable for monitoring the metal condensation process but independent means of assessing the actual state of metal dispersion (particle size and number density) must accompany these measurements.

3. Elevated substrate temperatures during metal deposition substantially increase the thermal stability of particles of a given size. The judicious combination of impingement flux and substrate temperature can control particle sizes.

4. Post-deposition annealing treatments can further improve particle size definitions but are limited by the amorphous/crystalline nature of some planar support materials and by chemical metal/support reactions. In-situ cleaning of partially oxidized Fe particles at elevated temperatures in H_2 is a definite possibility.

5. High metal impingement fluxes increase the maximum number density of particles and improve the signal/noise for particle property measurements. However, too high fluxes favor wider particle size distributions in the case of Fe on a-Al_2O_3.

Thanks are due to R.D. Moorhead of NASA/Ames who helped with the computer evaluation of the AES data. Part of this work was performed under NASA Grants # NCC2-323 and NCC2-394.

References

1. Mason, M.G.: Phys. Rev. B **27**, 748 (1983)
2. Schwab, G.M., Block, J., Mueller, W., Schultze, D.: Naturwissenschaften **44**, 582 (1958)
3. Szabo, Z.G., Solymosi, F., Batta, I.: Z. Phys. Chem. N.F. **17**, 125 (1958)
4. Boudart, M.: Proceedings of the 6th Congress of Catalysis. Vol. 1, p. 2. London: Chem. Soc. 1976
5. Ponec, V.: Metal Support and Metal Additive Effects in Catalysis, p. 63. Imelik, B., Naccache, C., Condurier, G., Praliaud, H., Meriandeau, P., Gallezot, P., Martin, G.A., Vedrine, J.C. (eds.) Amsterdam: Elsevier 1982
6. Tauster, S.J., Fung, S.C., Baker, R.T.K., Horsely, J.A.: Science **211**, 1121 (1981)
7. Bond, G. C.: Metal Support and Additive Effects in Catalysis. p. 1. Imelik, B., Naccache, C., Condurier, G., Praliaud, H., Meriandeau, P., Gallezot, P., Martin, G.A., Vedrine, J.C. (eds) Amsterdam: Elsevier 1982
8. Poppa, H.: Vacuum **34**, 1081 (1984)
9. Erdohelyi, A., Solymosi, F.: J. Catal. **84**, 446 (1983)
10. Hicks, R.F., Bell, A.T.: J. Catal. **90**, 205 (1984)
11. Huizinga, T., Van't Blik, H.F.J., Vis, J.C., Prins, R.: Surf. Sci. **135**, 580 (1983)

12. Cocke, D.L., Johnson, E.D., Merrill, R.P.: Catal. Rev.-Sci. Eng. **26**, 163 (1983)
13. Baker, R.T.K., Prestridge, E.B., McVicker, G.B.: J. Catal. **89**, 422 (1984)
14. Sushumna, I., Ruckenstein, E.: J. Catal. **90**, 241 (1984)
15. Tatarchuk, B.J., Dumesic, J.A.: J. Catal. **70**, 308 (1981)
16. Wang, T., Schmidt, L.D.: J. Catal. **70**, 187 (1981)
17. Kao, C.C., Tsai, S.C., Bahl, M.K., Chung, Y.W.: Surf. Sci **95**, 1 (1980)
18. Takasu, Y., Unwin, R., Tesche, B., Bradshaw, A.M., Grunze, M.: Surf. Sci. **77**, 219 (1978)
19. Brugniau, D., Parker, S.D., Rhead, G.E.: Thin Solid Films **121**, 247 (1984)
20. Chien, S., Shelimov, B.N., Resasco, D.E., Lee, E.H., Haller, G.L.: J. Catal. **77**, 301 (1982)
21. Poppa, H.: Ultramicroscopy **11**, 105 (1983)
22. Boudart, M., Topsoe, H., Dumesic, J.A.: The physical basis for heterogeneous catalysis. p. 337. Drauglis, E., Jaffee, R.I. (eds). New York, London: Plenum Press 1975
23. Poppa, H., Moorhead, D., Heinemann, K.: Thin Solid Films **128**, 251 (1985)
24. Poppa, H., Papageorgopoulos, C.A., Bauer, E.: Ultramicroscopy (in press)
25. These samples were prepared by Dr. Thomas Allen of the Optical Coating Laboratories in Santa Rosa, CA
26. Mittal, K.L.: Surface contamination. Vol. 2. New York, London: Plenum Press 1979
27. Farlow, G.C., White, C.W., McHargue, C.J., Appleton, B.R.: Mat. Res. Soc. Symp. Proc. Vol. 2, p. 395. Amsterdam, Oxford, New York: Elsevier 1984
28. French, T.M., Somorjai, G.A.: J. Phys. Chem. **74**, 2489 (1970)
29. Chang, C.C.: J. Appl. Phys. **39**, 5570 (1968)
30. Papageorgopoulos, C.A., Poppa, H.: (to be published)
31. Bauer, E., Poppa, H.: Thin Solid Films **121**, 159 (1984)
32. Doering, D.L., Poppa, H.: Proceedings of the 10th International Congress on Electron Microscopy. p. 499. Hamburg: Wissenschaftliche Verlagsgesellschaft 1982
33. Ladas, S., Poppa, H. Boudart: Surf. Sci. **U102**, 151 (1981)
34. Venables, J.A., Spiller, G.D.T.: Surface mobilities on solid materials. Binh, V.T. (ed.), p. 341. New York, London: Plenum Press 1983
35. Harsdorf, M.: Thin Solid Films **116**, 55 (1984)
36. Elliot, A.G.: Surf. Sci. **44**, 337 (1974)
37. Bauer, E., Poppa, H.: Thin Solid Films **12**, 167 (1972)
38. Biberian, J.P., Somorjai, G.A.: Appl. Surf. Sci. **2**, 352 (1979)
39. Rhead, G.E., Barthes, M.G., Argile, C.: Thin Solid Films **82**, 201 (1981)
40. Ossicini, S., Memeo, R., Ciccacci, F.: J. Vac. Sci. Technol. A **3**, 387 (1985)
41. Moorhead, R.D., of NASA/Ames, provided the computer plots
42. Seah, M.P., Dench, W.A.: Surf. Interf. Anal. **1**, 2 (1979)
43. Mathieu, H.J., Datta, M., Landolt, D.: J. Vac. Sci. Technol. A **3**, 331 (1985)
44. Brugniau, D., Parker, S.D., Rhead, G.E.: Thin Solid Films **121**, 247 (1984)
45. Methods of surface analysis. Czanderna, A.W. (ed.), p. 165. Amsterdam, Oxford, New York: Elsevier 1975
46. Heinemann, K., Osaka, T., Poppa, H., Avalos-Borja, M.: J. Catal. **83**, 61 (1983)
47. Brundle, C.R., Chuang, T.J., Wandelt, K.: Surf. Sci. **68**, 459 (1977).
48. Tatarchuk, B.J., Dumesic, J.A.: J. Catal. **70**, 323 (1981)
49. Wertheim, G.K., Dicenso, S.B., Youngquist, S.E.: Phys. Rev. Lett. **51**, 2310 (1983)
50. Parmigiani, F., Kay, E., Bagus, P.S., Nelin, C.: J. Electron Spectrosc. Relat. Phenom. **36**, 257 (1985)
51. Schreiffels, J.A., Belton, D.N., White, J.M.: Chem. Phys. Lett. **90**, 261 (1982)
52. Gonzales, L., Miranda, R., Ferrer, S.: Surf. Sci. **119**, 61 (1982)

H. Poppa
IBM Almaden Research Center
650 Harry Road
San José, CA 95120-6099
USA

C.A. Papageorgopoulos
Department of Physics
University of Ioannina
Greece

F. Marks
Department of Physics
Stanford University
Stanford, CA 94305
USA

E. Bauer
Physikalisches Institut
Technische Universität
D-3392 Clausthal Zellerfeld
Federal Republic of Germany

Atom Desorption Energies for Sodium Clusters

M. Vollmer and F. Träger
Physikalisches Institut der Universität Heidelberg, Federal Republic of Germany

Received July 28, 1986; final version August 19, 1986

The desorption energies of supported sodium clusters have been determined as a function of cluster size. Na_n clusters were formed by surface diffusion of sodium atoms adsorbed from a thermal atomic beam on a LiF(100) single crystal. Measurements have been performed by temperature programmed thermal desorption. The signals reflect fractional order desorption kinetics. The average cluster size could be controlled by varying the total number of sodium atoms on the surface. It was determined from scattering experiments. We find that the binding energies vary between 0.55 and 0.8 eV and only approach a constant value for clusters with diameters as large as 1,000 Å.

PACS: 36.40.+d; 68.45.Da; 82.65.−i

I. Introduction

Of particular importance for a detailed understanding of atomic clusters are the binding energies and their variation as the cluster size increases. In the past, binding energies have been reported in a number of theoretical papers [1–4]. Most calculations, however, were restricted to small clusters with at most forty atoms. Only few publications cited energies for larger aggregates the properties of which may already approach those of the bulk [5]. Systematic experimental determinations of cluster binding energies are not available at all.

In this paper we present measurements of atom desorption energies for sodium clusters, i.e. binding energies of individual atoms on the cluster surface. They have been determined by thermal desorption experiments, the clusters with radii of up to 1,000 Å being formed by surface diffusion on a LiF(100) single crystal. The cluster size distribution could be determined independently by scattering experiments. Besides the measurements of binding energies the goal of this work was to study the evaporation of small particles and to investigate fractional order thermal desorption processes systematically.

II. Experimental Methods

a) Cluster Formation and Cluster Size Determination

When atoms impinge on a cold insulator surface they are trapped in the surface potential (see e.g. [6]). Their residence time on the surface depends on the surface temperature and is given by $\tau_{res} = \tau_0 \exp(E_a/kT)$. Since the energy barrier for surface diffusion is typically only about one half of the binding energy E_a in the surface potential, the atoms can easily diffuse during their residence time which has been demonstrated e.g. for gold on alkali halide crystals [7].

For the present case of sodium atoms on LiF(100) the binding to the surface is initially accomplished by weak Van der Waals forces. However, if diffusing atoms occasionally collide with each other and form clusters or if they meet surface defects, the binding energy is increased. Therefore, the nucleation of atoms to clusters is energetically favourable. These clusters are likely bound to surface defects which act as nucleation centers. Metal clusters on surfaces and their size distributions have been studied extensively, in particular by electron microscopy (see e.g. [8–10]).

In the present experiment the growth of clusters during the exposure of the surface to a thermal atomic beam has been monitored by scattering measurements. The sodium beam is directed onto the surface and the rate of scattered atoms is measured as a function of time with a quadrupole mass spectrometer. With the reasonable assumption that only atoms desorb but clusters do not, the time dependence of the scattering signal reflects the increase of the area on the surface which is covered with clusters. Since atoms bind more strongly to these clusters than to the alkali halide crystal, the scattering signal decreases during the deposition. Within the framework of an atomistic model (see below) the time dependence of the scattering signal can be calculated. A fit to the experimental data then gives the mean cluster size for a given deposition time and a given constant flux of the atomic beam.

In order to extract the mean cluster radii from the scattering rates we have developed a simple model. It is based on the following assumptions:

1. The Na atoms which are initially adsorbed with 100% probability on the cooled LiF(100) surface may either desorb after a mean residence time $\tau_{res} = \tau_0 \exp(E_a/kT)$ or form clusters via surface diffusion.

2. Since clusters are bound more strongly to the surface only single atoms can desorb.

3. The surface diffusion of clusters is negligible.*

4. The surface diffusion of adatoms can be described as a random walk. During the diffusion the atoms cover an area R^2. According to Einstein [11] it is related to the time of diffusion τ_{diff} to the nearest capture site. Therefore, each nucleus or growing cluster is considered to be surrounded by a capture zone of radius R. This is plausible, since under our experimental conditions the probability that an adatom is captured by the nearest cluster [12] is of the order of 1.

5. The distribution of clusters on the surface can be approximated by a square lattice.

Under these assumptions the scattering of atoms from the surface, the diffusion of adatoms and the nucleation of clusters can be described quantitatively.

We first calculate the mean distance $R(t)$ which an atom has to migrate on the surface before it reaches a cluster. Half the distance of the centers of two neighbouring clusters is denoted by $R_{1/2}$, the number density of clusters by $1/(4R_{1/2}^2)$ and the mean cluster size by $\langle R(t)\rangle$. r_0 is the lattice constant. In a distance R' from the center of a cluster there are $2\pi R'\,dR'/r_0^2$ adsorption sites which serve as possible starting points for single atom diffusion. We obtain for $R(t)$:

$$R(t)=\frac{r_0^2}{\pi(R_{max}^2-\langle R\rangle^2)}\int_{\langle R\rangle}^{R_{max}}\frac{1}{r_0^2}\cdot 2\pi R'(R'-\langle R\rangle)\,\mathrm{d}R' \qquad (1)$$

R_{max} is the radius of the capture zone and is calculated from $\pi R_{max}^2=4R_{1/2}^2$, so that $R(t)$ finally becomes:

$$R(t)\simeq\frac{\left(\frac{3}{4}-y+\frac{\pi}{12}y^3\right)}{\left(1-\frac{\pi}{4}y^2\right)} \qquad (2)$$

with $y=\langle R\rangle/R_{1/2}$. y is the ratio of the mean cluster size and the mean half distance of neighbouring clusters. $y=1$ describes the situation where the clusters have grown to a size at which they start to touch each other, i.e. the onset of coalescence. In the present experiments y was always considerably smaller than 1.

From the mean distance the average time can be calculated which an atom needs to diffuse to a cluster and get bound more strongly. Einstein (see above) gives the following formula for random walk diffusion [11]:

$$R^2=D\times t \qquad (3)$$

R is the mean square distance that the particle diffuses in a time t; D is the diffusion coefficient. The mean diffusion time $\langle\tau_{diff}\rangle$ before reaching a cluster at the distance $R(t)$ is

$$\langle\tau_{diff}\rangle=(\langle R(t)\rangle/r_0)^2\,\tau_{diff,1} \qquad (4)$$

$\tau_{diff,1}$ is the average time required for one diffusion hop over the distance of the lattice constant r_0. Since the scattering signal is very weak, i.e. most Na atoms impinging on the surface are adsorbed, the diffusion time $\langle\tau_{diff}\rangle$ must be considerably smaller than the residence time τ_{res} on the surface.

Therefore, the steady state rate equation for the number of single adatoms on the surface is

$$n_1(t)=F A_{free}\langle\tau_{diff}\rangle \qquad (5)$$

where F is the sodium flux and $A_{free}=A_0[1-(\pi/4)y^2]$ is the surface area which is not covered with clusters. The scattering signal is therefore given by:

$$S(t)=\alpha\cdot\frac{n_1(t)}{\tau_{res}} \qquad (6)$$

where α stands for the detection sensitivity of the system. One finally obtains by combining Eqs. (2)–(6):

$$S(t)=C\cdot R_{1/2}^2(t)\cdot h(y(t)) \qquad (7)$$

with

$$C=\frac{\alpha\cdot F\cdot A_0}{r_0^2}\cdot\frac{\tau_{diff,1}}{\tau_{res}}$$

$$h(y)=\frac{\left(\frac{3}{4}-y+\frac{\pi}{12}y^3\right)^2}{\left(1-\frac{\pi}{4}y^2\right)}.$$

---

* Several contradictory observations about surface diffusion of clusters can be found in the literature. Most measurements have been made by electron microscopy. It seems that cluster diffusion, if at all, only occurs at relatively high temperatures and is possibly stimulated by electron bombardment [13, 14]. We therefore believe that surface diffusion of clusters is negligible under the conditions of the present experiment

This equation contains two variables, namely $\langle R(t)\rangle$ and $R_{1/2}$. Therefore, a second equation is needed for the calculation of $S(t)$. It is derived from the total coverage $N(t)$ which is determined by thermal desorption following the scattering experiment. In the above used model of hemispherical clusters, $N(t)$ is

$$N(t)=\frac{\pi}{6}\cdot\frac{A_0}{r_0^3}\cdot y^2(t)\cdot\langle R(t)\rangle. \quad (8)$$

From nucleation experiments it is known that for large deposition times, but well before coalescence occurs, the number density of clusters reaches a constant value, the saturation density of nuclei [15]. Then $R_{1/2}(t)$ becomes constant and only the number of atoms accumulated in a cluster increases. We can therefore write

$$y=g\cdot t^{1/3} \quad (9)$$

with

$$g=\sqrt[3]{\frac{6\cdot F\cdot r_0^3}{\pi\cdot R_{1/2}}}.$$

Combining Eq. (9) and Eq. (7) we finally obtain for $S(t)$

$$S(t)=C\cdot R_{1/2}^2\cdot h(g\,t^{1/3}) \quad (10)$$

which describes the scattering signal for long deposition times when the number density of clusters has reached the saturation value.

The factor C can be determined with a least square fit of the scattering rates to the analytical expression Eq. (10). In principle, $S(t)$ and $N(t)$ then are sufficient to extract $R_{1/2}(t)$ and $\langle R(t)\rangle$.

Under certain circumstances the number density of nuclei is reached so fast that the $y=g\cdot t^{1/3}$ law holds for almost the whole scattering signal. This can be verified by applying the $y\propto t^{1/3}$ dependence with the same parameters as obtained for long deposition times also to the initial decrease of the scattering signal.

It should be noted that in several nucleation theories the surface diffusion of clusters as well as decay processes of clusters have been taken into account. Especially the overlap of competing capture zones yields single adatom densities which are expressed by Bessel functions [16, 17]. We believe, however, that such a mathematical treatment is not necessary here. Also, our specific choice of the cluster distribution should not have a critical influence on the result. A different spatial distribution can only slightly modify the size of the capture zone. Nevertheless, Monte Carlo simulations of the diffusion and nucleation would be helpful.

b) *Thermal Desorption Experiments*

Thermal desorption [18] experiments are carried out by heating the sample with constant rate β and simultaneously recording the rate of desorbing particles as a function of temperature. The desorption process is described theoretically with the following Arrhenius equation:

$$-\mathrm{d}n(T)/\mathrm{d}t=n(T)^x\,\nu\exp(-E_a/kT) \quad (11)$$

with $T(t)=T_0+\beta t$.

$\mathrm{d}n(T)/\mathrm{d}t$ denotes the rate of desorbing particles, and $n(T)$ the remaining coverage. E_a is the activation energy for desorption. The frequency factor ν corresponds to the vibration of the adsorbed particle in the surface potential and is of the order of $10^{12}/s$. x is introduced as a parameter which describes the formal order and the kinetics of the desorption. $n(T)^x$ represents the number of particles on the surface which can participate in the desorption process.

III. Experiment

The experimental arrangement for thermal desorption experiments is shown schematically in Fig. 1. It consists of an ultrahigh vacuum system with a base pressure of typically 2×10^{-10} mbar. A quadrupole mass spectrometer serves to detect desorbing sodium atoms. A glass cell with high purity metallic sodium is attached to the system. Through heating sodium atoms diffuse out of a small orifice and pass a number of liquid nitrogen cooled diaphragms, which form a well collimated beam. It has a diameter of approximately 8 mm at the center of the UHV-chamber. Here, it impinges either on a quartz crystal microbalance or on the substrate. Both are attached to a manipulator and can be cooled to liquid nitrogen temperature. The substrate consists of a LiF(100) single crystal with a diameter of 10 mm and a thickness of about 4 mm. It is spring loaded against a small molybdenum oven that can be heated resistively up to 800 K. A NiCrNi thermocouple is integrated in the crystal holder and measures the surface temperature of the crystal. The sample is surrounded by a liquid nitrogen cooled copper shield (see Fig. 2). It serves as an additional aperture for the atomic beam and guarantees that sodium atoms only condense on the substrate. In addition, during thermal desorption experiments this shield prevents sodium from diffusion into the vacuum system. Only those atoms can pass that travel into the direction of the quadrupole mass spectrometer. Thus, the background signal is kept negligibly small. The frequency of the quartz crystal microbalance, the surface temperature of the LiF crys-

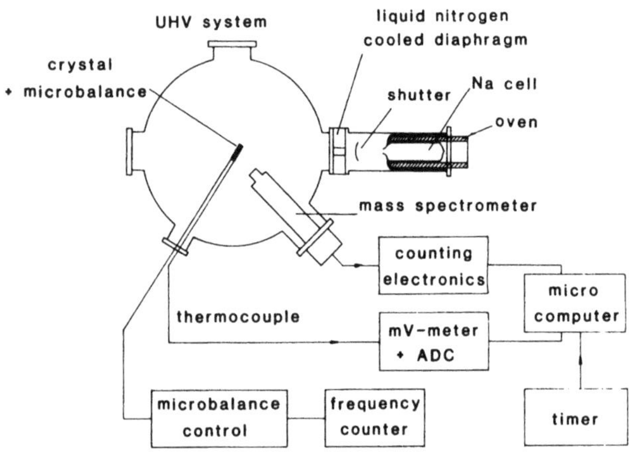

Fig. 1. Experimental arrangement for thermal desorption experiments with sodium clusters on a cold LiF(100) single crystal surface

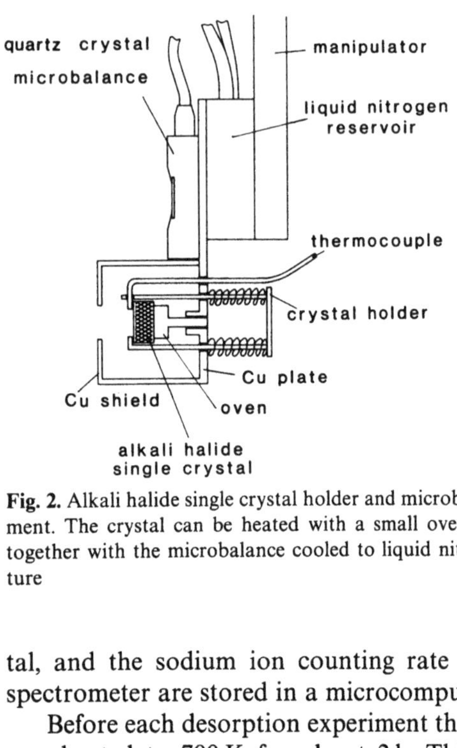

Fig. 2. Alkali halide single crystal holder and microbalance arrangement. The crystal can be heated with a small oven to 700 K and together with the microbalance cooled to liquid nitrogen temperature

tal, and the sodium ion counting rate of the mass spectrometer are stored in a microcomputer.

Before each desorption experiment the LiF crystal was heated to 700 K for about 2 h. This procedure serves for two purposes: firstly, contaminations on the surface are removed and, secondly, almost all active sites for subsequent adsorption of residual gases are annealed. As has been demonstrated for H_2O on alkali halide surfaces [19], the surface remains very clean for several hours in a UHV environment after such a heat treatment. This has also been verified by Auger electron spectroscopy [20]. Between two runs, the crystal was cleaned by heating to 700 K for more than 30 min.

The arrangement for the scattering experiments was nearly the same as for the thermal desorption studies. For the detection of scattered atoms during Na-deposition on the liquid nitrogen cooled LiF surface, a sample holder similar to that in Fig. 2 but without a coppershield was used.

The sodium coverage on the LiF-surface was determined by measurements with a quartz crystal microbalance [21]. The coverage of one (equivalent) monolayer with about 3.7×10^{14} atoms corresponds to a frequency change of 3.5 Hz. In the present experiment the flux of the atomic beam has been determined first. For this purpose the oven for heating the sodium cell was held at constant temperature and the gold coated quartz crystal surface was exposed to the sodium beam for typically 5 min. With the resulting change of the microbalance frequency the flux could then be determined accurately. Therefore, the coverage of the LiF-crystal can also be calculated precisely even for deposition times of only several seconds during the actual experiment. They result in coverages of fractions of a monolayer. The different sticking coefficients of the Na atoms on the gold surface of the microbalance and on the LiF-crystal have been investigated by scattering measurements and adequately taken into account.

IV. Measurements and Results

a) Scattering Experiments

The result of a measurement of the scattering rate as function of time is shown in Fig. 3. As expected, the signal gradually decreases. The solid line refers to the analytical dependence for $S(t)$ which is based on the atomistic model explained above. A least square fit of Eq. (10) for long deposition times (400 to 600 s), i.e. when the number density of clusters has reached the saturation value, was used to determine the factor C.

As can be seen from Fig. 3, the calculated scattering signal $S(t)$ fits the experimental data very well. Even the amplitude of the signal for the first few seconds of Na deposition is in excellent agreement with the theoretical fit. Thus, the $y \propto t^{1/3}$ approximation seems to describe the whole scattering signal for the case of Na_n clusters on a well annealed LiF(100) surface. This indicates (see above) that the number density of clusters saturates very rapidly. We obtain the value of 5×10^8 clusters/cm^2, which is comparable to the number of defects on well annealed alkali halide surfaces. Therefore, it is very likely that the defects act as nucleation centers. The second fit parameter, the amplitude C, gives the ratio of $\tau_{\mathrm{diff},1}/\tau_{\mathrm{res}}$. It is very small, i.e. only amounts to about 10^{-7}. Therefore, the surface diffusion to clusters proceeds very fast explaining the small rate of scattered atoms. Unfortunately, the number of clusters on the surface is relatively small so that the cluster diameters quickly reach values of the order of 100 Å. We find from the

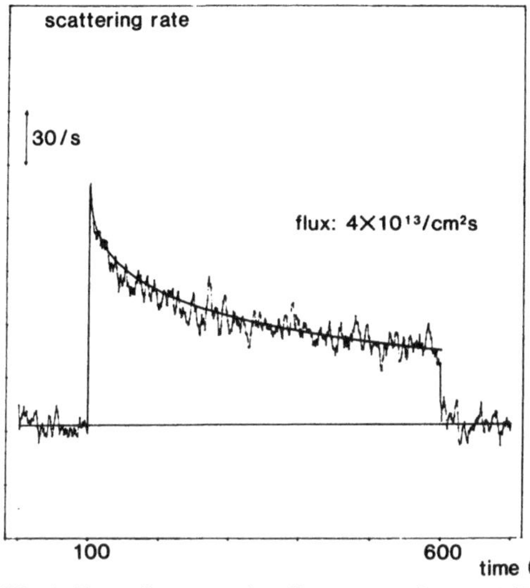

Fig. 3. Scattering rate of sodium atoms from a LiF(100) surface as a function of deposition time at a constant Na flux of 4×10^{13} atoms/cm^2 s. The solid line represents a least square fit for deposition times between 400 and 600 s according to the theory explained in the text. The experimental curve was obtained by adding up the results of four subsequent experiments

scattering experiments that clusters with radii between $\simeq 100$ Å and 1,500 Å were investigated.

b) Thermal Desorption Experiments

Thermal desorption spectra for different initial coverages are displayed in Fig. 4. The most characteristic feature is the shift to larger desorption temperatures with increasing coverage. According to the description of the desorption process with an Arrhenius equation (see above), this implies fractional order thermal desorption, i.e. for the order x holds $x<1$.

An Arrhenius plot with $\log(dn/dt)$ versus $1/T$ gives the energy E_a, if the change in $n(T)^x$ is negligible. This condition holds if the analysis is restricted to that part of the signal where the desorption rate gradually starts to increase, i.e. where the coverage decreases by at most 4% of its initial value [22]. Such an analysis is easily possible for the good signal-to-noise ratios associated with large coverages, i.e. large clusters. For low coverages, however, E_a can only be determined with reasonable accuracy if a 20% or even 40% decrease enters the analysis. This introduces an error in E_a. We account for this error in the following way. The order x is calculated from the previously derived energy value E_a with the relation:

$$x = \frac{\beta \cdot E \cdot n(T_p)}{(k \cdot T_p^2 \cdot \frac{dn}{dt}(T_p)}. \qquad (12)$$

It can be calculated from the analytical expression for the shape of the desorption signal (Eq. 11) and the condition that the derivative of the rate of desorbing particles is zero at the temperature T_p where the signal is maximum, i.e. $d/dT[dn/dt]=0$. The value for x is then used for a corrected Arrhenius diagram with $\log[(dn/dt)/n^x]$ versus $1/T$, from which a new E_a value is computed and so forth until selfconsistency is obtained. This procedure is actually equivalent to a fit of the experimental data to the lineshape of the signal. As a result of the thermal desorption spectra, we finally obtain the desorption energy E_a as a function of the average cluster size and the order of desorption. The dependence of these energies on the cluster radius is displayed in Fig. 5. The order of desorption is $x=0.79(8)$.

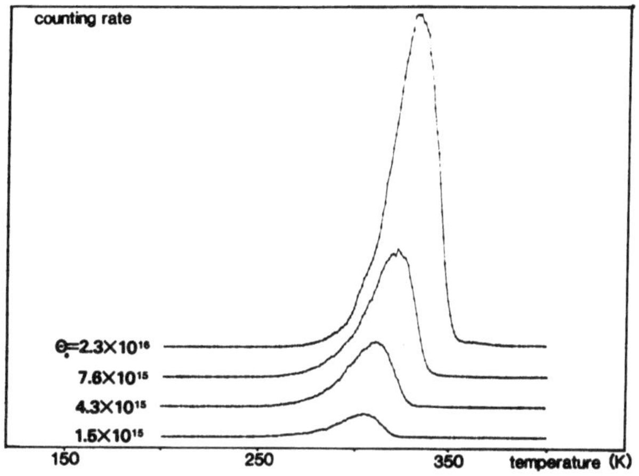

Fig. 4. Thermal desorption spectra of sodium atoms evaporating from Na_n clusters for different coverages θ on the surface, i.e. different cluster sizes

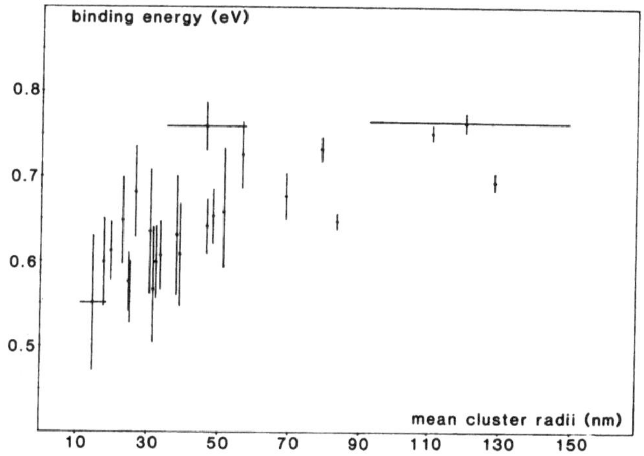

Fig. 5. Binding (desorption) energies of sodium clusters as a function of mean cluster size

V. Discussion

a) Order of Desorption

The order of $x=0.79(8)$ can be explained with desorption of atoms from clusters. In a simple picture of

f hemispherical clusters of only one size R, one finds for the total number of atoms n_{total}, the number of atoms at the cluster surface n_{surf}, and at the cluster perimeter n_{perim}:

$$n_{\text{total}} = (2\pi/3) f \times (R/r_0)^3$$

$$n_{\text{surf}} = 2\pi f \times (R/r_0)^2$$

$$n_{\text{perim}} = 2\pi f \times (R/r_0)$$

where r_0 denotes the lattice constant. If we assume that atoms can only desorb from the cluster surface, the order of desorption is given by

$$x = \log(n_{\text{surf}})/\log(n_{\text{total}}).$$

In the present case the number of clusters per square centimeter is approximately $f = 5 \times 10^8$ (see above). For an average cluster radius of $R = 100$ Å and the lattice constant of bulk sodium of $r_0 = 3.7$ Å we obtain $x_{\text{surf}} = 0.93$. Similarly, under the same conditions, the order is $x_{\text{perim}} = 0.82$ if atoms only desorb from the perimeters of the clusters. With a Gaussian cluster size distribution similar values for x are obtained. Neglecting for the moment the slight dependence of the order x on the cluster radius [23], it seems likely that the desorption primarily takes place from the cluster perimeters. In a more realistic picture of a cluster the atoms would then primarily desorb from edges of similar binding energy. The kinetics of fractional order desorption will be discussed in more detail elsewhere [23].

b) Binding Energies

As can be seen from Fig. 5 the binding energies slowly approach a saturation value of approximately 0.8 eV. A similar increase of the binding energies with coverage was also observed for Cu and Au on graphite substrates [24]. However, no attempt was made to relate the energies to certain cluster sizes. The saturation value of 0.8 eV observed here is smaller than the heat of sublimation of 1.113 eV [25] and the heat of vaporisation of 0.93 eV [26]. At a first glance one would expect that the energies measured here should approach one of these two values. We will therefore first discuss the physical meaning of the extracted energies.

In the present experiment the clusters are relatively large and cold so that the desorption of atoms should be governed by the heat of sublimation. On the other hand one has to keep in mind that the thermodynamic properties of small particles can be quite different from those of the bulk material. For example, the vapor pressure and the enthalpy of vaporisation are size dependent [27]. Also, the melting temperature decreases with decreasing particle size, an effect which is particularly pronounced for cluster diameters below 100 Å [28, 29]. Since the melting temperature of bulk sodium is 97.8° C, but the thermal desorption spectra peak around 0° C for the smallest clusters with $\langle R \rangle \simeq 100$ Å investigated here, it seems unlikely that the melting temperature is already reached at the initial part of the desorption spectra from which the energies are extracted. Consequently, the experimental values (at least for the largest clusters) have to be compared to the heat of sublimation of sodium. We therefore have to discuss the apparent difference of the sublimation energy of 1.113 eV and the binding energies of maximum 0.8 eV. Here, the kinetics of the desorption process is helpful. Since the order is in agreement with the assumption of evaporation from the perimeters of the Na-clusters, the desorption energies of these atoms may help to explain the observed saturation value (see Fig. 5). Clearly, the number of nearest neighbours influences the binding energy [5]. This effect becomes particularly important at the cluster perimeter where atoms are bound less strongly than in the interior of the cluster. In a simple geometrical model one expects the number of nearest neighbours for perimeter atoms of a large cluster to vary between about 0.65 and 0.7 of the corresponding number for atoms on a flat surface of bulk sodium. Thus, the bulk sublimation energy of 1.12 eV should decrease to $\simeq 0.8$ eV for perimeter atoms, which is in agreement with our data. These arguments are certainly qualitative only. Nevertheless, they explain the experimental results nicely and illustrate together with the experimental data that the coordination number is of primary importance for the understanding of cluster binding energies. However, more experimental and theoretical work is needed to understand in detail why the desorption energies of maximum 0.8 eV are substantially lower than the heat of sublimation. Unfortunately, most theoretical papers only give values averaged over the entire cluster so that a comparison to the experimental values cannot be made. Finally, one should keep in mind, that the substrate may influence the binding energy. We believe, however, that such effects are of minor importance here since they should show up only for rather small clusters. In any case, the general trend of the binding energies can indeed be regarded as caused by quantum size effects.

In conclusion, we have analysed fractional order desorption kinetics and determined binding energies of sodium clusters. The average size and the number density of the clusters could be derived from scattering experiments. The model developed here for surface diffusion and cluster formation is supported, for example, by the result of 5×10^8 clusters/cm^2 which very likely coincides with the number of defects on the surface. Measurements with clusters of smaller size and comparable signal-to-noise ratio should be possible by artificially increasing the number of defects to which the clusters stick. This can be accomplished e.g. by electron bombardment of the surface after the heat treatment [30]. For a given coverage the average cluster size will then be smaller than in the present work. Actually, it should be possible to study the evaporation of particles as small as 10 Å. For future measurements electron microscopy for direct cluster size determination would also be desirable.

References

1. Skala, L.: Phys. Status Solidi B **107**, 351 (1981); B **109**, 733 (1982)
2. Iniguez, M.P., Baladron, C., Alonso, J.A.: Surf. Sci. **127**, 367 (1983)
3. Rao, B.K., Jena, P., Shillady, D.D.: Phys. Rev. B **30**, 7293 (1984)
4. Koutecký, J., Fantucci, P.: Chem. Rev. **86**, 539 (1986)
5. Kadura, P., Künne, L.: Phys. Stat. Sol. B **88**, 537 (1978)
6. Hurkmans, A., Overbosch, E.G., Olander, D.R., Los, J.: Surf. Sci. **54**, 154 (1976)
7. Robinson, V.N.E., Robins, J.L.: Thin Solid Films **20**, 155 (1974)
8. Robins, J.L.: Surf. Sci. **86**, 1 (1979)
9. Venables, J.A., Spiller, G.D.T., Hanbücken, M.: Rep. Prog. Phys. **47**, 399 (1984)
10. Schmeisser, H.: Thin Solid Films **22**, 83 (1974)
11. Einstein, A.: Ann. Phys. **17**, 549 (1905)
12. Lewis, B.: Surf. Sci. **21**, 273 (1970)
13. Honjo, G., Takayanagi, K., Kobayashi, K., Yagi, K.: Phys. Status Solidi A **55**, 353 (1979)
14. Kinosita, K.: Thin Solid Films **85**, 223 (1981)
15. Poppa, H.: J. Appl. Phys. **38**, 3883 (1967)
16. Stowell, M.J.: Philos. Mag. **26**, 349 (1972)
17. Lewis, B., Rees, G.J.: Philos. Mag. **29**, 1253 (1974)
18. Menzel, D.: Thermal desorption. In: Springer Series in Chemical Physics. Vol. 20, Chemistry and Physics of Solid Surfaces IV. Berlin, Heidelberg, New York: Springer 1982
19. Estel, J., Hoinkes, H., Kaarmann, H., Nahr, H., Wilsch, H.: Surf. Sci. **54**, 393 (1976)
20. Weibler, W.: PhD Thesis, Heidelberg 1982 (unpublished)
21. Pulker, H.K.: Thin Solid Films **32**, 27 (1976)
22. Habenschaden, E., Küppers, J.: Surf. Sci. **138**, L 147 (1984)
23. Vollmer, M., Träger, F.: (to be published)
24. Arthur, J.R., Cho, A.Y.: Surf. Sci. **36**, 641 (1973)
25. Kittel, Ch.: Introduction to Solid State Physics. 5th Edn. New York: J. Wiley (1976)
26. CRC Handbook of Chemistry and Physics, 63th edn. (1983)
27. Sambles, J.R., Skinner, L.M., Lisgarten, N.D.: Proc. R. Soc. London Ser. A **318**, 507 (1970)
28. Hagesawa, M., Hoshino, K., Watabe, M.: J. Phys. F **10**, 619 (1980)
29. Buffat, Ph., Borel, J.-P.: Phys. Rev. A **13**, 2287 (1976)
30. Palmberg, P.W., Todd, C.J., Rhodin, T.N.: J. Appl. Phys. **39**, 4650 (1968)

M. Vollmer
F. Träger
Physikalisches Institut
Universität Heidelberg
Philosophenweg 12
D-6900 Heidelberg
Federal Republic of Germany

The Role of Small Silver Clusters in Photography

P. Fayet[1], F. Granzer[2], G. Hegenbart[2], E. Moisar[2, 3], B. Pischel[2], and L. Wöste[1]

[1] Ecole Polytechnique Fédérale de Lausanne, Institut de Physique Expérimentale, Lausanne, Switzerland

[2] J.W. Goethe-Universität, Institut für Angewandte Physik, Frankfurt, Federal Republic of Germany

[3] AGFA-GEVAERT AG, Forschung und Entwicklung, Leverkusen, Federal Republic of Germany

Received April 7, 1986; final version May 9, 1986

Size-selected silver cluster ions have been produced in a mass spectrometer arrangement and have been deposited onto binder-free AgBr microcrystals. Clusters containing less than four silver atoms did not catalyze development (reduction) of the microcrystals. However, application of clusters containing four or more silver atoms rendered the crystals immediately developable in a conventional photographic developer. This for the first time proves directly, that a so-called latent image speck under these conditions of development requires a minimum number of four silver atoms.

PACS: 82.50.−m; 07.68.+m; 36.40.+d

1. Introduction

Despite world-wide efforts to develop and improve new methods for storage of optical information (magnetic media, electronic photography etc.), the system most widely used and most frequently applied still uses the light sensitive properties of the silver halides. Its applications are almost universal: aside from the gigantic market of amateur photography, silver halide systems are used in the printing and reprographic domain, as X-ray photography in medicine and materials research, in documentation and in various fields of science, to mention just a few.

The basic principle is always the same: The light sensitive detectors, which also serve as the storage elements, are silver halide microcrystals, the size of which is usually in the range of 0.1 to several μm diametre. They are embedded in a binder matrix, which is usually gelatin, and which is situated on a support such as film base, paper or glass plates.

When a microcrystal absorbs a sufficient number of light quanta, a cluster of silver atoms – the so-called latent image speck – will be formed. This silver cluster changes fundamentally one property of the microcrystal: it becomes developable. By development is understood the reduction of the crystal to metallic silver in a reducing environment – the developer. A virgin silver halide crystal will not be immediately reduced, perhaps not before many minutes or even an hour have elapsed. At a crystal bearing a latent image silver cluster, however, reduction commences at once. Hence, the latent image cluster acts as a catalyst by greatly accelerating reduction. The sufficiently exposed and, hence, developed microcrystals form the information elements of which the photographic image is composed in black and white systems. Secondary chemical reactions of the oxidized developer with certain components in the layer lead to the formation of dyes in the vicinity of the developed microcrystals in colour materials.

Surprisingly, many details of this basic process are not precisely known. The minimum size of the latent image silver clusters has been estimated from various indirect experiments, a direct determination, however, was lacking so far. This minimum size is certainly larger than one or two atoms. Why such a "critical" size is required, is a second and not yet fully solved problem, and even the processes leading to the latent image cluster are not yet completely understood [1–4].

The present investigation is concerned with the problem mentioned first: the critical latent image size.

2. Latent Image Formation

Silver halides are almost insulators in the dark. A very low electrical conductivity is attributed to the presence of Frenkel type defects. In thermal equilibrium a few silver ions have left their lattice sites and form mobile interstitial ions. A corresponding number of vacancies is thus created. They bear a formal negative charge in respect to the bulk lattice; correspondingly, the interstitials are positively charged

$$Ag_L \rightleftharpoons Ag_i^+ + Ag_v^- \quad (1)$$

Here the subscript L denotes a lattice ion (the charge of which is compensated by the surrounding ions of opposite charge), i stands for interstitial and v for vacancy.

When the crystal is exposed to light within the absorption range, a significant increase in conductivity occurs. It is due to the liberation of mobile electrons from the valence band assigned to the halide ions which are lifted to the energy of the conduction band. A corresponding number of positive holes (identical with halogen atoms) is also formed

$$Br_L \overset{h\nu}{\rightarrow} e^- + h^+. \quad (2)$$

Here, Br_L is the apparently neutral halide ion on a lattice site. The interstitial ion and the photo electron could react, forming a silver atom

$$Ag_i^+ + e^- \rightarrow Ag^o. \quad (3)$$

The holes, in turn, could recombine with electrons

$$e^- + h^+ \rightarrow Br_L. \quad (4)$$

Thus a halide ion is restored, and absorption of a photon does not cause a permanent effect. In order to avoid such loss reactions, it is one of the main objects of optimization the performance of the microcrystals to eliminate the holes, before they could recombine with photo electrons.

A single silver atom (which many authors believe to be unstable) and even a cluster of perhaps two or three atoms are unable to catalyze development. This can be concluded from experiments, in which the presence of these so-called sub-size specks has been proved e.g. by decoration methods [5, 6]. Developable latent image silver clusters are formed under optimum conditions, when a microcrystal has absorbed approximately four photons. According to Eq. (3) this should lead to four silver atoms which, however, are required to form one aggregate of, hence, "critical" size. Aside from the "critical size" phenomenon, this concentration and aggregation process constitutes the major and not yet fully understood step of the photographic process.

There exist several theories trying to explain, why instead of a random and disperse deposition of silver atoms their aggregation under favourable conditions is directed to just one site. It has been assumed [7] that photo electrons are captured by existing electron traps. In sequence, these immobilized electrons will attract by Coulombic forces interstitial silver ions, thus forming a silver micro phase. In this model – the Gurney-Mott theory – the very essential removal of holes is not explicitly treated. This, however, is one of the central assumptions of the Mitchell model [8, 9]. Without the hazard of recombination, electrons are believed to combine in random and reversible events with interstitials. At a size starting at Ag_3 an interstitial can be adsorbed to the cluster. The positively charged entity then attracts an electron and grows. In this step Coulombic interactions are again involved. The Malinowski model [10] does not include any charge effects. Electrons and interstitials combine to isolated and mobile silver atoms. Holes associate with silver ion vacancies to neutral complexes. By random processes, preferentially at certain condensation sites, the silver atoms combine to clusters.

In addition to these models which speculate about assumed reaction paths, the formation of silver has been more generally treated as a process of phase formation [11, 12] in a solvent system – the silver halide – in which a supersaturation is created by the liberated photo electrons. This approach describes best the fact, that development requires a certain minimum size of a latent image cluster. It has been experi-

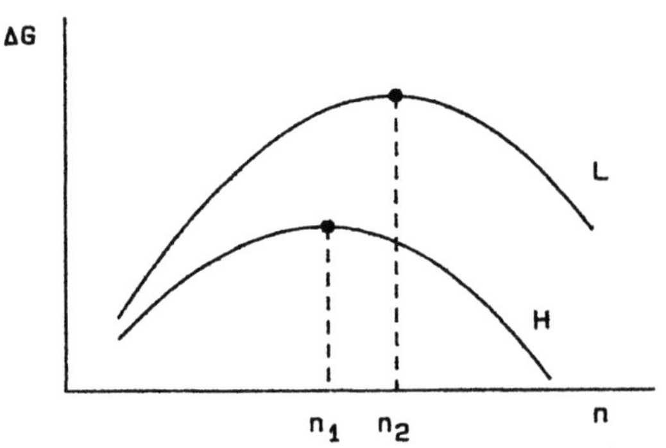

Fig. 1. Free energy of aggregates in dependence of their size for two levels of supersaturation (L: low, redox potential less negative; H: high, redox potential more negative). In case of a higher supersaturation the critical size will be smaller (schematic)

mentally found that this "critical" size is not a fixed and constant number of silver atoms. It seems to depend on the redox potential of the developer [13, 14]. In an immediately starting development process no or just a negligible activation energy is involved. An aggregate (of "critical" size) at which unrestricted growth proceeds, is situated on the maximum of the free energy vs. size curve (Fig. 1). The position of the maximum along the size axis depends on the prevailing supersaturation which, in a developer, is determined by the redox potential.

3. Experimental

Obviously many problems of latent image properties could be investigated, if silver clusters of precisely known size were available and could be attached to silver halide microcrystals. As reported elsewhere [15], a mass spectrometer arrangement is well suited to produce beams of single-size clusters (Fig. 2). Argon ions are produced in an ion source. They are directed onto a silver target, from which silver clusters of various sizes and charges are sputtered. By an energy filter and a mass spectrometer quadrupole cluster ions of desired mass Ag_n^+ (with $n=1 \dots 9$) can be selected. The beam of single-size cluster ions is directed to the specimen, which is a layer of binder-free monodisperse AgBr microcrystals on a glass plate, which is coated with a very thin adhesive layer of hardened gelatin. The binder-free microcrystals were prepared from a photographic emulsion containing cubic shaped AgBr microcrystals. The gelatin of the emulsion was destroyed by enzymatic degradation followed by an oxidizing treatment in order to remove residual traces of organic matter. After repeated sedimentation and redispersing steps the crystals were allowed to sediment onto the glass plates in less than a monolayer. The glass plates were also covered with an indium-tin oxide (ITO) layer which made them conductive. This was essential in order to eliminate repulsive forces caused by the impingement of charged clusters, which otherwise would badly spoil the geometry of the cluster beam (Fig. 3).

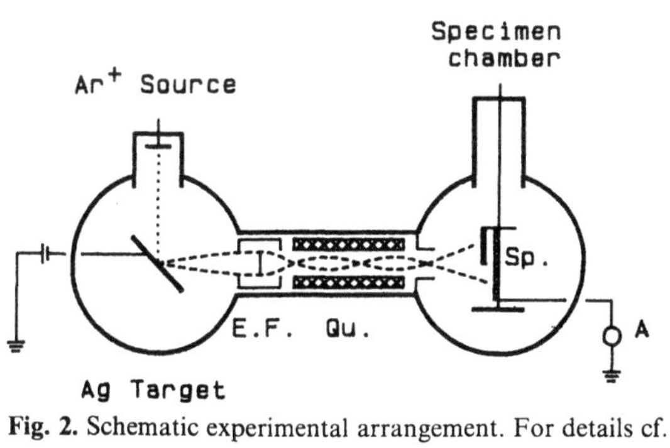

Fig. 2. Schematic experimental arrangement. For details cf. text

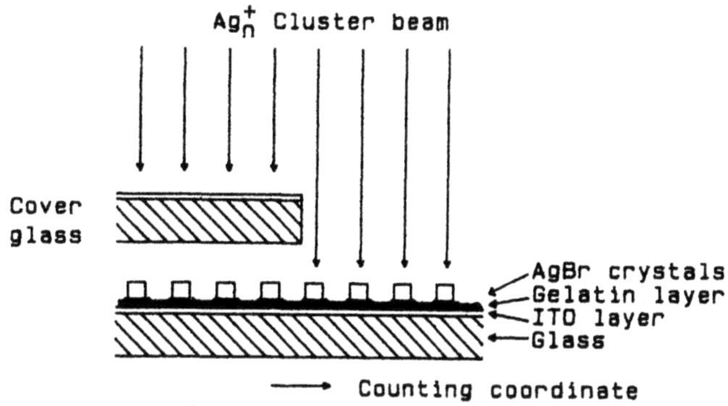

Fig. 3. Scheme of the specimen used for the exposure of AgBr to a beam of size-selected clusters

For exposure to the size-selected beam the specimen was moved out of the specimen chamber. About half of it was shielded by a cover glass. This shielded part of the specimen served as an internal blank; it received the same (if any) background radiation as the unshielded part, but it was not exposed to the beam. The cluster current could be measured by replacing the specimen with a Faraday cup detector. The exposure times were selected in order to let in the average ten clusters impinge on a crystal.

After exposure the specimen was returned to the specimen chamber, which could be removed and taken to a dark chamber for development. A commercial developer (AGFA Neutol) was used; the development time was 30 s. Without fixation the plates were rinsed, dried and examined under a microscope. The fraction of developed crystals both in the exposed and in the shielded area was obtained by counting several hundred fields of view.

The fraction of developed crystals in both parts of the specimen were identical, when beams with Ag^+ and Ag_3^+ had been used. However, as soon as the beam contained the next larger clusters Ag_4^+, almost all crystals in the exposed area had become developable (Fig. 4). Similarly, high degrees of development were observed, when larger clusters, up to Ag_9^+, had been used.

In order to obtain a sufficiently high cluster density in the beam and to avoid extensively long exposure times, a certain accelerating voltage of approximately 8 eV per atom had to be applied. This causes the cluster ions to impinge at the crystals with a velocity much higher than thermal energy. Artifacts resulting therefrom, like fragmentation of the impinging clusters or radiation damage released in the substrate, can be ruled out for the following reasons:

If some fraction of the impinging mono-sized Ag_n^+-clusters fragments into smaller clusters then the final size distribution of silver aggregates deposited on the AgBr-grains should range from the monomer to Ag_n.

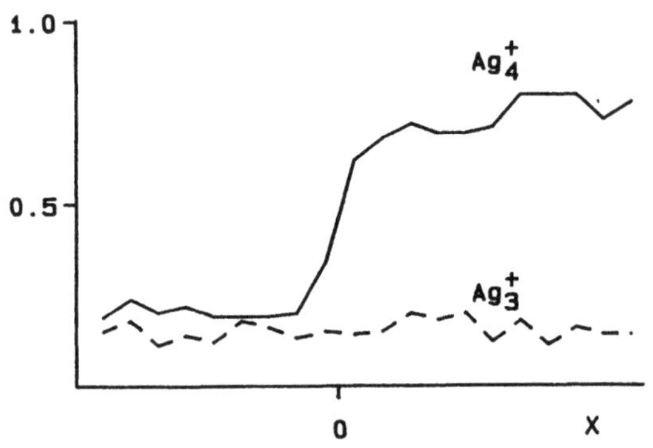

Fig. 4. Fraction of developed microcrystals (ordinate) vs. counting coordinate X. The region on the right side of the coordinate marked "0" has been exposed to the cluster beam; the region on the left side was shielded and served as internal blank. Curves for aggregates with 3 and 4 silver atoms are shown

Consequently at any given time, the *largest* silver cluster on the surface of an AgBr-grain will correspond to the size of the arriving unfragmented clusters in the single size cluster beam. Therefore, in order to interpret unambiguously the results shown in Fig. 4, one has to assume that it is always the fraction of the *un*fragmented Ag_4^+-clusters which renders the corresponding AgBr-grain developable.

Radiation damage, i.e. the spontaneous formation of silver aggregates by the impact of clusters onto the AgBr-microcrystals, has never been observed, even at energies as high as 75 eV for the Ag-monomers.

4. Discussion and Outlook

For the first time, direct evidence is provided that under normal development conditions the "critical" size required for developability corresponds to cluster sizes of four silver atoms. Since under optimum conditions four absorbed photons render a silver halide crystal developable, the latent image specks must be formed by a direct reaction of photo electrons and silver ions as indicated by (3). An alternative model [16] according to which photo electrons merely trigger the coagulation of a great number of pre-existing silver atoms to a huge silver cluster consisting of hundreds and more atoms, is now finally disproved by the present results.

The method described here can in future be used for a great number of exciting investigations. It should be possible, e.g. to study precisely the relationship between the critical size and the redox potential of developers, which is assumed to follow the well-known Gibbs-Thomson equation which relates the size of minute particles to their solubility or vapour pressure. An other interesting problem could be investigated as well: what are the properties of mixed clusters containing e.g. silver and gold? For nearly 50 years it is known that doping the crystal surface of silver halides increases their sensitivity. Several assumptions have been made on the mechanism of this widely used effect, but precise and unambiguous knowledge is again lacking. The present method could very soon deepen our understanding.

The authors wish to express their gratitude to AGFA-GEVAERT AG and to Deutsche Forschungsgemeinschaft for generous financial support of the work.

References

1. Frieser, H., Haase, G., Klein, E. (ed.): Die Grundlagen der photographischen Prozesse mit Silberhalogeniden. Frankfurt/M.: Akadem. Verlagsges. 1968
2. James, T.H. (ed.): The theory of the photographic process. 4th. Edn. New York, London: Macmillan Publ. Co. 1977
3. Granzer, F., Moisar, E.: Physik in unserer Zeit **12**, 22, 36 (1981)
4. Moisar, E.: Chemie in unserer Zeit **17**, 85 (1983)
5. James, T.H., Vanselow, W., Quirk, R.F.: PSA J. **14**, 349 (1948)
6. Hamm, F.A., Comer, J.J.: J. Appl. Phys. **24**, 1495 (1953)
7. Gurney, R.W., Mott, N.F.: Proc. R. Soc. **A 164**, 151 (1938)
8. Mitchell, J.W.: J. Phys. Chem. **66**, 2359 (1962)
9. Mitchell, J.W.: Photogr. Sci. Eng. **22**, 1, 249 (1978), **23**, 1 (1979)
10. Malinowski, J.: Photogr. Sci. Eng. **14**, 112 (1970)
11. Moisar, E., Granzer, F., Dautrich, D., Palm, E.: J. Photogr. Sci. **25**, 12 (1977), **28**, 71 (1980)
12. Moisar, E.: Photogr. Sci. Eng. **26**, 124 (1982)
13. Hillson, P.J.: J. Photogr. Sci. **22**, 31 (1974)
14. Konstantinov, I., Malinowski, J.: J. Photogr. Sci. **23**, 1, 145 (1975)
15. Fayet, P., Wöste, L.: Z. Phys. D – Atoms, Molecules and Clusters **3**, 177 (1986)
16. Galashin, E.A., Senchenkov, E.P., Chibisov, K.V.: Dokl. Akad. Nauk SSSR **181**, 124 (1969)

P. Fayet
L. Wöste
Ecole Polytechnique
Fédéral de Lausanne
Institut de Physique Expérimentale
CH-1015 Lausanne
Switzerland

F. Granzer
G. Hegenbart
B. Pischel
Institut für Angewandte Physik
Johann-Wolfgang-Goethe-Universität
Abt. f. wissenschaftliche Photographie
Robert-Mayer-Strasse 2–4
D-6000 Frankfurt am Main 11
Federal Republic of Germany

E. Moisar
AGFA-GEVAERT AG
Forschung und Entwicklung
D-5090 Leverkusen
Federal Republic of Germany

Magnetic Measurements on Stable Fe(0) Microclusters. Part 2*

A Mößbauer Spectroscopic Study

F. Schmidt, A. Quazi, A.X. Trautwein[1], G. Doppler[1], and H.M. Ziethen[1]
Institut für Physikalische Chemie, Universität Hamburg,
Federal Republic of Germany
[1] Institut für Physik, Medizinische Universität Lübeck,
Federal Republic of Germany

Received April 7, 1986; final version May 21, 1986

Zero valent iron microclusters were stabilized inside the pore structure of *A*-type zeolites. The cage diameter of the matrix was 1.1 nm. Mößbauer spectra show superparamagnetic behaviour down to at least 77 K. Below 10 K, slow relaxation of most of the superparamagnetic moments was observed. Information on the magnetic properties was obtained from the Mößbauer spectra below 4.2 K and from measurements in an external magnetic field applied parallel to the gamma beam. Evidence has been found for both bulk-like and non-crystalline iron clusters.

PACS: 75; 76; 81

Introduction

Part 1 of this series [1] reported on some results of the magnetic properties of Fe(0) microclusters, which have been stabilized up to room temperature by means of encapsulation inside a zeolite matrix. Measurements on a sample, which was reduced by means of a very small excess of sodium vapour, have shown that magnetization at 1.7 K with fields up to 7 T was higher than in bulk iron. However, after subtracting the calculated magnetization of a species with spin $S=0.5$, the saturation magnetization of the iron clusters was found to be 10% lower than in bulk iron. This decrease has been discussed in terms of a size effect of the smallest particles. The magnetization with $S=0.5$ has been attributed to defects formed upon reduction according to the equation

$$\mathrm{Na}(0)+[\mathrm{Na}(I)\text{-O-Zeolite}]\rightarrow \mathrm{Na}(I)+ +[\mathrm{Na}(I)\text{-O-Zeolite}].-$$

The size distribution of this sample was found to be bidisperse, with approximately 75 atomic % smaller than $(0.7\ \mathrm{nm})^3$, and the remainder with an average value of approximately $(2.5\ \mathrm{nm})^3$. A remarkable number of clusters was found to be built up by only a few iron atoms.

A second interpretation has been published recently [2], whereby the influence of reducing agent and matrix on the magnetic properties was studied.

The purpose of the present study is to get more information on the magnetic properties of Fe(0) microclusters using a different and independent method. The big advantage of Mößbauer spectroscopy is the possibility of determining spontaneous magnetization via hyperfine splitting. This method does not depend on determining sample weight. The prerequisite of this method, however, is the presence of a sufficiently large cluster size, or low temperature, or a high external magnetic field so that the magnetization is fixed within the time scale of the Mößbauer experiment. Because of the superparamagnetism of the very small clusters studied here, very low temperatures or very high external magnetic fields are necessary to meet these requirements.

Experimental

The samples were prepared by ion exchange of LTA zeolite $(\mathrm{Na_{12}Si_{12}Al_{12}O_{48}\cdot(H_2O)_{27}})$ by means of a

---
* Part 1: [1]

calculated amount of ferrous ions. A mnemotic code was used to identify the zeolite structure [3]. Typically 0.5 g of oxygen-free zeolite were exchanged by means of a calculated amount of Fe dissolved in a corresponding volume of 2N H_2SO_4. The filtrate and the combined washing waters were back-titrated with respect to iron. Further details of the ion exchange have been published in [4]. The ferrous zeolite thus obtained, $Na_{7.23}Fe(II)_{2.39}Al_{12}Si_{12}O_{48} \cdot (H_2O)$, $M=$ 1,727.53 g/mol corresponding to 7.72 wt.% iron on a dry weight basis, was dehydrated completely at 673 K and 10^{-5} torr and was reduced by means of a 100% excess of sodium vapour compared to the amount necessary for complete reduction. The sodium was placed at the bottom of a glass tube and the dehydrated sample was supported 3 cm above the bottom (on a small glass frit). At the top of the tube a Mößbauer cell was sealed to the glass tube. A half-length heating mantle was brought to 573 K, and the sodium vapour passed the sample zone, condensing at the top of the tube. The sodium mirror thus formed served as a trap for impurities such as oxygen, which could have entered the tube before the reduced sample was sealed off. After a heating period of 107 h, the mantle was removed and the tube was allowed to cool. Then the zeolite was placed in the Mößbauer cell, which was sealed off and mounted in the spectrometer. The kinetics of this reduction have been published in [2].

Details of sample preparation and measurement have been reported for magnetic [1] and Mößbauer spectroscopic measurements [2, 5] for various sodium excess samples. All Mößbauer spectra were obtained using constant acceleration standard Mößbauer equipment. Calibration was performed by using an α-iron foil: All isomer shifts are reported with respect to this absorber at room temperature. The source was a 50 mCi ^{57}Co in a Rh sample.

The spectra were analyzed by using conventional LSQ fit procedures with Lorentzian line profiles. The hyperfine field distribution was obtained via a procedure described by Le Caer and Dubois [6]. The use of their program is gratefully acknowledged.

Results and Discussion

The hyperfine (hf) structure of a Mößbauer spectrum is governed by three contributions. A change in the *s*-electron density of the iron nucleus which might arise from a change in valence will result in an altered Coulomb interaction, causing a shift of the nuclear levels. For this "electric monopole interaction" the term "isomer shift", δ, is used. For more detailed information, see Ref. 7. The second non-vanishing term of the electrostatic interaction of the iron nucleus with its surrounding electronic charge is the quadrupole coupling, ΔE_Q which results from the interaction of the nuclear quadrupole moment with the electric field gradient [8]. The third part of the hf structure of a Mößbauer spectrum is magnetic splitting arising from the interaction of the nuclear magnetic dipole moment with the magnetic field, either external or internal or both. For zero valent iron, this magnetic splitting B_{hf} is proportional to the atomic magnetic moment of the iron.

The parameters of the ferrous iron containing zeolite are in agreement with many previously published results on comparable samples indicating stoichiometric ion exchange. In [4], the isomer shift is $\delta=$ 0.81 mm/s, relative to α-iron, (in that paper the values were reported with respect to Co/Cu and not to α-iron), and the quadrupole splitting is $\Delta E_Q=0.47$ mm/s; or furthermore, in [9] the corresponding values are $\delta=0.83$ mm/s (relative to α-iron) and $\Delta E_Q=$ 0.47 mm/s. These Fe^{2+} ions are located at $S(II)$ sites ($=S_{1A}=A$, B sites [10], $3m[C_{3v}]$), where they are coordinated by three oxygen ions in a planar trigonal configuration similar to site $S(II)$ (site G [10]) in dehydrated Y zeolite [11].

First, we will discuss the Mößbauer spectra of the reduced samples taken in zero external magnetic field. Figure 1 shows the Mößbauer spectrum, taken at room temperature: the corresponding parameters are given in Table 1. We can distinguish at least four species, i.e.

(1) metallic iron with $\delta=0.0$ mm/s and $\Delta E_Q=$ 0.0 mm/s,

(2) ferrous ions with $\delta=0.74$ mm/s and $\Delta E_Q=$ 0.45 mm/s,

(3) ferrous ions with $\delta=0.83$ mm/s and $\Delta E_Q=$ 2.26 mm/s,

(4) ferric ions with $\delta=0.33$ mm/s and $\Delta E_Q=$ 0.77 mm/s.

The parameters of species (2) are close to those of the unreduced dehydrated high spin ferrous ions, indicating incomplete reduction in our sample. The decrease of δ (0.74 mm/s vs. 0.83 mm/s) reflects an increase of the electron density at the iron nucleus, probably caused by a decrease of the Fe–O bond distance. This shrinking may be produced by the space-consuming zero valent iron clusters.

Species (3) is assigned to high spin ferrous ions in a three-dimensional coordination. These sites are similar to those found in dehydrated A zeolites, too [4]. According to Rees [11], these ions are not coordinated by remaining OH groups and it is proposed that they are due to a local structural breakdown.

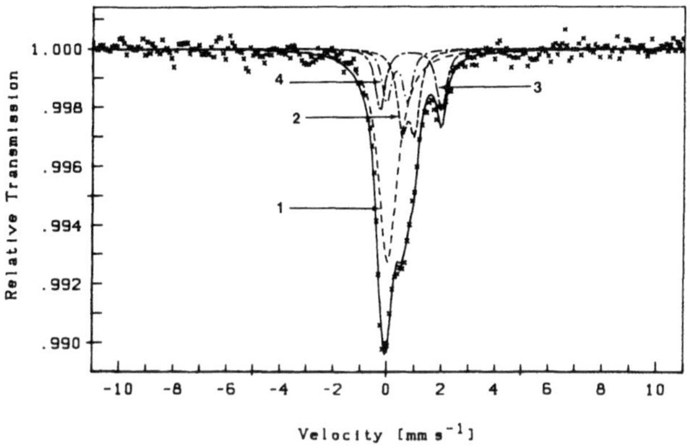

Fig. 1. Mößbauer spectrum of reduced Fe–LTA zeolite measured at room temperature: x indicates measured points, the line corresponds to an LSQ fit. For explanation of subspectra 1–4, see text

Table 1. Parameters of the Mößbauer spectrum measured at room temperature. The isomer shift δ is related to α-iron at room temperature. The errors of δ, ΔE_Q and Γ are ± 0.03 mm/s. The total absorption area is $A_{tot} = 1.814\% \cdot$ mm/s

Parameters	Species			
	1	2	3	4
δ[mm/s]	0	0.74	0.83	0.33
ΔE_Q[mm/s]	0	0.45	2.26	0.77
Γ[mm/s]	0.86	0.43	0.44	0.41
A[%]	53	18.8	15.9	12.3

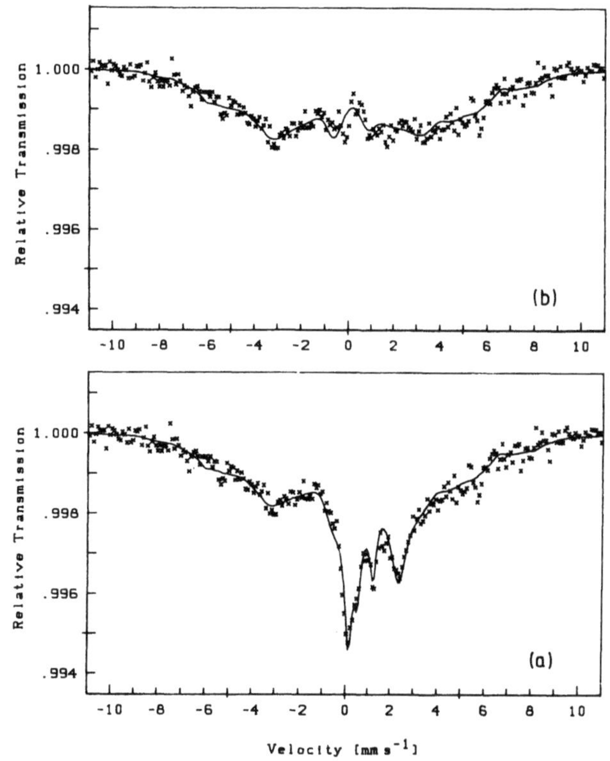

Fig. 2. **a** Mößbauer spectrum of reduced Fe–LTA zeolite measured at 4.2 K: x indicates measured points, the line corresponds to an LSQ fit of subspectra 1–3 plus magnetic hyperfine field distribution of Fig. 3. **b** Remaining spectrum after having been stripped of the non-magnetic lines listed in Table 2. The line corresponds to the magnetic hyperfine field distribution of Fig. 3

Because this species was not found in the dehydrated sample prior to reduction, local matrix destruction probably occurred on reduction.

Species (4) in Fig. 1 (approximately 12% relative absorption area) is due to high spin ferric ions formed upon internal oxidation of the reduced clusters. Due to this species, the magnetization isotherms of this sample measured at different temperatures do not superimpose, so differing from the behaviour reported in part 1 of this series.

External oxidation can be excluded because of the metallic sodium mirror, mentioned in the experimental section. The location of the ferric ions is not yet clear; further studies are needed to clarify this point.

Species (1) in Fig. 1 is due to metallic iron clusters and will be discussed in more detail below.

The values of the isomer shift and of the quadrupole splitting of the room temperature spectrum (Fig. 1) do not significantly change down to 77 K. These parameters show normal temperature dependence. However, below 10 K the spectrum is completely different and the features of this low temperature spectrum do not change down to 1.5 K! A typical spectrum measured at 4.2 K is shown in Fig. 2a. The two ferrous components are still not magnetically split, and their parameters are given in Table 2. The area fraction of the two magnetically split components is given in the note to Table 2. The parameters of the spectrum measured at 1.5 K are practically the same as those measured at 4.2 K.

Again, species (1) is assigned to metallic iron. The single line with only 3.4% of the total area is due to iron clusters in the fast relaxing superparamagnetic state. Details of the magnetic relaxation will be discussed below. The main part of species (1) is in a magnetically blocked state and shows a broad magnetic absorption of zero valent iron. In the following, we shall argue about this broad absorption and discuss

- the distribution of hyperfine fields due to dipolar cluster – cluster interaction;
- the effect of lattice vibration of cluster and matrix, and bound diffusion of clusters within the matrix;
- superparamagnetic relaxation of the clusters;
- the distribution of hyperfine fields due to different coordination and oxidation state of iron.

At first it is obvious that a very broad distribution of absorption lines is present in the sample, as indi-

Table 2. Parameters of species not magnetically split in the Mößbauer spectrum measured at 4.2 K. The isomer shift δ is related to α-iron at room temperature. The errors of δ, ΔE_Q and Γ are ± 0.03 mm/s. The total absorption area is $A_{tot} = 2.76\% \cdot$ mm/s at 4.2 K

Parameters	Species		
	1	2	3
δ[mm/s]	0.12	0.84	1.2
ΔE_Q[mm/s]	0	0.71	2.28
Γ[mm/s]	0.3	0.35	0.85
A[%]	3.4	6.8	20.4

The area fraction of the magnetically split components 1 and 4 amounts to 69.4%

cated in Fig. 2b, obtained from the experimental spectrum (Fig. 2a) when stripped of the non-magnetic species 1–3 (Table 2).

This distribution may be caused by dipolar cluster – cluster interaction. Assuming that two neighbouring cavities of the matrix are each occupied by clusters, with volume $V = (1.1\ \text{nm})^3$, the maximum value of the dipole field amounts to $B_{dip} = 1.4$ T. This value has to be added to and subtracted from the hyperfine fields. The resulting high field part, however, is much too small to cause the observed broad absorption pattern shown in Fig. 2b. Therefore, a distribution of hyperfine fields due to dipolar cluster – cluster interaction cannot be ruled out, but is certainly not the reason for the observed broad spectrum.

Other possible reasons for broadening of absorption lines are lattice vibration and Brownian diffusion in a limited space (bound diffusion of clusters within the zeolite cavities). Especially the latter of the two effects is known to result in an increase of line-broadening with increasing temperature [12], corresponding in magnitude to the overall broadening of the spectrum shown in Fig. 2b. The number of atoms within the clusters, which fit into the cavities of zeolites, is relatively small. Strong phonon coupling between cluster and zeolite is necessary to "absorb" the recoil momentum during nuclear γ absorption, thus yielding a measurable Mößbauer effect. Here, the nature of the atomic vibrations in small clusters must be discussed. The vibration pattern is sometimes derived on the basis of the variations in the Mößbauer effect probability. The variations in the recoil free fraction f for highly dispersed particles are associated with two effects, one of them being the surface effect which should result in an increased $\langle r^2 \rangle$. Since the specific surface increases with the decrease of the cluster size, the probability f should decrease for this reason alone. The second effect is related to the variations in the phonon spectrum of these particles and it can produce a contrary result, at least in unsupported clusters. Both effects depend on the amount of phonon coupling between cluster and matrix.

The f value changes remarkably between 1.5 K (total absorption area amounts to approximately 4.5) and 4.2 K (total absorption area amounts to 2.76). At higher temperatures the f values change, too, as can be calculated from the total absorption area of the spectra at room temperature (Table 1). However, the line width and the line shape of the stripped spectrum (Fig. 2b) do not change when the temperature changes from 1.5–4.2 K. Therefore, the observed changes of the f values are probably due to lattice vibrations, but the observed broadness of the stripped spectrum at low temperature cannot be due to bound diffusion of clusters.

An additional possible reason for the observation of broad absorption lines is superparamagnetic relaxation with low frequency.

In an assembly of non-interacting particles, each with volume V, the relaxation time for a spontaneous change of direction of the magnetization vector is given by the well-known Néel equation:

$$\tau^{-1} = f_0 \exp[-\Gamma K V / (k T)].$$

Here f_0 is a frequency factor usually estimated as $10^{+9}\ \text{s}^{-1}$ and Γ is a geometry factor. The frequency factor f_0 depends on the Larmor frequency of the magnetization vector in an effective magnetic field. If the relaxation time is larger than the time required for the measurement ($\tau \gg \nu_L^{-1}$), a snap shot is taken, and Zeeman splitting is observed: ν_L is the Larmor frequency of the nuclear spin within the effective magnetic field. For iron ν_L^{-1} is ca. 10^{-8} s. For $\tau \ll \nu_L^{-1}$, only the time average of the magnetic field is observed, i.e. the Zeeman splitting will disappear. This is the fast relaxation state, which is often called the superparamagnetic state.

In this model, the broadening of the slowly relaxing component could be caused by a distribution of volume and/or magnetic anisotropy of the clusters. This can be established either by applying lower temperatures or an external magnetic field. The latter will be discussed below. According to this explanation, only a small fraction of species (1) (area fraction ca. 3.4%) is in the superparamagnetic state below 10 K, whereas most of species (1) shows a magnetically split spectrum due to slow relaxation of superparamagnetic clusters (Fig. 2, Table 2). Above 10 K, the slowly relaxing component decreases at the expense of the intensity of the fast relaxing component.

At 77 K, a very small amount of the slowly relaxing superparamagnetic component is still present, whereas at 130 K and at 295 K the Mößbauer spec-

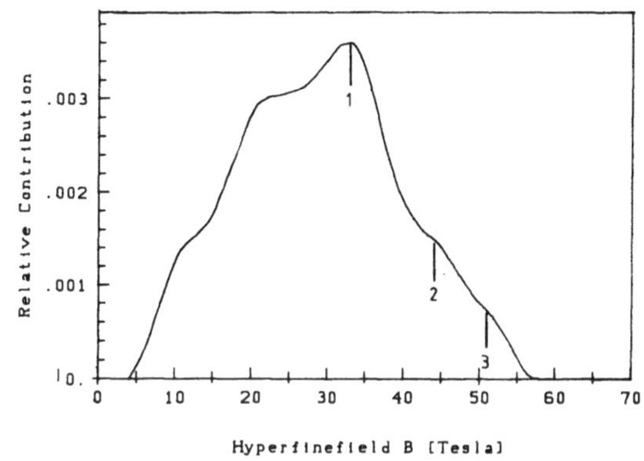

Fig. 3. Magnetic hyperfine field distribution of the "experimental" spectrum of Fig. 2b. For comparison: (1) crystalline zero valence iron clusters, (2) partially oxidized Fe clusters, (3) Fe(III)

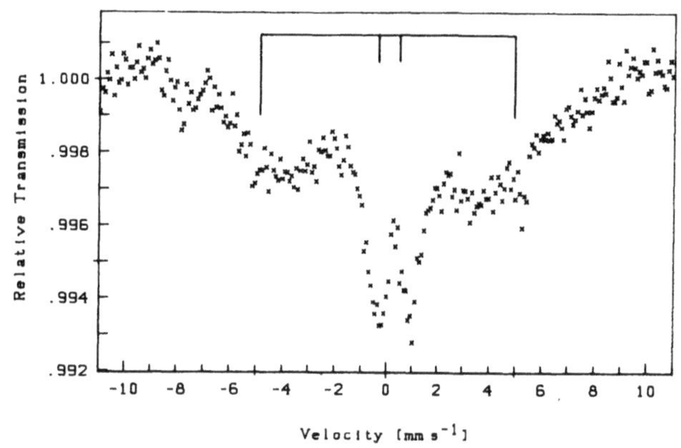

Fig. 4. Mößbauer spectrum of reduced Fe – LTA zeolite measured at 4.2 K in an external magnetic field of 2.47 T, applied parallel to the γ beam. The bars indicate the hyperfine field of the crystalline α-iron particles

trum of the zero valence iron clusters consists of a single absorption line only. For each temperature, the isomer shift of this single line is identical to the δ value of bulk iron at the corresponding temperatures with respect to α-iron at room temperature. It should be mentioned that this identity is clear evidence of the absence of any charge transfer between cluster and matrix. Below 10 K the intensity of the fast relaxing spectral component does not increase in favour of the intensity of the slowly relaxing spectral component when changing from 1.5–10 K. Therefore, changes in superparamagnetic relaxation play a role above 10 K only.

In the following, we assume that a hyperfine field distribution is responsible for the observed broad line pattern. Figure 3 shows the result obtained using Le Caer's method [6] applied to the stripped spectrum of Fig. 2b. In this distribution, several regions can be distinguished.

Ferric iron corresponds to hyperfine fields of ca. 50 T (approximately 12% of the total area). Hyperfine fields of about 40 T (approximately 20% of the total area) are probably due to surface effects caused either by adsorption of traces of oxygen or by interaction with the oxide matrix. This value of about 40 T is typical for partially oxidized iron clusters in zeolites. It is still being investigated whether this component is due to reversibly absorbed oxygen or to a chemically different surface phase.

The maximum at 34 T is due to crystalline zero valent iron clusters. The low field branch (<30 T) is attributed to non-crystalline zero valent iron because similar distributions were observed by Bjärman et al. [13] for sputtered non-crystalline iron. This similarity has lead us to propose the following model for the iron clusters: the clusters consist of a nucleus with a bulk-like structure, whereas the surface atoms show remarkable deviations from the iron – iron coordination of the infinite crystal. This model is confirmed by the observation of strong lattice vibrations of the small clusters even at very low temperatures (see above). Another tentative explanation is that both bulk-like clusters and non-crystalline clusters coexist. The non-crystalline clusters might consist of polyhedra such as trimers, icosahedra, etc., and are probably isolated from each other in the crystalline zeolite matrix. A third plausible explanation is that each cluster has the same structure as a small piece of bulk material, but because the iron – iron coordination of the edge and corner atoms differs from that of the bulk atoms, the corresponding hyperfine fields are different. Which of the three proposed models is actually valid needs further investigation.

In any of these three models, the observation of low values of the hyperfine field could explain a possibile decrease of the spontaneous magnetization, which was reported in [1].

Our interpretation of Fig. 3 on the basis of a hyperfine field distribution is confirmed by an additional experiment, i.e. by applying an external magnetic field to the sample. Figure 4 shows the Mößbauer spectrum, taken at 4.2 K in an external magnetic field of 2.47 T, applied parallel to the gamma beam. The spectrum can be attributed partly to bulk-like iron clusters (marked by bars in Fig. 4), partly to ferric iron and iron oxides (absorption area left of −5 mm/s and right of +5 mm/s), and partly to non-crystalline behaviour of iron clusters (absorption area between the outer bars at ca. ±5 mm/s).

Conclusion

By using a zeolite matrix it is possible to stabilize zero valent iron clusters up to room temperature. The size of most of the clusters is limited by the cage dimension of the matrix to less than approximately 1.1 nm in diameter. Above 10 K these clusters are superparamagnetic under the conditions of a Mößbauer experiment. Therefore it is clear that the reason for the fluctuation of the magnetization is the very low anisotropy energy barrier between neighbouring directions due to the very small volume of the clusters. Furthermore part of the iron was found to exhibit the same magnetic hyperfine field as bulk α-iron, indicating the presence of very small crystalline structures. However, it is also evident that part of the iron is located in non-crystalline surroundings.

Financial support from the "Deutsche Forschungsgemeinschaft" and from the "Fonds der Chemischen Industrie" is gratefully acknowledged.

References

1. Schmidt, F., Stapel, U., Walther, H.: Ber. Bunsenges. Phys. Chem. **88**, 310 (1984)
2. Schmidt, F.: In: Industrial applications of the Mößbauer effect. New York, London: Plenum Press (1986; in press)
3. Olson, D.H., Meier, W.M.: In: Atlas of zeolite structure types. p. 57, Pittsburgh: IZA Polycrystal Book Service 1978
4. Schmidt, F., Gunßer, W., Adolph, J.: ACS Symp. Ser. **40**, 291 (1977)
5. Schmidt, F., Adolph, J.: In: Proceedings of the International Conference on Mößbauer Spectroscopy, p. 454, Jaipur (1981)
6. Le Caer, G., Dubois, J.M.: J. Phys. E: Sci. Instrum. **12** 1083, (1979)
7. Shenoy, G.K., Wagner, F.E.: In: Mößbauer isomer shifts. Amsterdam: North-Holland 1978
8. Gütlich, P., Link, R., Trautwein, A.X.: Mößbauer spectroscopy in transition metal chemistry. In: Inorganic Chemistry Concepts. Vol. 3, p. 69. Berlin, Heidelberg, New York: Springer-Verlag 1978
9. Rees, L.V.C.: In: Structure and reactivity of modified zeolites. p. 1. Amsterdam: Elsevier 1984
10. Mortier, W.J.: In: Compilation of extra framework sites in zeolites. p. 19. London: Butterworth SC. Ltd. 1982
11. Rees, L.V.C.: In: Metal microstructures in zeolites. p. 33. Amsterdam: Elsevier 1982
12. Parak, F., Trautwein, A.X.: In: Brussels hemoglobin symposium, Schnek, A.G., Paul, C. (eds.), p. 299. Edition de l'Université de Bruxelles (1983)
13. Bjärman, S., Wäppling, R.: J. Magn. Magn. Mater. **40**, 219 (1983)

F. Schmidt
A. Quazi
Institut für Physikalische Chemie
Universität Hamburg
Laufgraben 24
D-2000 Hamburg 13
Federal Republic of Germany

A.X. Trautwein
G. Doppler
H.M. Ziethen
Institut für Physik
Medizinische Universität Lübeck
Ratzeburger Allee 160
D-2400 Lübeck
Federal Republic of Germany

Photofragmentation of Mass Resolved Carbon Cluster Ions

M.E. Geusic, M.F. Jarrold, T.J. McIlrath*, L.A. Bloomfield, R.R. Freeman, and W.L. Brown
AT & T Bell Laboratories, Murray Hill, New Jersey, USA

Received April 7, 1986; final version May 12, 1986

The results of a detailed study of the photodissociation of carbon cluster ions, C_3^+ to C_{20}^+, are presented and discussed. The experiments were performed using internally cold cluster ions derived from pulsed laser evaporation of a graphite target rod in a helium buffer gas followed by supersonic expansion. The mass selected clusters were photodissociated using 248 nm and 351 nm light from an excimer laser. Photofragment branching ratios, photodissociation cross sections and data on the laser fluence dependence of photodissociation are reported. For almost all initial clusters, C_n^+, the dominant photodissociation pathway was observed to be loss of a C_3 unit to give a C_{n-3}^+ ion. This observation is interpreted as indicating that dissociation occurs by a statistical unimolecular process rather than by direct photodissociation. The photodissociation was found to be linear with laser fluence for $n>5$ with 248 nm and 351 nm light; quadratic for $n=5$ for 248 nm and 351 nm; and linear for $n=4$ at 248 nm. Dissociation energies for the carbon cluster ions implied by these results are discussed. The photodissociation cross sections were found to change dramatically with cluster size and with the wavelength of the photodissociating light.

PACS: 36.40+d; 33.8.Gj

I. Introduction

Over the past five years activity in the field of cluster science has increased rapidly [1]. The principal motivation for studying clusters is interest in how their physical and chemical properties change from molecular to bulk as the cluster size increases. In addition, clusters are technologically important, the best known examples being in catalysis.

The recent advances in the studies of clusters have been made possible by the development of experimental techniques which permit the generation of clusters of virtually any element in the periodic table [2]. However, investigating the properties of these clusters is not straightforward. The cluster sources currently in use generate clusters with a broad range of sizes, making it difficult to investigate the properties as a function of size. Furthermore, detecting the clusters has several associated problems: pulsed laser ionization, which is generally used, introduces uncertainties due to multiphoton dissociation. Investigating the properties of cluster ions rather than their neutral analogs avoids the detection problems and permits selection of a particular cluster before it is probed. This is the approach we have employed in the present work.

The application of conventional spectroscopic techniques, such as resonant two photon ionization and laser induced fluorescence, to the larger clusters has generally not met with success because of the large number of excited states and their extremely short lifetimes. However, photodissociation, which has been used for several years to investigate the spectroscopy and dissociation dynamics of ions [3], can be used to probe the properties of the larger clusters. Unfortunately, photodissociation does not generally yield the detailed spectroscopic information available from techniques such as resonant two photon ionization and laser induced fluorescence. On the other hand, photodissociation does provide useful information about the dissociation processes themselves.

* On leave from Institute of Physical Sciences and Technology, University of Maryland, College Park, MD 20742, USA

In this paper we report the results of a study of the photodissociation of mass selected carbon cluster cations containing between 3 and 20 atoms using 248 nm and 351 nm light. Carbon clusters are believed to be important astronomical molecules, having been identified in stars and comet tails, and they also probably play a role in soot formation. Carbon clusters have been studied on several occasions in the past by both experimental [4–6] and theoretical [7–9] means. C_2 and C_3 are now well characterized species, however, very little is known about the larger clusters. Kaldor and coworkers [10] recently reported some studies of the larger clusters. They investigated the distribution of carbon clusters generated by pulsed laser evaporation and coursely bracketed their ionization potentials. More recently Smalley and coworkers [11] reported some additional studies of the larger carbon clusters and suggested that C_{60} was a particularly stable species with the structure of a soccer-ball, although this structure is the subject of controversy [12].

II. Experimental Methods

A schematic diagram of the experimental apparatus is shown in Fig. 1. The experiment is pulsed and run at a repetition rate of 10 Hz. An experimental sequence starts with the production of the cluster ions. A pulsed valve (Lasertechnics) produces a 100–400 μs helium pulse which flows down a 0.1 cm diameter channel over a rotating graphite target rod. A Q-switched Nd:YAG laser (second harmonic) focused to a 0.1–0.05 cm diameter spot (approximately 1×10^8 W cm^{-2}) vaporizes the target producing a plasma entrained in the helium carrier gas. The gas flows along a 0.5 cm diameter channel of variable length and then expands into the main chamber. The beam is skimmed and enters a differentially pumped inner chamber containing a three plate acceleration region. High voltage pulses, 3 μs long, of 2.25 kV and 1.8 kV are applied to the back and middle plates respectively, accelerating the cluster ions out of the acceleration region and into the first time-of-flight region. The ion beam passes through deflection regions, an einzel lens and then enters the pulsed mass isolator which consists of two parallel plates biased at 0 and 100 V. When the cluster of interest enters the mass isolator the 100 V side is pulsed to ground, permitting a mass selected packet of clusters (around 10^3) to pass through. The packet is deccelerated to 1 kV and crossed with a pulsed excimer laser beam which is focused to approximately 0.4 cm^2. The unfragmented ions and product ions are then reaccelerated to 5 kV and enter a second time-of-flight region at the end

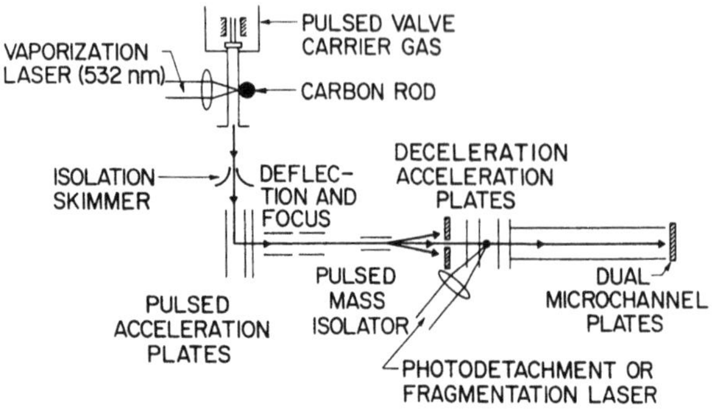

Fig. 1. Schematic diagram of the experimental apparatus

of which they are detected using dual microchannel plates. The signal is amplified, and analyzed by a 100 MHz transient recorder. The digitized signal is then transferred along a camac dataway to a PDP 11/34 computer for analysis. The photodissociating laser intensity is monitored by a UV diode and boxcar averager and is fed to the computer through a camac A to D converter.

III. Results

The carbon cluster ions for these studies were derived directly from the source rather than from laser photoionization of neutral clusters. Work from this laboratory was the first to show the feasibility of this direct production technique [13]. The distribution of carbon cluster ions measured with the pulsed mass isolator turned off is shown in Fig. 2. Intense peaks in the mass spectrum ("magic" numbers) can be seen with a periodicity of four at $n=7$, 11, 15, 19 and 23. The distribution of cluster ions from the source can be quite different from the distribution of ionized neutral clusters. For example, for photoionized carbon clusters only species with an even number of atoms are observed for $n>36$; however, for positive and negative cluster ions derived directly from the source both even and odd species occur [13]. Photoionized neutral clusters are likely to be internally excited in this photoionization process. The main advantage of deriving the cluster ions directly from the source is that they should be internally cooled in the expansion. We expect substantial rotational cooling and some vibrational cooling to have occurred in the expansion. This should lead to internal temperatures below the temperature of the buffer gas which is close to room temperature, although it has not yet been possible to measure this internal temperature directly.

III. A. Photofragment Branching Ratios

An example of a time-of-flight mass spectrum of the photofragments from C_{18}^+ photodissociation is shown

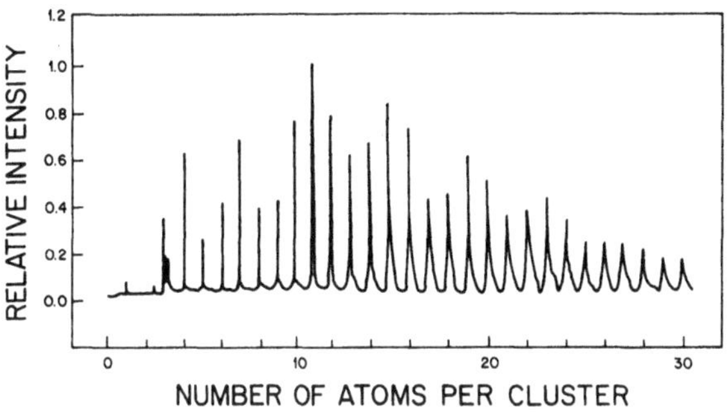

Fig. 2. Mass Spectrum of the carbon cluster cations derived directly from the source

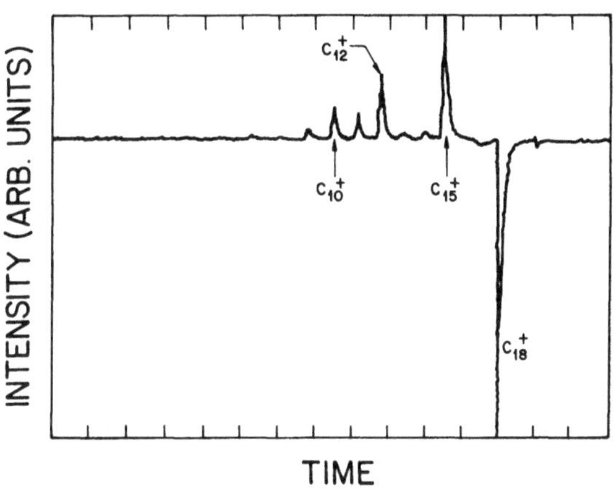

Fig. 3. Time-of-flight mass spectrum of the photofragments from C_{18}^+ photodissociation measured with 248 nm light and a photodissociating laser fluence of approximately 12 mJ cm^{-2}

in Fig. 3. The spectrum was measured with 248 nm light and a photodissociating laser fluence of approximately 12 mJ cm^{-2}. The main peaks in the product ion mass spectrum are due to C_{10}^+, C_{11}^+, C_{12}^+, and C_{15}^+. The C_{18}^+ peak is negative because of the background subtraction techniques used. Data were accumulated in the computer with the photodissociating laser alternately on and off. The signals are normalized and the laser-off signal is subtracted from the laser-on signal. Thus the negative C_{18}^+ signal gives a measure of the depletion of the C_{18}^+ by photodissociation.

The photofragment branching ratios were found to vary quite dramatically with photodissociating laser fluence. We will illustrate this with photodissociation of C_{18}^+ with 248 nm light. The laser fluence dependence of the intensity of the two most intense fragments, C_{15}^+ and C_{12}^+, are shown in Fig. 4. With low laser fluences, <4 mJ cm^{-2}, the C_{15}^+ intensity increases linearly with laser fluence. For higher laser fluences the C_{15}^+ intensity stops increasing and for fluences greater than 10 mJ cm^{-2} starts to decrease. At the same time, for fluences greater than 4 mJ cm^{-2} the C_{12}^+ product increases rapidly. The increase in

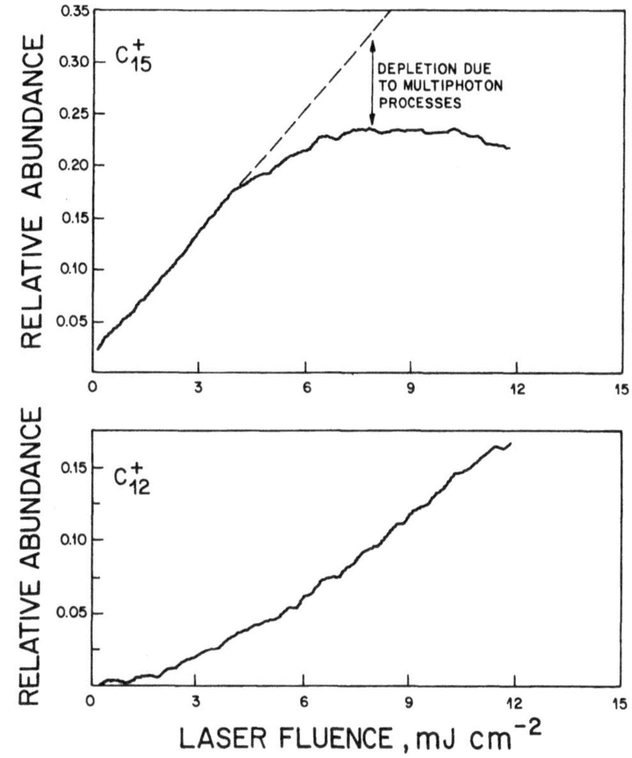

Fig. 4. Laser fluence dependence of the intensity of the C_{15}^+ and C_{12}^+ photofragments from C_{18}^+ photodissociation with 248 nm light

the C_{12}^+ intensity with laser fluence appears quadratic (rather than linear) suggesting the C_{12}^+ arises from a multiphoton process. We believe that the quadratic behavior for C_{12}^+ and the fall-off in the rate of increase of C_{15}^+ for laser fluences above 4 mJ cm^{-2} are both due to multiphoton processes. For example, C_{18}^+ might absorb a photon and lose a C_3 unit to give C_{15}^+ which can subsequently absorb a second photon and lose a second C_3 unit to give C_{12}^+:

$$C_{18}^+ \xrightarrow{+h\nu} C_{18}^{+*} \longrightarrow C_{15}^+ + C_3$$
$$C_{15}^+ \xrightarrow{+h\nu} C_{15}^{+*} \longrightarrow C_{12}^+ + C_3. \tag{1}$$

This sequential photon absorption process requires that the first dissociation occurs within the period of the photodissociating laser pulse. Nanosecond dissociation lifetimes are not unreasonable, however, we cannot distinguish this two step absorption process from a process in which the two photons are absorbed in the parent C_{18}^+ followed by two sequential C_3 loses:

$$C_{18}^+ \xrightarrow{+h\nu} C_{18}^{+*} \xrightarrow{+h\nu} C_{18}^{+**} \longrightarrow C_{15}^{+*} + C_3$$
$$C_{15}^{+*} \longrightarrow C_{12}^+ + C_3. \tag{2}$$

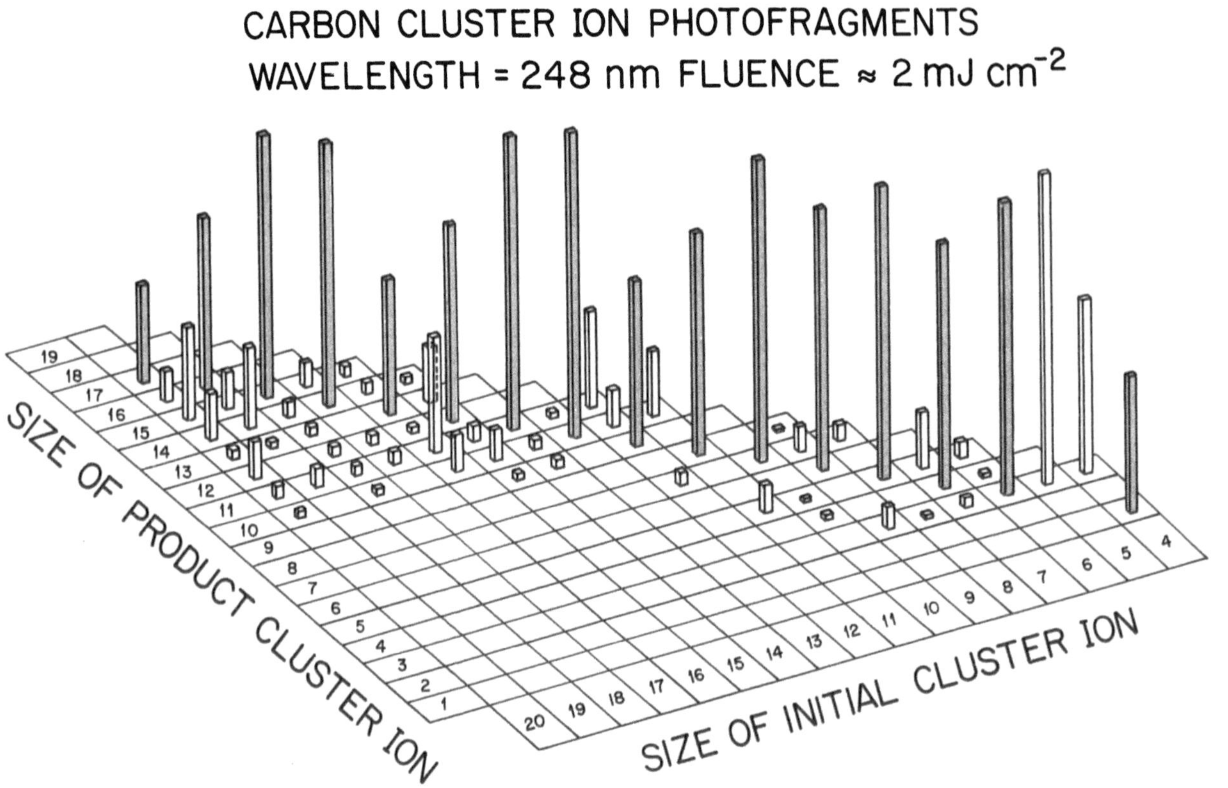

Fig. 5. Three dimensional histogram of the photofragments from carbon cluster ion photodissociation with 248 nm light at a fluence of approximately 2 mJ cm^{-2}. The peaks in the histogram arising from C_3 loss have been shaded

Experiments using two lasers, with one of them delayed, would help to resolve this question.

In some cases (but not for C_{18}^+) the amount of depletion of the first generation daughter (C_{15}^+ for C_{18}^+ photodissociation, see Fig. 4) matches the depletion expected from the directly measured photodissociation cross section of such a cluster ion as a parent. This supports the two step absorption process (Eq. 1 above). However, in other cases cross sections several times larger than the measured cross sections are required to account for the depletion of the first generation daughters. These larger cross sections could arise because the daughters are produced internally excited with a larger density of states.

In Fig. 5 we show a three dimensional histogram of the photofragments from carbon cluster ion photodissociation with 248 nm light using low laser fluences ($\approx$2 mJ cm^{-2}). With this low laser fluence, processes with intensity greater than 10% of the total are generally linear with laser fluence. It is clear from the results shown in Fig. 5 that the loss of a C_3 unit dominates the fragmentation of these carbon cluster cations. For some of the larger cluster ions, loss of C_5 is also an important fragmentation pathway. The photofragment branching ratios measured with 248 nm and 351 nm light are tabulated in Fig. 6. The results shown in Fig. 6 were obtained with a laser fluence of approximately 2 mJ cm^{-2} at both wavelengths. The photofragment branching ratios at 351 nm are very similar to those obtained at 248 nm. Differences of less than 5% of the total should not be considered significant. Similar results were obtained using a doubled Nd:YAG laser (532 nm). However, these results will not be discussed here in more detail because the poor beam quality of the Nd:YAG laser made fluence dependence studies difficult.

III. B. Laser Fluence Dependence of Photodissociation

The laser fluence dependence of photodissociation can be used to bracket the dissociation energies of the cluster ions by determining if photodissociation is one photon (depletion of parent ion linear with laser fluence) or multiphoton (quadratic depletion). For these studies it is critical that the photodissociating laser beam is homogeneous and has no hot spots. We used an excimer laser which has good beam quality.

The major problem with this method of determining dissociation energies is that if either of the cross sections for a two photon process is very much larger than the other the depletion will be controlled by

Size of initial cluster ion \ Size of product cluster ion	1	2	3	4	5	6	7	8	9	10	11	12	13	14	15	16	17	18	19	20
4	44 (62)		56 (37)																	
5			100 (100)																	
6			94 (96)	1	5 (4)															
7			3	78 (82)	18 (13)	(6)														
8			1 (1)		94 (92)	(3)	5 (4)													
9			7		(1)	84 (89)	8 (10)	1												
10				2 (6)	1 (2)		97 (92)													
11					9	(3)	(4)	71 (79)		21 (14)										
12					(2)		4 (5)		53 (67)	12 (10)	31 (16)									
13										98 (90)	2 (10)									
14								(8)	3	3	94 (92)									
15									2 (10)	10 (15)	5 (8)	63 (67)	18	2						
16										12 (8)	37 (34)	2	43 (58)	4	3					
17											4 (5)	3 (4)		84 (92)	8					
18										2	3	3 (3)	3 (3)	5 (12)	84 (83)					
19										(3)	6 (8)	(3)	2 (6)	26 (15)	12 (7)	55 (57)				
20										2 (1)	4 (3)	12 (2)	3	15 (7)	30 (39)	10	31 (47)			

Fig. 6. Table showing photofragment branching ratios for photodissociation of carbon cluster ions with 248 nm and 351 nm light at a fluence of approximately 2 mJ cm^{-2}. The 351 nm data is in the brackets. Differences between the branching ratios at 248 nm and 351 nm of less than 5% of the total should not be considered significant

Table 1. Summary of laser fluence dependence of the depletion of the parent ion for the photodissociation of carbon cluster ions with 248 nm and 351 nm light

Cluster Size	248 nm, 4.98 eV	351 nm, 3.53 eV
C_3^+	not enough[a] depletion	not enough[a] depletion
C_4^+	appears one photon	not enough[a] depletion
C_5^+	multiphoton	multiphoton
$C_6^+ + C_{20}^+$	appears one photon	appears one photon

[a] Probably multiphoton, see text

the smaller cross section and appear linear (i.e., one photon). One way to alleviate this problem is to work with very small laser fluences and very low depletions which of course results in signal to noise problems. The results we obtained for carbon cluster ions with 248 nm and 351 nm light are summarized in Table 1. At 248 nm (4.98 eV) for cluster ions with $n=4$ and $n=6$-20 photodissociation appears one photon and for $n=5$ photodissociation is multiphoton. At 351 nm (3.53 eV) the photodissociation of clusters with $n=6$-20 appears one photon. For C_3^+ we were unable to obtain sufficient depletion with either 248 nm or 353 nm light to determine the laser fluence dependence. As will be discussed in more detail below the dissociation energy of C_3^+ is known to be greater than 5 eV so the low depletions we observed are due to the fact that photofragmentation must be multiphoton. Similarly, for C_4^+ with 351 nm light we were unable to obtain sufficient depletion to reliably determine the laser fluence dependence and again this is probably a multiphoton process.

III. C. Photodissociation Cross Sections

Total photodissociation cross sections are derived from measurements of the depletion of the parent cluster ion as a function of laser fluence. To attempt to avoid multiphoton processes we work with low depletions (generally less than 10%). The cross sections we derive are one photon cross sections. The results are shown in Fig. 7. For clusters for which

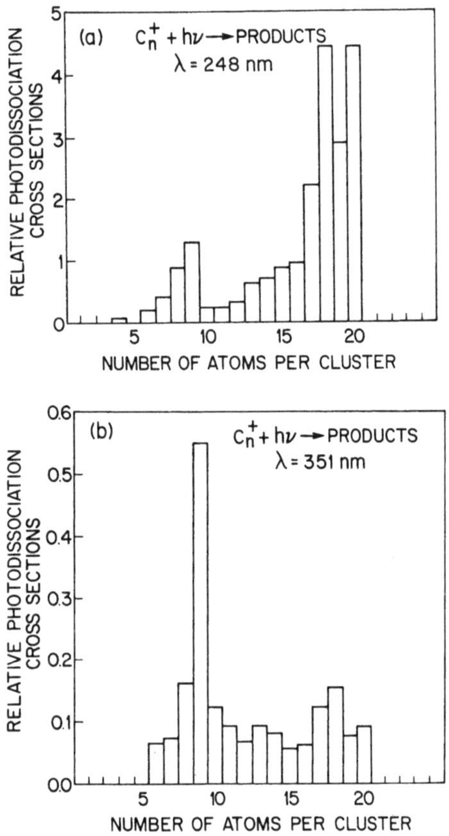

Fig. 7 a and b. Photodissociation cross sections for the photofragmentation of carbon cluster ions at **a** 248 nm and **b** 351 nm. Approximate absolute values for the cross sections can be derived from the relative values by multiplying the scale by 10^{-17} cm^2

only multiphoton photodissociation was observed (C_3^+ and C_5^+ at 248 nm and 351 nm and C_4^+ at 351 nm) the cross sections shown in Fig. 7 are zero. The cross sections measured with 248 nm light are shown in Fig. 7a and those measured with 351 nm light are shown in Fig. 7b. Approximate absolute values for the cross sections can be derived by multiplying the relative values plotted in Fig. 7 by 10^{-17} cm^2 [14]. The relative cross sections for clusters which are close neighbors are probably reliable to within $\pm 30\%$. However, systematic errors, such as variation in the cluster ion packet size, make the relative cross sections for clusters that are well separated less reliable.

With 248 nm light the cross sections increase to a maximum at $n=9$, fall dramatically and then rise again as the cluster size increases. With 351 nm light the cross sections also increase to a maximum at $n=9$ and then fall but they do not increase substantially again as the cluster size increases. The cross sections measured at 351 nm are much smaller (by roughly an order of magnitude) than the cross sections measured with 248 nm light. The photodissociation cross sections measured for the carbon cluster ions with 248 nm light have absolute values in the 10^{-18} cm^2 to 10^{-17} cm^2 range. Cross sections of this magnitude are significantly larger than typical photoionization cross sections. Thus if carbon cluster ions are generated by photoionization of neutral species substantial photodissociation will also occur.

IV. Discussion

Perhaps the most striking result from these studies is the dominance of the C_3 loss channel in the photofragmentation of carbon cluster ions. These results differ from those for silicon and germanium clusters studied previously in this laboratory [15, 16]. The photofragment branching ratios for silicon and germanium cluster ions are quite similar to each other and are characterized by preferential loss of a single atom for initial cluster ions with less than seven atoms and formation of a Si_6^+ or Ge_6^+ species for initial cluster ions with more than six atoms.

What makes the C_3 loss channel "magic" depends on the dissociation mechanism. One possibility is that we are observing direct photodissociation where the building blocks of the cluster, C_3, are removed. We believe this explanation to be unlikely. The dominance of the C_3 loss channel probably arises because this is the lowest energy fragmentation pathway. The C_3 species is known to be very stable. Thus the photodissociation of the carbon cluster ions is probably not direct (i.e., a direct transition to a repulsive state) but involves a transition to a bound excited state which then internally converts to the ground state. Dissociation then occurs statistically (i.e., by a unimolecular reaction) on the ground state potential surface to give preferentially the lowest energy products. Further support for this conclusion is contained in the very similar branching ratios for the two photodissociating wavelengths. For a direct dissociation mechanism the branching ratios are expected to be strongly dependent on the excitation energy. However, the similarity in branching ratios cannot be regarded as proof of statistical dissociation as product branching ratios for unimolecular reactions are not generally independent of energy [17].

An apparent anomaly in the photofragment branching ratios occurs for $n=5$ where the only products observed are $C_3^+ + C_2$. The charge is on the C_3 product. This observation suggests that the ionization energy of C_2 is larger than that of C_3, making it energetically favorable to place the charge on the C_3 fragment. For the fragmentation of the larger clusters the ionization energy of the larger fragment must be smaller than that of C_3, for the neutral C_3 loss channel to be so dominant.

Table 2. Dissociation energies for the carbon cluster ions

Carbon cluster ion	Dissociation energies, eV	
	Laser fluence dependence	Literature[a]
C_3^+	(>4.98)	7.0
C_4^+	<4.98	4.3
C_5^+	>4.98	5.9
C_6^+	<3.53	<3.8
C_7^+	<3.53	<6.8[b]
$C_8^+-C_{20}^+$	<3.53	

[a] From References [4] and [15]. These values can only be considered reliable to ± 1 eV

[b] Ionization energy for this cluster estimated from IE (C_6) and IE (C_5)

Dissociation energies for the carbon cluster ions determined from the laser fluence dependence of photodissociation are summarized in Table 2. Our data suggests that $C_6^+-C_{20}^+$ have binding energies less than 3.53 eV, C_5^+ is bound by greater than 4.98 eV and C_4^+ bound by less than 4.98 eV. The binding energies thus oscillate quite substantially for the smaller clusters but for $n>6$ they drop to be less than 3.53 eV. It is interesting that the binding energy drops at C_6^+ where the C_3 loss channel starts to dominate the photodissociation.

We noted above that the most serious drawback with determining dissociation energies from laser fluence dependences is that the depletion can appear linear for a two photon process if one of the two cross sections is much larger than the other. There was some evidence, mentioned in the Results section, that the photodissociation cross sections increase with internal energy so apparently linear two photon processes could be a problem for the carbon cluster ions studied here. There are several other potential sources of error with this method of determining dissociation energies which will be briefly mentioned below. Internal energy in the cluster ions will lead to an underestimate of the clusters binding energy. As discussed above, we believe that the clusters sampled in this experiment have internal temperatures below room temperature. Activation barriers along the reaction coordinate for dissociation are another source of potential error leading to an overestimate in the cluster binding energy. For the carbon cluster ions studied here the product time-of-flight peaks are narrow, indicating low product relative kinetic energies, which suggests that there are no substantial activation barriers.

There is very little other experimental data available for the binding energies of these cluster ions for us to compare our measurements with. In column 2 of Table 2 we have combined the measured heat of formation of some of the smaller neutral clusters [4] with ionization energies [18] and derived values for the dissociation energies of the cluster ions with $n=3$ to 7. These derived values are not very accurate, mainly due to uncertainties in the ionization energies. However, there is reasonable agreement between the two columns in the Table, suggesting that the dissociation energies derived from the laser fluence dependence of photodissociation may be reliable. However, some recent *ab initio* calculations by Raghavachari [19] give us a reason to be concerned. Generally there is reasonable agreement between his calculated dissociation energies and the values we derive from the laser fluence dependences. However, his calculated dissociation energy for C_7^+ is more than 2 eV larger than the limit (<3.53 eV) we derive from our measurements. If the calculated value is correct then the photodissociation of C_7^+ should be non-linear at both 248 nm and 351 nm. This could be an example of a two photon process which is apparently linear. However, some recent work by McIlvany, Creasy and O'Keefe [20] suggests another possible explanation of this discrepancy for C_7^+. They investigated the reactivity of carbon cluster cations in an FT-ICR (fourier transform ion cyclotron resonance) spectrometer, generating their clusters by pulsed laser vaporization of a graphite target just outside the ICR cell. C_7^+ was observed to be unique in that it was only partially converted to products with some reactant gases. This result was interpreted as evidence for the presence of structural isomers of C_7^+. These isomers may not be present in significant amounts in our experiment because of the presence of the buffer gas in our source. However, if structural isomers are present in the C_7^+ clusters sampled in our photodissociation experiment we would determine the dissociation energy of the least stable isomer. Thus isomers could account for the discrepancy between our measurements and the theoretical results, and together these results could be interpreted as indicating that the higher energy isomer is more than 2 eV less stable than the most stable structure.

It is interesting to speculate on the origin of the dramatic variation of the photodissociation cross sections with cluster size. In an earlier communication [21] we noted that the drop in cross section at $n=10$ might reflect a shift in the absorption due to a change in structure from linear chains to monocyclic rings. There have been several calculations of the structures of carbon clusters and some predict this change in structure to occur at around $n=10$ [7]. However, the reported calculations are generally low level. Some recent higher level calculations for C_4 indicate that the rhombus structure is slightly lower in energy than the linear one [9], suggesting that the transition to a cyclic structure occurs for cluster sizes less than $n=10$, at least for the even clusters. A definitive answer to this question of the origin of the cross section variations requires higher level structure calculations and measurements of the photodissociation spectra to fully map out the shifts in the absorption which must be occurring.

V. Conclusions

In this paper we have reported the results of a detailed study of the photodissociation of carbon cluster ions containing between 3 and 20 atoms. The most striking result from these studies is the dominance of the C_3 loss channel in the photofragment branching ratios. This fragmentation pathway is "magic" probably because of the stability of the C_3 species. The photofragment cross sections were observed to change dramatically with cluster size. Clearly the absorptions must shift quite substantially and it would be interesting to measure photodissociation spectra, using a tunable laser, to map out the features in the absorption spectra in more detail. In this paper we reported limits on the dissociation energies of the carbon cluster ions derived from the laser fluence dependences of photodissociation. We rather critically reviewed this method of determining dissociation energies. The most serious problem is that two photon processes can appear linear if one of the cross sections is much larger than the other. The presence of significant quantities of structural isomers, as perhaps for C_7^+, also introduces uncertainties. If structural isomers are present the dissociation energy determined from our photodissociation experiment will be that of the least stable isomer. Despite all these problems the method of determining dissociation energies of these clusters from laser fluence dependences has one rather compelling advantage at the moment, it is the only experimental technique that will give this sort of information for the larger clusters.

We would like to thank K. Raghavachari for discussing the results of his *ab initio* calculations with us, and A. O'Keefe for providing preprints of his work on the reactivity of the carbon cluster cations.

References

1. See, for example, Powers, D.E., Hansen, S.G., Geusic, M.E., Michalopoulos, D.L., Smalley, R.E.: J. Chem. Phys. **78**, 2866 (1983) Bondybey, V.E., Heaven, M., Miller, T.A.: J. Chem. Phys. **78**, 3593 (1983); Rohlfing, E.A., Cox, D.M., Kaldor, A.: J. Phys. Chem. **88**, 4497 (1984); Martin, T.P.: J. Chem. Phys. **81**, 4426 (1984); Richtsmeier, S.C., Parks, E.K., Liu, K., Pobo, L.G., Riley, S.J.: J. Chem. Phys. **82**, 3659 (1985); Preuss, D.R., Pace, S.A., Gole, J.L.: J. Chem. Phys. **71**, 3553 (1979); Jacobson, D.B., Frieser, B.S.: J. Am. Chem. Soc. **107**, 1581 (1985); Peterson, K.I., Dao, P.D., Farley, R.W., Castleman, A.W.: J. Chem. Phys. **80** 1780 (1984); Johnson, M.A., Alexander, M.L., Lineberger, W.C.: Chem. Phys. Lett. **112**, 285 (1984); Meckstroth, W.K., Ridge, D.P., Reents, W.D.: J. Phys. Chem. **89**, 612 (1985); Jarrold, M.F., Illies, A.J., Bowers, M.T.: J. Am. Chem. Soc. **107**, 7339 (1985); Knight, W.D., Klemenger, K., DeHeer, W.A.: Phys. Rev. B **31**, 2539 (1985).
2. Dietz, T.G., Duncan, M.A., Powers, D.E., Smalley, R.E.: J. Chem. Phys. **74**, 6511 (1981); Bondybey, V.E., English, J.H.: J. Chem. Phys. **74**, 6978 (1981)
3. See, for example, Dunbar, R.C.: Molecular ions: spectroscopy structure and chemistry. Miller, T.A., Bondybey, V.E. (ed.) Amsterdam: North Holland 1983; Moseley, J., Durup, J.: J. Chim. Phys. **77**, 673 (1980)
4. Drowart, J., Burns, R.P., DeMaria, G., Ingram, M.G.: J. Chem. Phys. **31**, 1131 (1959)
5. Honig, R.E.: J. Chem. Phys. **22**, 126 (1954); Berkowitz, J., Chupka, W.A.: J. Chem. Phys. **40**, 2735 (1964); Furstenau, N., Hillenkamp, F.: Int. J. Mass Spectrom. Ion Phys. **37**, 135 (1981)
6. Huber, K.P., Herzberg, G.: "Molecular spectra and molecular structure. Vol. IV Constants of diatomic molecules. New York: Van Nostrand 1979; Gausset, L., Herzberg, G., Lagerguist, A., Rosen, B.: Disc. Faraday Soc. **35**, 113 (1963); Thompson, K.R., DeKock, R.I., Weltner, W.: J. Am. Chem. Soc. **93**, 4688 (1971)
7. Pitzer, K.S., Clementi, E.: J. Am. Chem. Soc. **81** 4477 (1959); Strickler, S.J., Pitzer, K.S.: Molecular orbitals in chemistry physics and biology. Pullman, B., Lowden, P.O. (eds.). New York: Academic Press 1964; Hoffman, R.: Tetrahedron **22**, 521 (1966)
8. Fougere, P.F., Nesbit, R.K.: J. Chem. Phys. **44**, 285 (1966); Peric-Radic, J., Romelt, J., Peyerimhoff, S.D., Buenker, R.J.: Chem. Phys. Lett. **50**, 344 (1977); Romelt, J., Peyerimhoff, S.D.,Buenker, R.J.: Chem. Phys. Lett. **58**, 1 (1978); Ewing, D.W., Pfeiffer, G.V.: Chem. Phys. Lett. **86**, 365 (1982)
9. Whiteside, R.A., Krishnan, R., DeFrees, D.J., Pople, J.A., Schleyer, P. von R.: Chem. Phys. Lett. **78**, 538 (1981)
10. Rohlfing, E.A., Cox, D.M., Kaldor, A.: J. Chem. Phys. **81**, 3322 (1984)
11. Kroto, H.W., Heath, J.R., O'Brian, S.C., Curl, R.F., Smalley, R.E.: Nature **318**, 162 (1985); Heath, J.R., O'Brian, S.C., Zhang, Q., Liu, Y., Curl, R.F., Kroto, H.W., Tittel, F.K., Smalley, R.E.: J. Am. Chem. Soc. **107**, 7779 (1985); Zhang, Q.L., O'Brian, S.C., Heath, J.R., Liu, Y., Curl, R.F., Kroto, H.W., Smalley, R.E.: J. Phys. Chem. **90**, 525 (1986)
12. Cox, D.M., Trevor, D.J., Reichmann, K.C., Kaldor, A.: J. Am. Chem. Soc. (in press)
13. Bloomfield, L.A., Geusic, M.E., Freeman, R.R., Brown, W.L.: Chem. Phys. Lett. **121**, 33 (1985)
14. The absolute scale is derived from the dimensions of the photodissociating laser beam at the point of photodissociation. The laser beam goes through a small aperature just before entering the vacuum system and appears homogeneous at the point of photodissociation. For clusters with large photodissociation cross sections we were able to deplete the parent by >80% with large laser fluences which indicates that there is good overlap between the cluster ion packet and the photodissociating laser beam. The absolute scale is probably reliable to within a factor of two
15. Bloomfield, L.A., Freeman, R.R., Brown, W.L.: Phys. Rev. Lett. **54**, 2246 (1985)
16. Bloomfield, L.A., Freeman, R.R., Brown, W.L.: unpublished data
17. See, for example, Jarrold, M.F., Bass, L.M., Kemper, P.R., Koppen, P.A.M. van, Bowers, M.T.: J. Chem. Phys. **78**, 3756 (1983)

18. Rosenstock, H.M., Draxl, K., Steiner, B.W., Herron, J.T.: J. Phys. Chem. Ref. Data **6**, Supplement No. 1 (1977), (Energetics of Gaseous Ions)
19. Raghavachari, K.: Private communication
20. McElvany, S.W., Creasy, W.R., O'Keefe, A.: J. Chem. Phys. (in press)
21. Geusic, M.E., McIlrath, T.J., Jarrold, M.F., Bloomfield, L.A., Freeman, R.R., Brown, W.L.: J. Chem. Phys. **84**, 2421 (1986)

M.E. Geusic
M.F. Jarrold
T.J. McIlrath
L.A. Bloomfield
R.R. Freeman
W.L. Brown
AT & T Bell Laboratories
Murray Hill, NJ 07974
USA

Decomposition Channels for Multiply Charged Ammonia Clusters

D. Kreisle★, K. Leiter, O. Echt★, and T.D. Märk
Institut für Experimentalphysik, Leopold Franzens Universität, Innsbruck, Austria

Received April 17, 1986; final version June 18, 1986

Multiply charged ammonia cluster ions are produced by adiabatic nozzle expansion and subsequent ionization by electron impact. They are analyzed in a double focussing sector field mass spectrometer (reversed geometry). Doubly charged clusters are only detected above a critical size of 51 and triply charged clusters above 121. Some of these multiply charged ions decay via metastable dissociation processes in the experimental time window accessible. Doubly charged ammonia clusters with sizes of $n \geqq 51$ lose one neutral monomer or, roughly ten times less probable, two neutral monomers. Conversely, triply charged ammonia clusters with sizes $110 \leqq n \leqq 120$ show an extremely asymmetric Coulomb dissociation resulting in doubly charged cluster ions of about 90% of the initial mass

PACS: 36.40.+d; 34.80.Gs; 35.20.Wg

1. Introduction

A particularly interesting subject in the growing field of cluster physics [1] is the production and stability of multiply charged cluster cations. These ions X_n^{z+} are usually only observed in a mass spectrometer if they are exceeding a critical size $n_c(z)$, because the repulsive forces between the separate charges in the cluster cause rapid ejection of charged fragments ("Coulomb explosion"). Critical sizes have been determined experimentally for several weakly bound clusters for charge states up to $z=4$ [2-6]. This subject has also been treated theoretically [7-11]. Little information, however, is available about *(i)* the dissociation mechanism of multiply charged clusters below n_c (i.e. the fragment size distribution) and *(ii)* the stability of multiply charged clusters above n_c (i.e. see [11]).

In this study we have analyzed metastable dissociations of doubly and triply charged ammonia clusters on a time scale of about 5 to 25 μs with respect to the ionization event. Doubly charged ammonia clusters with sizes slightly larger than the critical size of 51 mainly lose one or, roughly ten times less probable, two neutral monomers. No Coulomb explosions are detected for these ions. Conversely, triply charged ammonia clusters with sizes slightly below the critical size of 121 show an extremely asymmetric Coulomb dissociation resulting in doubly charged fragment ions of about 90% of the precursor ion mass. These results are in accordance with similar studies of multiply charged CO_2 clusters [11].

2. Experimental

The molecular beam electron impact mass spectrometer system and the general experimental technique have been described previously [12-14]. In short, the experimental setup consists of a modified [12] three electrode open-type electron impact ionization source and a high resolution double focussing reversed geometry sector field mass spectrometer. Neutral ammonia clusters are typically produced by expanding a mixture of 30% ammonia and 70% argon (with pressures up to 7 bar) under constant

★ Permanent address: Fakultät für Physik, Universität Konstanz, D-7750 Konstanz, FRG

temperature (typically between -20 and $-50\,°C$) through a 20 μm nozzle.

The expanding supersonic gas beam passes a skimmer (differential pumping stage [15]) and is crossed at right angles by an electron beam of variable energy. Ions produced are extracted at right angles from the ionization regime by a weak electric field, analyzed in the magnetic and electric sector fields and detected with a CuBe conversion dynode followed by a CEM used as analog ion detector.

Fragment ions produced by unimolecular dissociation in the first field free region of the mass spectrometer (between earth slit and magnetic sector field) are analyzed by decoupling the acceleration and analyzer fields [16]. With this method it is possible to investigate unimolecular dissociations occurring in a specific time window after formation caused either by metastable or by collision induced dissociation. The time to reach the sampling window after formation (i.e. earth slit) is for instance ~ 6 μs for $(NH_3)_{49}(NH_4^+)_2$ and ~ 7.5 μs for $(NH_3)_{117} \cdot (NH_4^+)_3$, whereas the sampling time (flight time in the field free region) was calculated to be ~ 10.7 and ~ 13.4 μs, respectively. The total time of these ions to reach the detector without decay is in the order of ~ 46 and ~ 58 μs, respectively. Prior to the present investigation, the general reliability of this setup for the detection of metastable dissociations was tested successfully by studying the metastable decay of $C_3H_8^+$ [17] and argon cluster ions [14, 15].

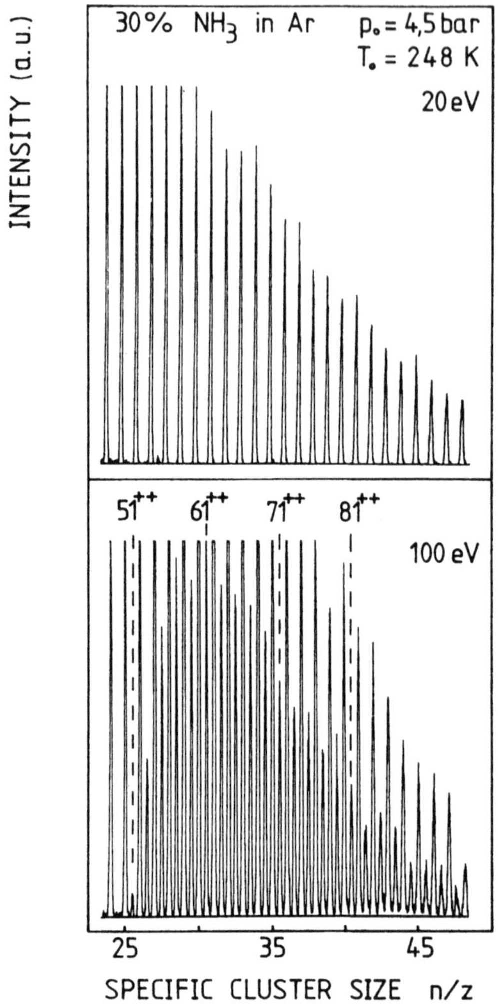

Fig. 1. Section of direct mass spectrum recorded following the electron impact ionization of ammonia clusters with two different electron energies, respectively. For the singly charged cluster peaks, $(NH_3)_{n-1} \cdot NH_4^+$ (for an identification see [18]), the values of n are plotted on the x-axis. The size of the doubly charged cluster ions, $(NH_3)_{n-2}(NH_4^+)_2$, is given above the respective peaks, starting with $n=51$

3. Results and Discussion

In the direct electron impact ionization mass spectrum of an ammonia cluster beam doubly charged cations can be observed for sizes, $n \geqq 51$. Figure 1 shows as an example the section of the mass spectrum around and above this critical size $n_c(2)=51$ for two different electron energies. It can be seen that at the higher electron energy additional peaks appear between two consecutive singly charged cluster ions. These peaks can be clearly identified as doubly charged ammonia cluster ions, due to their mass to charge ratio and due to their appearance potential and electron energy dependence. According to a previous study [5] their composition is $(NH_3)_{n-2}(NH_4^+)_2$. Their abundance in the mass spectrum, depending on the expansion conditions, reaches typically up to 50% of that of the adjacent singly charged cluster ions.

Moreover, triply charged clusters are also present in the electron impact ionization mass spectrum of an ammonia cluster beam, however, in this case we were not able to resolve the mass spectra completely. From an apparent step in the mass spectrum it can be concluded that the minimum existing size is $n_c(3)=121$. Their composition is tentatively assigned as $(NH_3)_{n-3}(NH_4^+)_3$.

In the case of multiply charged cluster ions, a knowledge of unimolecular dissociation mechanism could help to locate the position of the positive charges within the cluster system [5]. For doubly charged cluster cations two dissociation routes are possible, i.e.

$$(NH_3)_{n-2}(NH_4^+)_2 \rightarrow (NH_3)_{n-2-x}(NH_4^+)_2 + x(NH_3), \quad (1)$$

$$(NH_3)_{n-2}(NH_4^+)_2 \rightarrow (NH_3)_{n-2-x}NH_4^+ + (NH_3)_x \cdot NH_4^+. \quad (2)$$

Both, process (1), sometimes called evaporative dissociation, and process (2), sometimes called Coulomb explosion, have not been observed previously, despite an extensive search by Shukla et al. [5]. In the present study we were able to detect and analyse unimolecular dissociation of doubly charged ammonia cluster anions. The main dissociation channel observed is the loss of one neutral monomer, i.e.

$$(NH_3)_{n-2}(NH_4^+)_2 \rightarrow (NH_3)_{n-3}(NH_4^+)_2 + NH_3. \quad (3)$$

The loss of two neutral monomers also occurs (see Fig. 2), however, this mechanism is roughly ten times less probable than reaction (3) in the present time window. Both dissociation processes are observable only for cluster sizes above the critical size $n_c(2)=51$ (see Fig. 2). It could be shown (variation of background gas pressure [14]) that the dissociations observed are purely metastable [14]. The ratio of product ion intensity ($(NH_3)_{n-3}(NH_4^+)_2$ in reaction (3)) and precursor ion intensity ($(NH_3)_{n-2}(NH_4^+)_2$ in reaction (3)) increases with cluster size, i.e. from ~0.5 for a precursor ion cluster size of 53 to ~1.3 for a precursor ion cluster size of 63.

Conversely, no singly charged fragment ions were apparently produced via reaction (2) in our time window for precursor cluster ions with sizes ranging from below to above the critical size $n_c(2)=51$. Thus symmetric fission as predicted from the theory of Tomanek et al. [8] could not be detected. The absence of any detectable Coulomb explosion suggests, that unstable doubly charged cluster ions with sizes below the critical size decay very fast after production (i.e. before they reach the time window starting ~6 μs after production), whereas unstable doubly charged cluster ions with sizes above the critical size stabilize by evaporative processes (reaction (1)).

Finally, a possible loss of one neutral monomer in the case of triply charged ammonia cluster ions could not be analyzed because of other interfering decay channels. On the other hand, turning the voltage of the electric sector analyzer to a value of 1.34 U_0 we were able to discover a series of dissociations leading to doubly charged fragment ions via process:

$$(NH_3)_{n-3}(NH_4^+)_3 \rightarrow (NH_3)_{x-2}(NH_4^+)_2 + (NH_3)_{y-1}(NH_4^+) + z(NH_3) \quad (4)$$

with $110 \leqq n \leqq 120$, $n-14 < x < n-11$ and $n-3 = x + y + z$.

Thus in contrast to the case of doubly charged cluster ions, unstable triply charged ions with sizes below the critical size live long enough to allow detection of the respective "Coulomb" explosion. As can be seen, the doubly charged fragments have a size

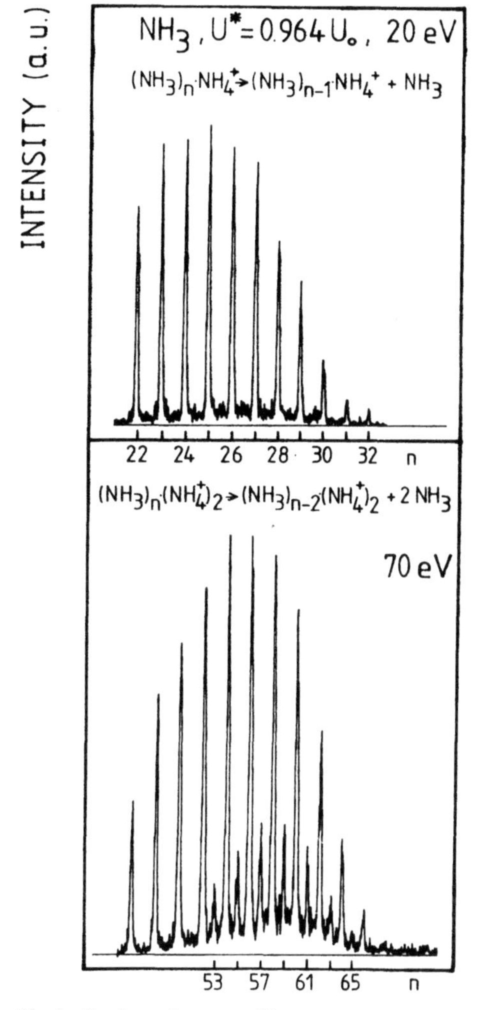

Fig. 2. Section of metastable mass spectrum recorded with an electric deflection voltage in the 90° sector field of 0.964 U_0, with U_0 deflection voltage for direct spectrum. The upper spectrum taken at an electron energy of 20 eV shows a series of decays from singly charged cluster ions, whereas the lower spectrum taken at an electron energy of 70 eV shows additional peaks due to the loss of 2 neutral monomers from a doubly charged cluster ion

of about 90% of the precursor ion size with a very narrow size distribution.

This very asymmetric fission of triply charged ammonia cluster ions contradicts the above-mentioned theoretical prediction [8], but agrees well with a similar study of CO_2 clusters [11]. One can rationalize the size distribution of fission fragments in a model based on the liquid drop approximation [11]. In short, for a given precursor size, the energy barrier for fission is lowest if a small singly charged fragment is ejected from the triply charged cluster. Fission of clusters being larger than the critical size cannot proceed without thermal excitation, even if the overall reaction would be exothermic. The excitation is, of course, introduced into the cluster by the ionization and subsequent structural rearrangement. At the critical size, the energy barrier for the most favorable fission channel is slightly positive such that the lifetime of these ions becomes comparable to the time scale of the mass analysis.

Work partially supported by the Österreichischer Fonds zur Förderung der Wissenschaftlichen Forschung under Project Nr. 5692 and by the Deutsche Forschungsgemeinschaft. It is a pleasure to acknowledge discussions with Prof. A.W. Castleman, Jr., Penn State University, USA, in the frame of the United State-Austria Cooperative Science Program.

References

1. Märk, T.D., Castleman, A.W. Jr.: Adv. At. Mol. Phys. **20**, 65 (1985)
2. Sattler, K., Mühlbach, J., Echt, O. Pfau, P., Recknagel, E.: Phys. Rev. Lett. **47**, **160** (1981)
3. Echt, O., Sattler, K., Recknagel, E.: Phys. Lett. **90**A, 185 (1982)
4. Ding, A., Hesslich, J.: Chem. Phys. Lett. **94**, 54 (1983)
5. Shukla, A.K., Moore, C., Stace, A.J.: Chem. Phys. Lett. **109**, 324 (1984)
6. See also: Echt, O.: Proceedings of the 5th Symposium on Atomic Surface Physics, Obertraun, Howorka, F., Lindinger, W., Märk, T.D. (eds.), p. 272 (1986)
7. Gay, J.G., Berne, B.J.: Phys. Rev. Lett. **49**, 194 (1982)
8. Tomanek, D., Mukherjee, S., Bennemann, K.H.: Phys. Rev. B**28**, 665 (1983)
9. Delley, B.: J. Phys. C**17**, L 551 (1984)
10. Echt, O.: J. Phys. C**18**, L 663 (1985)
11. Kreisle, D., Echt, O., Knapp, M., Recknagel, E., Leiter, K., Märk, T.D., Saenz, J.J., Soler, J.M.: Phys. Rev. Lett. **56**, 1551 (1986)
12. Stephan, K., Helm, H., Märk, T.D.: J. Chem. Phys. **73**, 3763 (1980)
13. Stephan, K., Märk, T.D., Helm, H.: Phys. Rev. A**26**, 2981 (1982)
14. Stephan, K., Märk, T.D.: Phys. Rev. A**32**, 1447 (1985)
15. Märk, T.D., Scheier, P., Leiter, K., Ritter, W., Stephan, K., Stamatovic, A.: Int. J. Mass Spectrom, Ion Proc. (in press)
16. Cooks, R.G., Beynon, J.H., Caprioli, R.M., Lester, G.R.: Metastable ions. Amsterdam: Elsevier 1973
17. Futrell, J.H., Stephan, K., Märk, T.D.: J. Chem. Phys. **76**, 5893 (1982)
18. Stephan, K., Futrell, J.H., Peterson, K.I., Castleman, A.W., Jr., Wagner, H.E., Djuric, N., Märk, T.D.: Int. J. Mass Spectrom. Ion Phys. **44**, 167 (1982)

D. Kreisle
O. Echt
Fakultät für Physik
Universität Konstanz
Postfach 7733
D-7750 Konstanz
Federal Republic of Germany

K. Leiter
T.D. Märk
Institut für Experimentalphysik
Leopold Franzens Universität
Technikerstrasse 15
A-6020 Innsbruck
Austria

An Improved Clusterion-Photoelectron Coincidence Technique for the Investigation of the Ionisation Dynamics of Clusters

L. Cordis, G. Ganteför, J. Heßlich, and A. Ding
Hahn-Meitner-Institut für Kernforschung, Bereich Strahlenchemie, Berlin, Germany

Received April 7, 1986; final version June 18, 1986

The introduction of photoion-photoelectron coincidence techniques has made it possible to investigate photoionisation properties of heavy clusters, which are not accessible by conventional mass spectrometry. This technique has been further developed in combination with a zero-volt electron energy analyser and greatly improved in performance.

The method has been applied to the investigation of different homogeneous and heterogeneous clusters. This type of cluster experiment requires both a very high resolution and a large dynamic range in order to identify also clusters present in low abundance.

As an example, a series of coincidence mass spectra of Xe clusters has been recorded at different wavelengths. Below a photon energy of 11.1 eV, the range of observable clusters shifts to higher cluster sizes with decreasing energy. Appearance potentials and the binding energy of different cluster ions were obtained. Intensity fluctuations, already observed in spectra with electron bombardment ionisation (magic numbers), have also been detected in the coincidence spectra and become most pronounced near the ionisation threshold. This indicates that these fluctuations are caused by the size-dependent stability of the ionic and not the neutral cluster. Furthermore, the threshold size does not change linearly with cluster size. The binding energy per particle seems to change drastically around $n=13$ which indicates the existence of a shell structure in the cluster ion.

PACS: 33.65; 33.80E; 36.40

Introduction

Interest in the properties of neutral and ionic clusters has grown rapidly in recent times. Besides investigating intermolecular forces [1] and chemical dynamics [2], it is particularly interesting to study the transition between the gas phase and the condensed state of matter. Recently, the study of surface phenomena like chemisorption [3] in isolated systems and surface reactivity [4] of clusters has become feasible through the development of many new experimental techniques for the investigation of free cluster beams of very different composition.

A major obstacle in the understanding of cluster properties is the lack of a genuinely size-sensitive detection procedure. Methods such as spectroscopy or electron scattering fail to provide direct insight into the size distribution of clusters while mass spectroscopy requires ionisation of the cluster.

Extensive fragmentation occurs during the ionisation process due to the energy released as the cluster changes from the neutral to the ionic equilibrium configuration. A prerequisite for the understanding of the fragmentation process is knowing the amount of energy dissipated in the cluster. For this reason a detailed energy balance of the system before and after ionisation is needed. For photoionisation, the energy dissipated in the system is the energy of the photon. It is partly released as internal energy in the cluster with the remaining energy appearing as kinetic energy of the ejected electron. In order to collect more exact information, it is necessary to measure the kinetic energy of the ejected electron.

Electron bombardment ionisation is even more

complex, as the energy of the system is distributed between 3 particles – as internal energy of the cluster and as kinetic energy of both electrons.

Experimental Method

On the basis of previously developed PIPECO and TEPICO [5, 6] methods we have devised an improved method for investigating the energy released into a cluster after photoionisation: pairs of electrons and cluster ions are produced by irradiating a molecular beam containing a small percentage (1% to 2%) of clusters with monochromatic VUV light from the Berlin Synchrotron Source BESSY. A weak continuous electric field separates the ion-electron pairs produced by photoionisation. Electrons below a certain energy determined by the electric field are extracted into a drift region consisting of an axial cylinder with a channeltron multiplier attached to an aperture on the side of the cylinder (Fig. 1). The field penetration caused by the high electric potential of the multiplier serves to guide the electrons towards the multiplier. The majority of the electrons with energies higher than the threshold is emitted perpendicular to the axis and lost. Only those electrons emitted in the direction of the axis will be collected and eventually limit the energy resolution.

After amplification and pulse shaping, the multiplier signal is used to start a time-to-amplitude converter. It also triggers a pulse generator which applies a high voltage pulse to a repeller electrode inside the primary ionisation chamber, thus accelerating the ion towards the detector (c.f. Fig. 1). The ion then undergoes continuous acceleration in a section of constant electric field and passes through a field-free drift region before reaching the ion detector. The pulse height of the repeller pulse, the length of the acceleration region, the length of the drift region and the final acceleration potential were chosen so as to achieve spatial focussing, thus optimising the mass resolution. By using a low extraction field for the electron and subsequently a pulsed accelerating field for the ion, both the energy resolution of the electron and the time resolution of the ion can be optimised independently. The coincidence methods used without the extraction pulse yielded high energy resolution of the ejected electrons but had an inferior time resolution since the ions remained for a longer time in the low extraction field. The reason for this is that the time-of-flight depends on the location where the ions are produced and shows large variations due to the finite diameter of the photon beam.

Furthermore, the coincidence method shares all the advantages of the time-of-flight techniques without requiring installation of pulsed ion sources. It is therefore possible to use it in connection with continuous or quasi continuous light sources. This is of particular interest for the use of synchrotron radiation for time-of-flight photoionisation mass spectrometry, as the pulse structure of a synchrotron is usually too fast to be used with conventional time-of-flight techniques. A disadvantage of the coincidence technique is, however, that no other electron ion pair may be generated during the travelling time of the ion as the technique relies on correlation between the electron and ion formed from the same particle. This limits the maximum achievable counting rate to approximately one third of the inverse maximum flight time. Higher rates will result in additional statistical coincidences, whose number increases quadratically with the average counting rates.

Using high acceleration voltages (1 kV) and short drift tubes (10 cm), maximum rates of approximately 10 kHz are feasible. In the previously used setup only 1 to 2 kHz had been obtained.

The coincidence method combines the following advantages of conventional time-of-flight mass spectrometry and photoelectron spectroscopy:

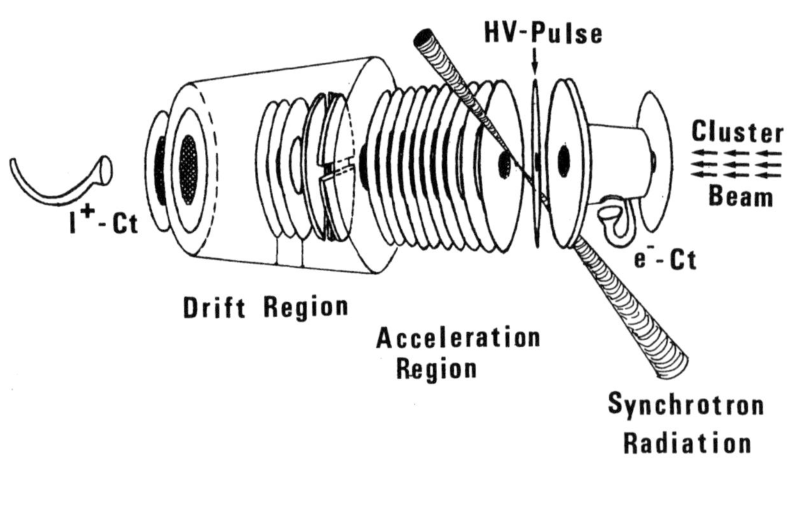

Fig. 1. Schematic diagram of the experimental setup

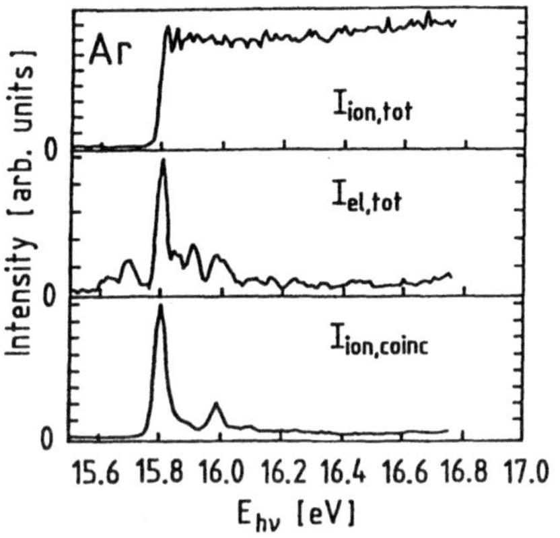

Fig. 2. Photoionisation efficiency curve (top), photoelectron spectrum (middle) and photoion-photoelectron coinicidence spectrum (bottom) of a supersonic argon beam

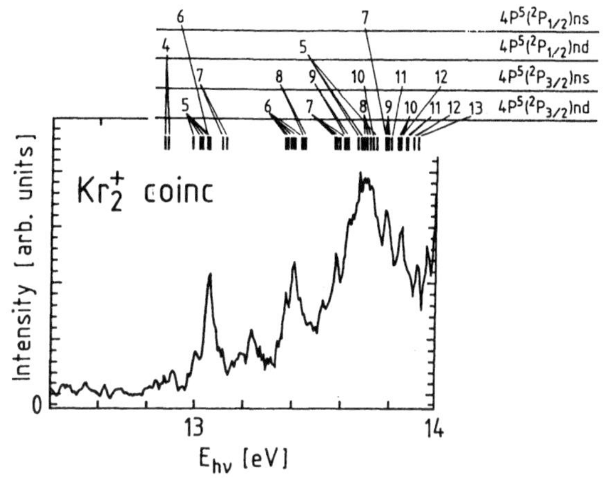

Fig. 3. Photoion-photoelectron coincidence spectrum of the krypton dimer ion

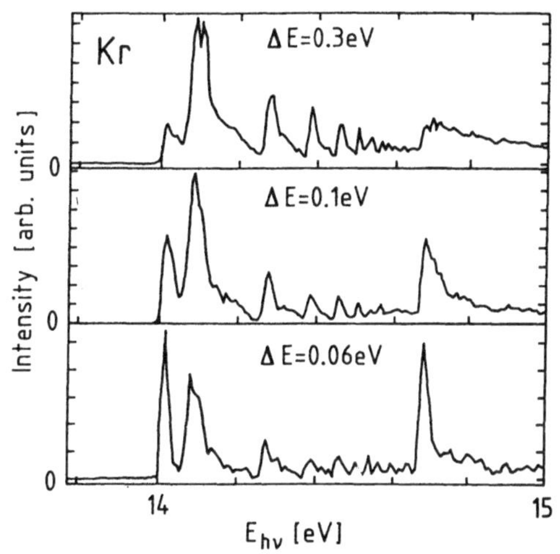

Fig. 4. The effect of the resolution of the zero-volt electron energy analyser on the spectrum of krypton atoms

1) It is possible to record mass spectra over an exceptionally large mass range. This is a particular advantage in conjunction with cluster experiments as large clusters cannot be accessed using conventional quadrupole or magnetic mass spectrometers.

2) By using electronic gates, it is possible to scan several masses simultaneously as a function of photon energy. The number of channels is limited only by the electronic equipment available. This multiplex gain constitutes a great advantage compared with the standard techniques and makes the method one of the most sensitive ones. The absence of narrow slits increases the sensitivity further.

3) It is possible to perform mass selective electron spectroscopy. This is particularly useful when dealing with systems containing a large number of constituents of different masses as in cluster experiments. Figure 2 compares the different methods of photoionisation mass spectrometry, conventional electron spectroscopy and the coincidence method applied to the analysis of supersonic expansion of argon gas. The supersonic argon beam contains a small fraction of dimers and larger clusters. In spite of the photoelectron spectrum (middle) exhibiting many more details than the standard photoionisation efficiency curve (top), interpretation is very difficult if more than one mass contributes to the spectrum. The bottom curve shows the photoion-photoelectron coincidence spectrum for a selected mass range (argon monomer). The direct ionisation channels leading to the $^2P_{3/2}$ and the $^2P_{1/2}$ ionic states of the argon ion are clearly visible.

Similar spectra have been taken for the masses of the higher multimers. An example of a Kr_2^+ coincidence spectrum is shown in Fig. 3 together with the positions of the molecular Rydberg states of the excited krypton dimer. Compared with the results of Ng et al. [6], the spectrum shows a higher degree of modulation and different relative peak heights. This is not surprising as the spectrum in Fig. 3 displays the contribution of only those ions generated together with threshold electrons.

The resolution of the zero-volt energy analyser can be changed by variation of the extraction field. Figure 4 shows electron-ion coincidence spectra of Kr atoms using different resolution of the electron analyser. The direct ionisation of Kr into Kr^+ $^2P_{3/2}$ and Kr^+ $^2P_{1/2}$ generates electrons which carry the remainder of the energy, i.e. low-energy electrons are ejected near the respective thresholds only. The electrons ejected as a consequence of an autoionisation process carry energy corresponding to the finite energy difference between this neutral level and the lower lying ionisation threshold. Therefore the coincidence method is rather insensitive to autoionising states if the resolution of the energy analyser is sufficiently high.

Results and Discussion

The method has been applied to a number of systems of homogeneous clusters. Illustrative examples are the mass spectra of Xe_n^+ taken at different photon energies, which are shown in Fig. 5. Spectra produced with high-energy photons are very similar to those generated by electron bombardment ionisation, showing prominent mass peaks at Xe_{13}^+, Xe_{19}^+ and Xe_{25}^+ [7]. For a given photon energy there is a minimum cluster size for which photoionisation can occur. The clusters disappear altogether below a photon energy of 10.60 eV. It has to be noted, however, that the present experimental setup is not sensitive to clusters larger than $n=50$ for reasons of detector sensitivity. It is therefore not surprising that the ionisation energy of liquid Xe (9.87 eV as calculated from data contained in Ref. 8) is not reached.

Ionisation efficiency curves of clusters exhibit similar shapes: a slow rise above the ionisation threshold followed by a steep increase to a maximum onto which resonant structures are superimposed. Above a certain energy – which usually coincides with the appearance of fragment ions – the efficiency decreases again. These features are usually interpreted as indirect and direct ionisation channels. While the latter correspond to direct Franck-Condon transitions into the ionic state, the slowly rising part is thought to be caused by transitions via highly excited neutral states which lie close to the curves of the appropriate ion [9]. This process, which is somewhat similar to associative ionisation [10] is usually not efficient, so explaining the low intensity observed near threshold. The latter corresponds more to an adiabatic transition into a cluster geometry of minimum energy while the direct ionisation process corresponds to a vertical transition.

It is therefore useful to note both the threshold energy for photoionisation and the onset of the direct ionisation, which is shown as a function of cluster size in Table 1. The values represent the difference in binding energy between the neutral and the ionic species. The accuracy of the experimental values is approximately ±0.05 eV.

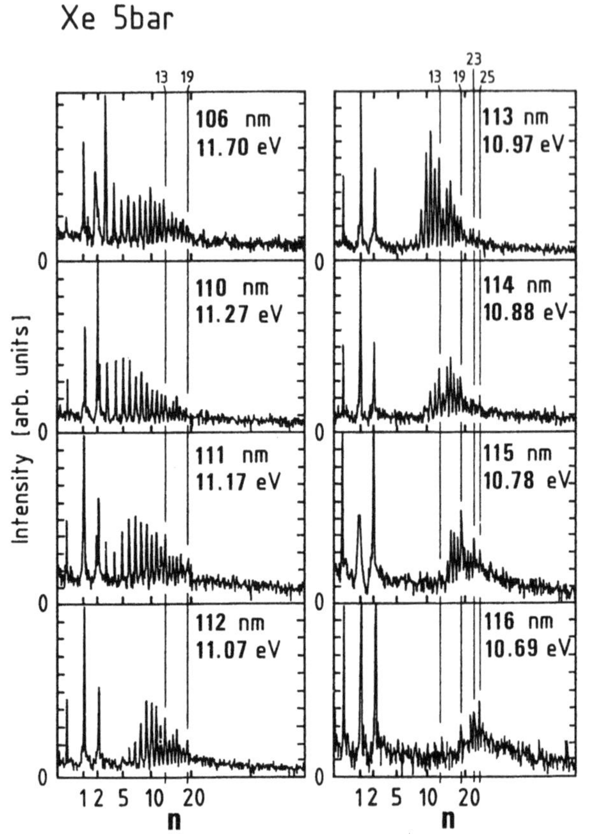

Fig. 5. Photoion-threshold electron coincidence mass spectra of Xe clusters recorded at different wavelengths. The abscissa shows cluster size. The monomer peak is due to the finite intensity of the second order of the spectrometer used

The small cluster ions up to Xe_4^+ all appear at the same energy, indicating that the binding energy does not change significantly from the neutral to the ionic cluster. From $n=5$ to $n=13$ the bond energy per particle (i.e. the difference of appearance energy between two successive clusters) is significantly larger, for $n>13$ it decreases again but does not disappear. This is interpreted as due to the shell structure of the ionic cluster [11]: above $n=13$ a second shell is formed with much less binding energy per particle than found for the first shell. From this it is clearly seen that Xe_{13}^+, Xe_{19}^+ and Xe_{25}^+ have an appearance energy lower than that of their neighbouring clusters,

Table 1. Photoionisation thresholds of Xe_n clusters in eV[a] (in brackets: onset of the direct ionisation process)

n		*n*		*n*		
2	11.00 (11.12)	8	10.90 (10.97)	14	10.77	(10.82)
3	11.00 (11.12)	9	10.85 (10.92)	15	10.65	(10.75)
4	11.00 (11.12)	10	10.80 (10.85)	16	10.65	(10.72)
5	10.95 (11.12)	11	10.75 (10.80)	17, 18	10.70	(10.75)
6	10.92 (11.05)	12	10.75 (10.80)	19, >23	<10.62	(<10.65)[b]
7	10.92 (11.00)	13	10.70 (10.80)	20, 21, 22	10.65	
				liquid Xe	9.87 [8]	

[a] The threshold for Xe is 12.13 eV

[b] No clusters $n<40$ were ionised at 10.60 eV

stressing the exceptionally high stability of these ions. The measured threshold values for the dimer ion are slightly lower than those given by Ng et al. [6]. Our values for Xe_2^+ and Xe_3^+ are considerably lower than the ones given by Poliakoff et al. [12], which we explain by the higher sensitivity of our experimental setup.

Parallel to the appearance of a lower threshold, the upper limit of the cluster distribution is also found to depend on the photon energy. It is known from experiments with electron bombardment ionisation that large cluster ions are easily formed under the present experimental conditions [13]. This leads us to the conclusion that the surplus energy is transferred to the ejected electron. The particular stability of large cluster ions could be explained by the effect whereby the electron carries away a great fraction of the absorbed energy.

The technical assistance of Mr. J. Lehmann and of the staff of the Berlin synchrotron facility BESSY is gratefully acknowledged.

References

1. Hutson, J.M., Howard, B.J.: Mol. Phys. **45**, 769 (1982); **45**, 791 (1982)
2. King, D.L., Dixon, D.A., Herschbach, D.R.: J. Am. Chem. Soc. **96** 3328 (1974)
3. Geusic, M.E., Morse, D.M., Smalley, R.E.: J. Chem. Phys. **82**, 590 (1984)
4. Whetton, R.L., Cox, D.M., Trevor, D.J., Kaldor, A.: J. Phys. Chem. **89**, 566 (1985)
5. Baer, T.: State selection by photoion photoelectron coincidence. In: Gas phase ion chemistry. Vol. 1. New York: Academic Press 1979; Tanaka, K., Kato, T., Guyon, P.M., Koyano, I.: J. Chem. Phys. **77**, 4441 (1982)
   Guyon, P.M.: Threshold photoion-photoelectron coincidences. In: Electronic and atomic collisions. Oda, N., Takayanagi, K. (eds). Amsterdam, Oxford, New York: North Holland Publ. Co. 1980
6. Ng, C.Y., Trevor, D.J., Mahan, B.H., Lee, Y.T.: J. Chem. Phys. **66**, 446 (1977)
7. Echt, O., Sattler, K., Recknagel, E.: Phys. Rev. Lett. **47**, 1121 (1981)
8. Schmidt, W.F.: IEEE Trans. Electron. Ins., **EI-19**, 389 (1984)
9. Runge, S., Pesnelle, A., Perdrix, M., Watel, G.: Phys. Rev. A **32**, 1412 (1985)
10. Carlson, N.W., Taylor, A.J., Schawlow, A.L.: Phys. Rev. Lett. **45**, 18 (1980)
    Carlson, N.W., Taylor, A.J., Jones, A.J. Schawlow, A.L.: Phys. Rev. **A24**, 822 (1981)
11. Saenz, J.J., Soler, J.M., Garcia, N.: Chem. Phys. Lett. **114**, 15 (1985)
12. Poliakoff, E.D., Dehmer, P.M., Dehmer, J.L., Stockbauer, R.: J. Chem. Phys. **76**, 5214 (1982)
13. Heßlich, J., Ding, A.: Unpublished results

L. Cordis
G. Ganteför
J. Hesslich
A. Ding
Hahn-Meitner-Institut
für Kernforschung Berlin GmbH
Bereich Strahlenchemie
Postfach 390128
D-1000 Berlin 39
Germany

Study of the Fragmentation of Small Sulphur Clusters

M. Arnold, J. Kowalski, G. zu Putlitz, T. Stehlin, and F. Träger
Physikalisches Institut der Universität Heidelberg, Federal Republic of Germany

Received July 1, 1986; final version July 22, 1986

The fragmentation of sulphur clusters caused by electron impact ionization was studied. For this purpose, a beam of S_n-clusters with $n \leqq 8$ was generated in a gas aggregation source and ionized by electrons of variable energy. Special care was taken to maintain constant nucleation conditions so that the neutral cluster composition remained unchanged. It was found that the cluster ion mass spectra drastically depend on the electron energy. Even near threshold fragmentation processes contribute significantly to the dependence of the ion intensities on the electron energy.

PACS: 36.40.+d; 34.80.G

Introduction

In many aspects clusters represent a still unknown state of matter intermediate between atoms and molecules on the one hand and the solid state on the other. More refined experiments are certainly needed to improve our knowledge on these particles. Of particular importance are measurements which permit an unambiguous determination of the size dependences of cluster properties. Many efforts were made to derive information on the size distribution of the neutral clusters and on their properties from ion mass spectra. However, it is now generally accepted that these spectra do not necessarily reflect the neutral cluster composition. Fragmentation associated with the ionization process can be such a dominating effect that the mass distributions of the neutral and the ionized clusters become substantially different. Therefore, the detailed study of electron-cluster [1] and photon-cluster interactions with particular emphasis to fragmentation is indispensable.

Here, we report on measurements of the fragmentation of small sulphur clusters with $n \leqq 8$ caused by electron impact ionization. The experiments have been performed in the course of charge exchange studies with sulphur cluster ions [2]. The final goal is to generate a beam that only contains neutral clusters of a single size. Sulphur lends itself naturally to fragmentation studies, since this element tends to form aggregates that range in size up to the well known eight-membered rings [3, 4]. Therefore, if the cluster source is operated such that larger species do not grow, the upper ($n=8$) and the lower ($n=1$) mass limit of the neutral cluster beam are well-defined. This facilitates the interpretation of the measurements.

Experiment

The experimental set-up is arranged as follows: An intense beam of S_n-clusters with $1 \leqq n \leqq 8$ is generated in a gas aggregation source [5, 6]. The helium pressure is chosen such that the nucleation of aggregates larger than S_8 is prevented. After passing a differential pumping stage to remove the buffer gas, the clusters are ionized with an electron beam. The cluster ions are mass-selected in an *rf*-quadrupole filter, accelerated, and detected with a channeltron.

To study the influence of the electron energy on the cluster ion spectra all other experimental parameters must be kept constant. In particular, the sulphur clusters are generated under fixed nucleation conditions, e.g. the crucible temperature ($T=210°$ C), and the pressure of the helium gas ($p=2.8$ Torr) remain unchanged to ensure an identical neutral cluster composition. Also, the operating conditons of components like the quadrupole filter, two electrostatic lens systems, and the detector are fixed. Only the electron energy is systematically varied from ionization threshold of about 10 eV to 120 eV. In this range the cluster ion spectra are measured as a function of electron energy.

Results

Measurements with electron energies ranging from 40 eV to 120 eV yielded almost identical ion mass spectra. For electron energies below 40 eV, however, the mass distribution changes dramatically. This is illustrated in Fig. 1a and Fig. 1b which display ion mass spectra of S_n-clusters for two different electron energies. For ionization with an electron energy of 41 eV, S_3^+ and S_4^+ play a dominant role in the mass spectrum. In contrast, if the clusters are ionized with 11 eV, which is slightly above the ionization threshold (see below), S_3^+ and S_4^+ have completely vanished. Under the chosen experimental conditions neither doubly charged sulphur clusters, which occur only for aggregates with $n \geqq 20$ [7], nor singly charged cluster ions larger than S_8^+ were detected. It is known from studies of sulphur clusters up to S_{56}^+ [7] that ions of all masses with $n \leqq 56$ appear in the spectra after the ionization event. We therefore conclude that fragmentation of species larger than S_8 does not contribute here to the observed intensities of cluster ions with $n \leqq 8$.

The pronounced dependence of the cluster ion intensities on the electron impact energy can be analyzed as follows: The ion intensity of one particular

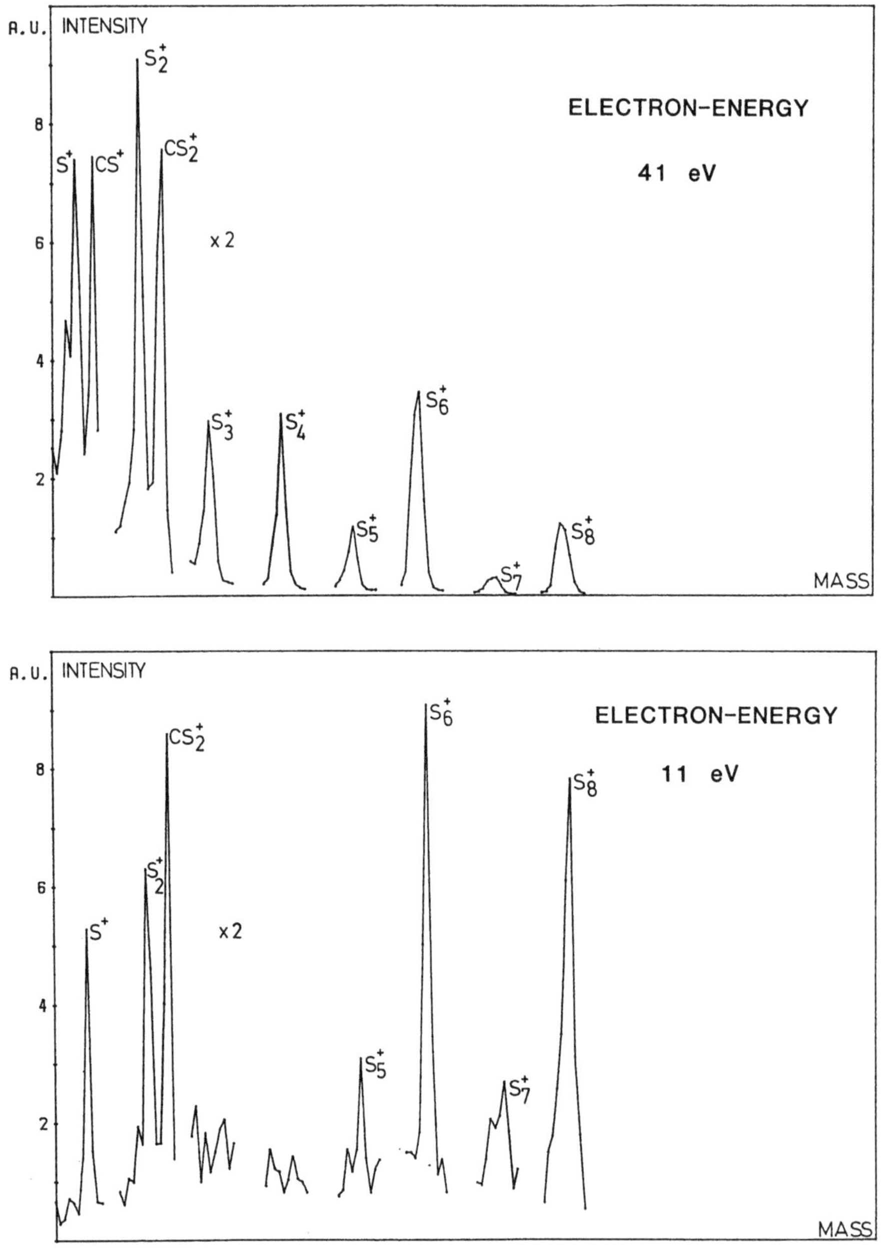

Fig. 1a and b. Mass spectrum of sulphur clusters ionized with an electron energy of 41 eV **(a)** and 11 eV **(b)**

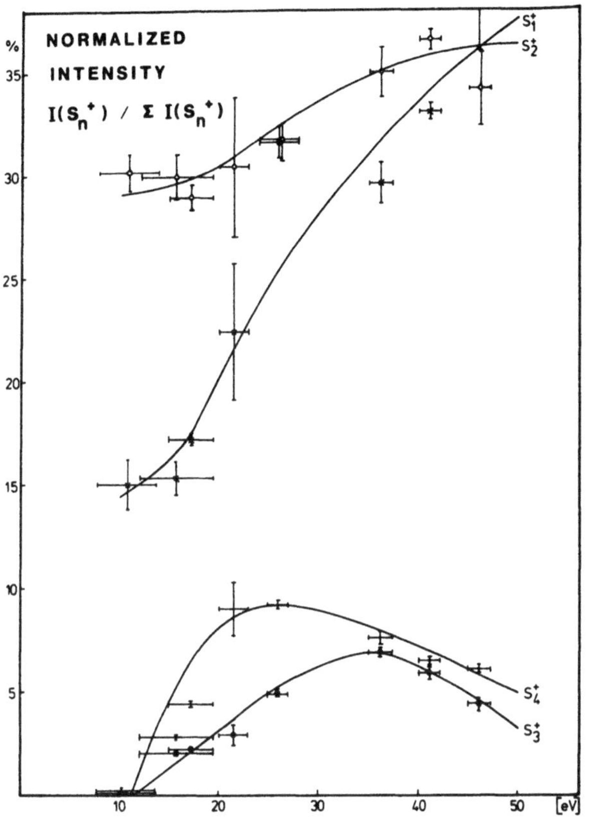

Fig. 2. Normalized intensity (see text) of sulphur cluster ions from S_1^+ to S_4^+ as a function of electron energy

cluster size n is not exclusively determined by direct ionization of S_n into S_n^+. The cluster ions S_n^+ may also be charged fragments originating from larger aggregates of size m. Furthermore, fragmentation into smaller products of size k must be considered. Therefore, the ion intensity $I(S_n^+)$ is given by:

$$I(S_n^+) = n(e^-) \times d \times [\sigma_n I(S_n) - \sum \sigma_{nk} I(S_n) + \sum \sigma_{mn} I(S_m)] \quad (1)$$

with $1 \leqq n \leqq 8$, $k<n$, and $m>n$. $n(e^-)$ stands for the electron density in the ionization region, and d gives the interaction length of the cluster beam and the electron beam. σ_n is the partial cross section for ionization, σ_{nk}, and σ_{mn} are the partial cross sections for ionization with subsequent fragmentation. Double and multiple ionization events, which are strongly suppressed at energies slightly above the threshold for single ionization, modify this simple relation for electron energies above approximately 30 eV. Since mass dependent variations of the transmission of the quadrupole filter and the detection system can hardly alter the general trend of the cluster ion intensities as a function of the electron energy, they are omitted in equation (1).

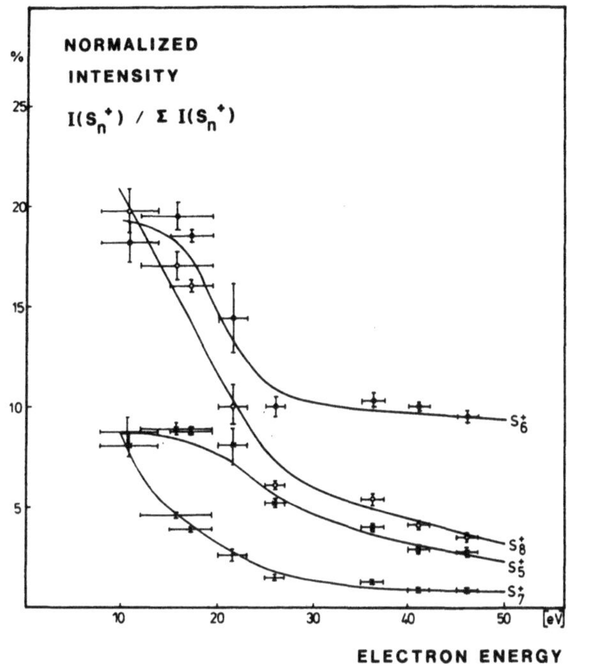

Fig. 3. Normalized ion intensity of S_n-clusters with $n \geqq 5$ as a function of electron energy

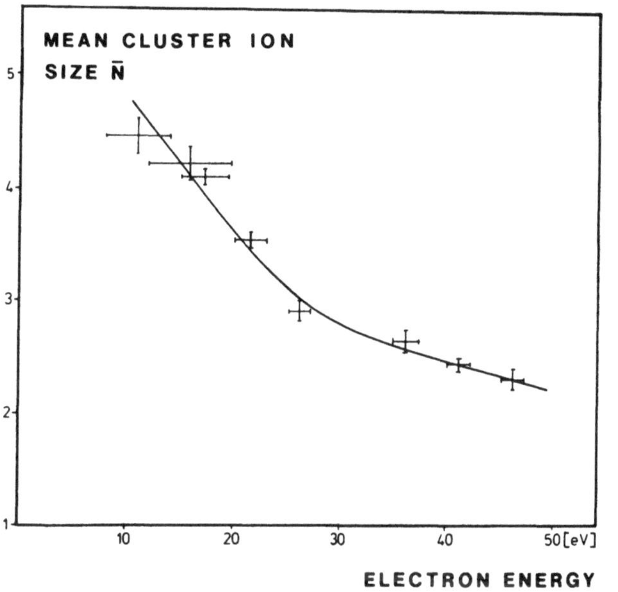

Fig. 4. Mean cluster ion size $\bar{N} = \sum n I(S_n^+) / \sum I(S_n^+)$ as a function of electron energy

By summarizing the ion intensities of all sulphur aggregates from S_1 to S_8, the fragmentation terms cancel. The total ion intensity $\sum I(S_n^+)$ therefore directly reflects the sum of the neutral cluster rates only modified by the respective ionization cross sections:

$$\sum I(S_n^+) = n(e^-) \times d \times \sum \sigma_n I(S_n) \quad (2)$$

with $1 \leqq n \leqq 8$. Experimental parameters like the electron density can be eliminated by normalizing the ion intensity of each particular cluster size $I(S_n^+)$ with respect to the sum of all ion intensities $\sum I(S_n^+)$ for each of the different electron energies:

$$\frac{I(S_n^+)}{\sum I(S_n^+)} = \frac{\sigma_n I(S_n) - \sum \sigma_{nk} I(S_n) + \sum \sigma_{mn} I(S_m)}{\sum \sigma_n I(S_n)}. \quad (3)$$

These normalized ion intensities are shown as a function of the electron energy in Fig. 2 and Fig. 3. The values for S_1^+ and S_2^+ grow with increasing electron energy, whereas those for S_3^+ and S_4^+ increase for small energies, reach a flat maximum, and then slowly fall off. Simultaneously, the intensities of the larger species with $n \geqq 5$ decrease very quickly. In addition, the mean cluster ion size defined by:

$$\bar{N} = \sum n I(S_n^+) / \sum I(S_n^+) \quad (4)$$

becomes smaller as the electron energy rises (see Fig. 4).

Discussion

As can be seen from equation (1), the variation of the normalized ion intensities and of the average cluster ion size can be caused by two effects: fragmentation processes and changes of the ionization cross sections as a function of the electron energy. Since the appearance potentials of all sulphur aggregates up to S_8 lie in a small energy range around 10 eV [8], the dependence of the ionization cross sections σ_n near threshold on the kinetic energy T of the electrons can be calculated from a simple Wannier type relation [9]:

$$\sigma_n(T) \propto [T - E_I(n)]^\alpha, \quad (5)$$

where $E_I(n)$ is the appearance potential of a cluster of size n. In our case, where only singly charged ions are considered, the exponent α is about 1 [9, 10]. This implies that σ_n increases as a function of the electron energy for each cluster size. The absolute values of σ_n at a fixed electron energy, however, are different and depend on the number of electrons of each sulphur species. It is generally assumed that the ionization cross section is proportional to the number of valence electrons in the cluster. The ion rate of S_8^+, for example, should therefore increase more rapidly than that of S_6^+ *. Consequently, the intensity of S_6^+ normalized with respect to the S_8^+-signal should decrease as a function of the electron energy. Experimentally, however, the opposite is observed. This does not only hold for $I(S_6^+)/I(S_8^+)$ but for all other ratios $I(S_n^+)/I(S_8^+)$ with $n < 8$ as well.

Since the term $\sigma_n I(S_n)$ in Eq. (1) obviously does not explain the experimental results, only the fragmentation terms $\sum \sigma_{nk} I(S_n)$ and $\sum \sigma_{mn} I(S_m)$ can give rise to the dependence of the normalized intensities on the electron energy. Obviously, the larger sulphur aggregates with $n \geqq 5$ can dissociate during the ionization event even for small electron energies. Consequently, the ion intensities of the smaller clusters are strongly enhanced by fragments of the larger species. The fact that the normalized intensities of S_3^+ and S_4^+ do not generally increase with increasing electron energy but even slowly decrease for higher energies can be due to several reasons: fragmentation channels may change, double ionization followed by Coulomb explosion may occur so that the relation for the ion intensity of one particular cluster size n given above becomes incorrect. Furthermore, the Wannier type dependence of the ionization cross sections $\sigma_n(T)$ is not valid for larger energies.

The results of our measurements are in agreement with studies on sulphur clusters generated by vaporization of red HgS in a Knudsen cell and ionized by electron impact with energies from threshold up to about 20 eV [11]. This agreement demonstrates that different nucleation conditions for the sulphur clusters which may result in a different population of the rotational-vibrational energy levels do not necessarily cause a different fragmentation behaviour. It seems that the clusters are heated up so much during the ionization process that differences in the original temperature distribution of the neutrals become insignificant.

In summary, we conclude that neither the variation of the ionization cross sections with the electron energy nor the different numbers of electrons of each S_n-cluster give rise to the measured behaviour. Fragmentation processes obviously contribute significantly to the dependence of the cluster ion intensities on the electron energy. As outlined above, the normalized ion intensities provide information which cluster sizes dissociate predominantly and which appear as charged fragments of larger aggregates. In addition, the decrease of the mean cluster ion size $\bar{N}$ shows that more and more clusters fragment as the electron energy becomes larger.

The results presented here constitute an example, which demonstrates that fragmentation occurs even near threshold. Therefore, the measured cluster ion spectra do not reflect the neutral cluster composition. This elucidates that extrapolations to small electron energies [12] can hardly reveal the "true" cluster abundance in the neutral beam. Consequently, a neutral beam which only contains clusters of a single size [2] is urgently needed for an unambiguous determination of cluster parameters.

* Strictly speaking this argument only holds for identical ionization potentials

This work was supported by the Deutsche Forschungsgemeinschaft.

References

1. Keesee, R.G., Castleman, A.W., Jr., Märk, T.D.: In: Proceedings of the "Fourth International Swarm Seminar", Lake Tahoe, VCH Publishers (1985)
2. Arnold, M., Kowalski, J., Putlitz, G. zu, Stehlin, T., Träger, F.: Z. Phys. A – Atoms and Nuclei **322**, 179 (1985)
3. Berkowitz, J.: In: Elemental sulphur. Meyer, B. (ed.), p. 125. New York: Wiley and Sons 1965
4. Berkowitz, J., Chupka, W.A.: J. Chem. Phys. **40**, 287 (1964)
5. Granqvist, C.G., Buhrmann, R.A.: J. Appl. Phys. **47**, 2200 (1976)
6. Echt, O., Sattler, K., Recknagel, E.: Phys. Rev. Lett. **47**, 1121 (1981)
7. Martin, T.P.: J. Chem. Phys. **81**, 4426 (1984)
8. Drowart, J., Goldfinger, P., Detry, D., Rickert, H., Keller, H.: Adv. Mass Spectrom. **4**, 499 (1967)
9. Wannier, G.H.: Phys. Rev. **90**, 817 (1953)
10. Read, F.H.: In: Electron impact ionization. Märk, T.D., Dunn, G.H. (eds.), p. 42. Wien, New York: Springer 1985
11. Rosinger, W., Grade, M., Hirschwald, W.: Ber. Bunsenges. Phys. Chem. **87**, 536 (1983)
12. Pfau, P., Sattler, K., Mühlbach, J., Recknagel, E.: Phys. Lett. **91**, 316 (1982)

M. Arnold
J. Kowalski
G. zu Putlitz
T. Stehlin
F. Träger
Physikalisches Institut
Universität Heidelberg
Philosophenweg 12
D-6900 Heidelberg 1
Federal Republic of Germany

Metal Clusters and Particles

A Few Concluding Theoretical Remarks

B. Mühlschlegel
Institut für Theoretische Physik, Universität zu Köln, Federal Republic of Germany

Received April 30, 1986

Basic research on "Small Particles and Inorganic Clusters" has flourished amazingly, and results have been rapidly communicated between scientists since the first international conference with this title was held in Lyon 10 years ago (at the strong insistence of J. Friedel). Four years later, in 1980 in Lausanne, the cluster beams of the Konstanz group were at the center of interest and have subsequently stimulated many other researchers – both experimentally and theoretically. The results of this stimulation were clearly perceived at the third international meeting in 1984 in Berlin [1]. The fourth conference of the series is scheduled to take place in 1988 – again in France. The present „International Symposium on Metal Clusters" is thus in the middle of these major European activities.

A symposium on small systems of variable size organized in Heidelberg brings our attention quite naturally to nuclear physics for which this city was and is a leading center. Also nuclear physics deals with objects of variable size. The scales are here fm and MeV instead of nm and eV. In spite of obvious differences between nuclear physics and cluster physics due to *1)* different forces and *2)* nucleons with isospin versus light electrons and heavy nuclei (or core ions) there are quite a number of similarities in handling the underlying many-body problem.

The Brueckner-Hartree-Fock method leads from bare to effective nuclear forces, explains ground state properties of spherical nuclei and offers a microscopic justification of the phenomenological shell model for such nuclei. An extension (Hartree-Fock-Bogoliubov) accounts for pairing effects for incomplete shells and for deformed nuclei. Quasiparticle excitations calculated this way explain nuclear spectra but the wealth of these spectra gives at the same time a clear indication of collective excitations. Nuclear fission is certainly a collective process, too. Again, phenomenological descriptions (liquid drop model) are supported by a microscopic approach. Here, both static and dynamic methods (such as time-dependent Hartree-Fock, path integral) are employed.

In order to treat structure, excitations, decay and reactions nuclear physics has developed several important schemes of the "ab initio" type as well as phenomenological models and semiclassical and statistical descriptions. It appears that experience with these methods might be quite useful also for the cluster theorist [2]. Even better it would be if also nuclear theorists could become seriously interested in cluster physics which is, as we are all convinced, a young and booming science.

Cluster- and small-particle physics – in spite of some similarities with nuclear physics – covers nevertheless a much broader and more complex field since, due to charge compensation, there is a natural bulk limit. Further, clusters can be coupled in various ways to their surroundings. They are embedded in a medium, they can interact with outside atoms and molecules, they can be placed on a surface, and they can be coupled to each other forming a granular structure. We should also see clusters and small particles in the context with enormous advances being made by nanolithographic techniques for producing artificial small electronic devices. Many familiar equilibrium and non-equilibrium properties will undergo basic changes in such super-small structures [3].

Experiments and theory discussed at the present symposium were mainly devoted to clusters of very small size. Coming from the atomic side the "truth" of the many-body system lies in the Schrödinger equation of electrons and nuclei. Global properties of stability are the subject of rigorous studies in mathematical physics (E. Lieb, B. Simon, W. Thirring). But quantum chemistry works hard to come near this "truth" concerning geometric and electronic structure. Beginning "ab initio" means here to write the Born-Oppenheimer many-electron wave function as a linear combination of Slater determinants. The Hartree-Fock and Configuration-interation equations following from the variation principle are solved as well as possible. The pseudopotential procedure treats only the "essential" electrons in the field of the remaining core ions. An alternative and widely used approach is given by the local electron-density functional theory which was originally developed for the inhomogeneous electron gas in bulk solids. Concerning all the different methods it is helpful to refer here to a review just completed by one of the participants of this symposium [4]. According to this review it appears that the biggest cluster calculated so far is Cu_{79}.

The theoretical contributions by Koutecký, Hermann, Buttet, Roseń and Jena use the above methods to obtain results described in detail in their papers. We learned that some of these results are very relevant for the experiments of the Bern group, of the Lausanne-Lyon-Paris collaboration, etc. Concerning the work of Hermann one should realize that he actually develops the microscopic theory of adsorption at a metallic surface which in turn produces important input parameters for surface physics.

Cluster physics covers such a large domain of different phenomena that there is ample room for new models and speculations. We might consider under this aspect the nucleation theory for formation and kinetics of clusters as presented by Yamada, and also Gspann's interesting arguments which lead him to the conclusion that metal clusters – in contrast to rare-gas clusters – should be liquid. Further, the efforts to explain the Konstanz results for the fission of multiply charged clusters have to be mentioned here.

As objects of variable size clusters have, unlike nuclei, a definite bulk limit. From a theoretical view point it is therefore interesting to study the leading deviations from bulk behavior due to finite size of the system. Since this approach was not discussed at the present meeting we briefly sketch some aspects of such small metal particles [5].

Stable elementary excitations around the Fermi energy ε_F are well established for the bulk system by Fermi liquid theory. In a system of finite size N (N = number of ions or "conduction" electrons, resp.) these continuous quasiparticle excitations are expected to become discrete with a mean energy-level separation of order $\delta = \varepsilon_F/N$ (Kubo gap, which is of order 0.1meV for $N \approx 10^5$). This simple observation will lead immediately to the prediction of various deviations from bulk behavior (quantum-size effects) when other energies such as thermal $k_B T$ or electromagnetic $\hbar\omega$ become as small as δ. For the example chosen, this would mean the infrared, and temperatures as low as 1K.

Bulk properties are often influenced by some characteristic length, and it is then of considerable interest to consider the case where the size of the system is decreased and becomes comparable or even smaller than this length. We mention here only the phase transition in superconductors at the transition temperature T_c which is substantially effected by increasing fluctuations of the order parameter with decreasing size. The result (also experimentally supported) is that superconductivity survives down to sizes small compared with the characteristic (pair coherence)-length but will cease to exist when δ becomes of order $k_B T_c$.

So far the surface of the particle was solely needed to create level discreteness via boundary conditions. In order to see a real surface influence one has to consider a specific shape. The sphere is here an obvious choice. It was used, as we all know, in many studies of different problems with classical methods (thermodynamics, electrodynamics, ranging from surface energy to Mie scattering). For itinerant electrons, however, the challenge is to translate to the spherical situation all the modern quantum mechanical knowledge obtained previously for electrons confined by a plane surface. This model problem was recently attacked in a selfconsistent manner by Ekardt with applications to dynamical response, etc. [6]. Such studies are, of course, very important for a deeper understanding of how various phenomena of local classical electrodynamics are modified with decreasing size of the metallic system. In addition, the sphere allows an extrapolation of radius $\to 0$, and one arrives here at a jellium shell model of the small cluster proposed by the Berkeley group.

We have said above that clusters or small particles can be placed in a medium and can be coupled together. They become thereby the elements of larger heterogeneous structures with unusual properties. Here we have seen very interesting examples in the contributions from Saarbrücken and from IBM-San Jose. – With the last remarks we are actually back to bulk systems. Let me therefore finally draw the attention to recent achievements in the field of bulk metals. There the new phenomena (mixed valence, heavy fermions, etc.) all occur in compounds (mostly rare earth) and are characterized by very strong local correlations. The cluster theorist should then in the future also pay more attention to such correlations; at the same time it is very nice that increasingly compound clusters are being produced and investigated.

This has been an exciting and high-quality meeting. Very interesting experimental results were presented, several of them even unknown a few months ago. I think it is in the name of all participants to extend cordial thanks to both of the organizers and their local helpers for their work. It was a great pleasure and unique experience that cluster science could be placed at the very beginning of activities in the beautiful new Internationale Wissenschaftsforum Heidelberg.

References

1. The proceedings of the three international meetings were published Lyon: J. Phys. (Paris) Colloq. C2 (1977), Lausanne: Surf. Sci. **106** (1981), Berlin: Surf. Sci. **156** (1985). In addition to these conferences several workshops, schools and small meetings on clusters have taken place in various countries. We mention here only the proceedings of a Nato School in France on "Impact of Cluster Physics in Material Science and Technology", Kluwer Academic 1986
2. An excellent source of information is: The nuclear many-body problem. In: Texts and Monographs in Physics. Ring, P., Schuck, P. (eds.). Berlin, Heidelberg, New York: Springer 1980
3. A lively written review is: Physics of mesoscopic systems by Y. Imry, to be published in 1986
4. Koutecký, J., Fantucci, P.: Theoretical aspects of metal atom clusters. Chem. Rev. (to appear)
5. Kubo, R., Kawabata, A., Kobayashi, S.: Electronic properties of small particles. Annu. Rev. Mater. Sci. **14**, 49 (1984). Mühlschlegel, B.: In: Percolation, localization and superconductivity. Goldman, A.M., Wolf, S.A. (eds.). Nato Adv. Study Inst. Ser. B. 109. New York: Plenum Press, 1984
6. Ekardt, W.: Phys. Rev. Lett. **52**, 1925 (1984); (to be published)

B. Mühlschlegel
Institut für Theoretische Physik
Universität zu Köln
Zülpicher Strasse 77
D-5000 Köln 41
Federal Republic of Germany

Index of Contributors

MIX
Papier aus verantwortungsvollen Quellen
Paper from responsible sources
FSC® C105338

If you have any concerns about our products,
you can contact us on
ProductSafety@springernature.com

In case Publisher is established outside the EU,
the EU authorized representative is:
Springer Nature Customer Service Center GmbH
Europaplatz 3, 69115 Heidelberg, Germany

Printed by Libri Plureos GmbH
in Hamburg, Germany